Lecture Notes in Mathematics 2188

More information about this series at http://www.springer.com/series/304

Friedrich Wehrung

Refinement Monoids, Equidecomposability Types, and Boolean Inverse Semigroups

Springer

Friedrich Wehrung
Département de Mathématiques
Université de Caen Normandie
Caen, France

ISSN 0075-8434 ISSN 1617-9692 (electronic)
Lecture Notes in Mathematics
ISBN 978-3-319-61598-1 ISBN 978-3-319-61599-8 (eBook)
DOI 10.1007/978-3-319-61599-8

Library of Congress Control Number: 2017950673

Mathematics Subject Classification (2010): 20M18; 20M14; 16E20; 06F05; 20M25; 28B10; 06E15; 08A30; 08A35; 08A55; 08B10; 08C05; 16E50; 18A30; 19A31; 19A49; 43A07; 46L80

Printed on acid-free paper

This Springer imprint is published by Springer Nature
The registered company is Springer International Publishing AG
The registered company address is: Gewerbestrasse 11, 6330 Cham, Switzerland

Contents

Chapter 1
Background

1.1 Introduction

Generally speaking, this book deals with arithmetical systems that arise naturally as invariants of various mathematical structures. By "arithmetical systems" we mean here (usually commutative[1]) *monoids* $(M,+,0)$. Depending on the context, the elements of M are usually thought of as *measures* or *dimensions*.

A prototype of such a monoid invariant is given by (additive) *cardinal arithmetic*. We say that two sets X and Y have the same cardinality, in notation $\operatorname{card} X = \operatorname{card} Y$, if X and Y are *equinumerous*, that is, there exists a bijection from X onto Y. This could serve as a definition of the cardinal "number" $\operatorname{card} X$, by setting $\operatorname{card} X$ equal to the class of all sets equinumerous to X; which, of course, raises the following foundational problem: if X is nonempty, then $\operatorname{card} X$ is a proper class. Although the problem of being able to define a suitable "function" $X \mapsto \operatorname{card} X$ cannot be solved in the general framework of Zermelo-Fraenkel set theory with neither the Axiom of Choice nor the Axiom of Foundation (cf. Jech [61, Theorem 11.2]), this has no impact on "practical" questions: most definitions and statements on cardinal numbers make sense even though those numbers are not themselves defined. For example, the sum of two cardinal numbers can be defined by the rule

$$\operatorname{card} X + \operatorname{card} Y = \operatorname{card}(X \cup Y)\,, \quad \text{for any disjoint sets } X \text{ and } Y\,, \qquad (1.1.1)$$

and it is easy to verify that the addition thus defined is both commutative and associative. Cardinal arithmetic is the study of cardinal numbers under the addition defined in (1.1.1).

[1] A notable exception to commutativity occurs with *ordinal algebras* (cf. Tarski [110]).

F. Wehrung, *Refinement Monoids, Equidecomposability Types, and Boolean Inverse Semigroups*, Lecture Notes in Mathematics 2188,
DOI 10.1007/978-3-319-61599-8_1

In the presence of the Axiom of Choice, the above-mentioned foundational problem disappears (define card X as the smallest ordinal equinumerous to X) and cardinal addition takes a very simple form. Namely, a cardinal number is either a nonnegative integer n or a transfinite cardinal $\aleph_\alpha$; cardinal addition extends the addition of nonnegative integers; further, $\kappa + \aleph_\alpha = \aleph_\alpha$ whenever $\kappa \leq \aleph_\alpha$.

Without the Axiom of Choice, the situation becomes far more interesting, as it enables us to extend results of cardinal arithmetic to much more general mathematical structures. For example, even without the Axiom of Choice, the Cantor–Schröder–Bernstein Theorem on cardinal numbers remains valid (i.e., if $\beta = \alpha + \xi$ and $\alpha = \beta + \eta$, then $\alpha = \beta$), and so do the multiplicative cancellation laws $m\alpha = m\beta \Rightarrow \alpha = \beta$ for positive integers m (cf. Tarski [108, 109]). None of those results is trivial, especially the second one; their proper algebraic settings were coined by Tarski in his monograph [109], under the name *cardinal algebras*. A cardinal algebra is a commutative monoid, endowed with an operation of addition of countably infinite sequences, subjected to a list of axioms valid for the addition of sequences of (possibly infinite) nonnegative real numbers.

Leaving aside the above-mentioned foundational problem, cardinal numbers form a (proper class) cardinal algebra. So do isomorphism classes of countably complete Boolean algebras under direct product (cf. Tarski [109, Theorem 15.27]). Hence, isomorphism classes of countably complete Boolean algebras satisfy both the Cantor–Schröder–Bernstein Theorem and multiplicative cancellation. Removing the countable completeness assumption from the Boolean algebras no longer yields a cardinal algebra; in fact *every countable commutative semigroup embeds into the monoid* **B** *of isomorphism types of all countable Boolean algebras* (cf. Ketonen [70]). In particular, the Cantor–Schröder–Bernstein Theorem and multiplicative cancellation both fail for **B**. The monoid-theoretical structure that survives the deletion of countable completeness is the one of *refinement monoid*. Isomorphism classes of Boolean algebras form a (proper class) refinement monoid.

Building on ideas of Vaught [114], Dobbertin investigated the connection between refinement monoids and back-and-forth families on Boolean algebras (cf. [32–35]). He proved, in particular, that many refinement monoids are ranges of certain abstract measures, which he calls *V-measures*, on Boolean algebras (see Theorem 4.6.7 in the present book). Those V-measures are, in turn, closely related to monoids arising from the *Banach-Tarski paradox* (cf. Banach and Tarski [18]), often called *equidecomposability types monoids*. Those monoids are also refinement monoids. The investigations on Dobbertin's V-measures and equidecomposability types monoids both involve the study of finitely additive, monoid-valued measures on Boolean algebras.

A third important source of refinement monoids is provided by *nonstable K-theory of rings*, also discussed in the present book. For any ring R, the monoid $\mathrm{V}(R)$ of all Murray–von Neumann equivalence classes of idempotent matrices over R is not, in general, a refinement monoid. The observation that $\mathrm{V}(R)$ is a refinement monoid whenever R is a von Neumann regular ring brought considerable development to the study of those rings (cf. Goodearl [51]). This observation got further extended to the larger class of *exchange rings* (cf. Ara [5]).

The present book is largely an account of the refinement monoids that arise from the class of all *Boolean inverse semigroups*. Boolean inverse semigroups can be described as inverse semigroups of partial isomorphisms, closed under finite disjoint union of functions, with Boolean semilattices of idempotents (cf. Lawson [74, 75], Lawson and Lenz [76]). The associated commutative monoids, called *type monoids of Boolean inverse semigroups*, were introduced in Wallis [116] and Kudryavtseva et al. [71]. Elements x and y in a Boolean inverse semigroup S have the same type—in notation, $\operatorname{typ}(x) = \operatorname{typ}(y)$—if they are related under Green's relation $\mathscr{D}$. We show, in the present book, that type monoids are identical to equidecomposability types monoids. Our in-depth study of those monoids raises close connections with Dobbertin's above-cited work, highlighting the crucial importance of V-measures. We also relate the type monoid $\operatorname{Typ} S$ of a Boolean inverse semigroup S to the nonstable K-theory of a certain ring, denoted by $K\langle S\rangle$ (where K is an arbitrary unital ring). The present work thus establishes connections between the three above-cited sources of refinement monoids (viz., cardinal arithmetic and generalizations; Dobbertin's V-measures and equidecomposability types monoids; nonstable K-theory of rings).

In preparation for the sections on type monoids, we also investigate Boolean inverse semigroups from an algebraic viewpoint. Our investigation starts with the crucial observation that Boolean inverse semigroups, with a suitable class of maps called *additive homomorphisms*, form a variety of algebras (in the sense of universal algebra) that we call *biases*. As biases turn out to be a congruence-permutable variety, Boolean inverse semigroups are much closer to rings than to semigroups.

1.2 Origin and Motivation

The present book is an outgrowth of a few lines scribbled by the author on a result, obtained in Ara and Exel (cf. [7, Theorem 7.11]), about monoids of equidecomposability types. Due to many ramifications to other topics, more came out than originally expected.

1.2.1 First Short Motivation

Let G be a group acting on a set Ω, and let $\mathcal{B}$ be a ring of subsets of Ω (viz., a nonempty set of subsets of Ω, closed under finite union and set difference). The *monoid of G-equidecomposability types of elements of* $\mathcal{B}$, denoted by $\mathbb{Z}^+\langle\mathcal{B}\rangle /\!\!/ G$ (explanations about that notation will follow), is defined as the commutative monoid defined by generators $[a]_G$, where $a \in \mathcal{B}$, and relations $[\varnothing]_G = 0$, $[ga]_G = [a]_G$, and $[a \sqcup b]_G = [a]_G + [b]_G$, where $\sqcup$ denotes disjoint union.

A large part of the present work arises from the question asking which monoids can be represented as monoids of equidecomposability types, see for example Kerr

[68, Question 3.10], Kerr and Nowak [69], Rørdam and Sierakowski [99, p. 285]. A partial result in that direction, due to Ara and Exel (cf. [7, Theorem 7.11]), states that every finitely generated conical commutative monoid embeds into the equidecomposability types monoid of a Boolean algebra under a free action of a free group. We prove a stronger result, namely:

Every countable[2] *conical refinement monoid is isomorphic to some* $\mathbb{Z}^{+}\langle\mathcal{B}\rangle /\!\!/ G$.

(The requirements that G is free and its action is free both come for free: just replace G by any free group preimage F of G and let any element of F act as its image in G; then replace Ω by $\Omega \times F$, so the action of F becomes free.) This result, which we state in Theorem 4.8.9 (in a different, but equivalent, form), turns out to be an easy consequence of known results, mostly by Hans Dobbertin. Nonetheless, this is only the tip of the iceberg. The present work is mostly devoted to show what lies underneath.

1.2.2 Second Short Motivation

It is known since Tarski [109] that in the context of Sect. 1.2.1 above, the assumption that G be *exponentially bounded* implies that the monoid $\mathbb{Z}^{+}\langle\mathcal{B}\rangle /\!\!/ G$ satisfies the monoid implication[3] $\mathsf{x}+\mathsf{z}=\mathsf{y}+2\mathsf{z} \Rightarrow \mathsf{x}=\mathsf{y}+\mathsf{z}$ (such commutative monoids are nowadays called *strongly separative*).

Alexander Pruss raised on http://mathoverflow.net/questions/140693 the question whether the *supramenability* of G was sufficient, and this in the context where $\mathcal{B}$ is a powerset algebra. By invoking a measure-theoretical result established in Armstrong [15, Proposition 1.7], Pruss obtained there a positive answer.

However, a direct attempt at extending Pruss' argument to arbitrary Boolean rings $\mathcal{B}$ (and using the results of Moreira Dos Santos [84] instead of Armstrong's) could not lead us further than proving that $\boldsymbol{a}+\boldsymbol{c}=\boldsymbol{b}+2\boldsymbol{c}$ implies that $\boldsymbol{a}$ and $\boldsymbol{b}+\boldsymbol{c}$ are equivalent modulo the least congruence of $\mathbb{Z}^{+}\langle\mathcal{B}\rangle /\!\!/ G$ with separative quotient.

In this work we solve completely the extended form of Pruss' question (cf. Theorem 5.3.8). Once again, this is only the tip of the iceberg.

As to Wagon's question whether every supramenable group is exponentially bounded, Theorem 5.3.8 seems to be in danger of getting some day stripped of its content by its specialization to exponentially bounded groups, which we state in Theorem 5.3.6 (originating in Tarski [109, Theorem 16.10]). On the other hand, the general belief seems to be currently in favor of a negative solution to Wagon's question (see, in particular, Examples 71 and 74 in Ceccherini-Silberstein et al.

---

[2]Throughout this work, "countable" will always mean "at most countable".

[3]We will always write (syntactical) objects in sans serif fonts, such as x, y, z, p, ... while keeping the notation x, y, z, p, ... for the objects that they interpret, denizens of a given mathematical structure.

[26]); thus it is plausible that the inclusion in the present work of Theorem 5.3.8 is a reasonable anticipation.

1.2.3 Third (Not So Short) Motivation

The Murray–von Neumann equivalence classes of square matrices over a ring R form a commutative monoid $\mathrm{V}(R)$, encoding the *nonstable K-theory of* R (cf. Sect. 1.3.4). The problem of which monoids appear as $\mathrm{V}(R)$, for various types of rings R, has been, for decades, an active field of research. For example, every conical commutative monoid with order-unit is isomorphic to $\mathrm{V}(R)$ for some unital hereditary ring R: this is proved in Theorems 6.2 and 6.4 of Bergman [20] for the finitely generated, unital case, and in Bergman and Dicks [21, p. 315] for the general, unital case. The general, non-unital case is established in Ara and Goodearl [9, Proposition 4.4].

The problem above, restricted to various classes of rings, yields fascinating open problems, especially for rings that are either regular (in von Neumann's sense[4]), exchange rings, or C*-algebras of real rank zero. The problem, as to whether every conical conical refinement monoid appears as $\mathrm{V}(R)$ for a regular ring R, first appeared in print in Goodearl [51]. Counterexamples, in any cardinality beyond $\aleph_2$, were constructed in Wehrung [124]. The cases of cardinality either countable or $\aleph_1$ are still open. For an interesting survey on the regular case, see Ara [6].

The intuition underlying the arguments involved in Wehrung [124] was relying heavily on the concept of equidecomposability briefly discussed in Sects. 1.2.1 and 1.2.2 above, laced with some basic infinite combinatorics. A crucial concept used there was the one of a *measure*. Here, measures are finitely additive measures, defined on Boolean algebras, with values in commutative monoids.

The class of measures relevant to our matters, introduced in Dobbertin [33], are called *V-measures*. Elements in a countable Boolean algebra, with the same measure, are related by a measure-preserving partial automorphism. The semigroups of partial automorphisms thus considered are *inverse semigroups*, as opposed to groups. In addition, those inverse semigroups are closed under finite orthogonal join (which, for partial functions, coincides with the least common extension). Such semigroups, called *Boolean inverse semigroups*, have been over the last few decades an active topic of research, see for example[5] Lawson [74, 75], Lawson and Lenz [76], Kudryavtseva et al. [71]. By definition, an inverse semigroup S with zero

[4] A ring R is *regular* if its multiplicative semigroup is regular, that is, for all $x \in R$ there exists $y \in R$ such that $x = xyx$.

[5] Exel [41] also uses the term "Boolean inverse semigroup", but in a weaker sense than the one used here.

is Boolean if its semilattice of idempotents is (generalized) Boolean and S has finite orthogonal joins.

1.2.4 A Statement of Purpose

Various experiments, about the questions raised in Sects. 1.2.1–1.2.3, suggested that a number of problems pertaining to those questions could be successfully handled via an appropriate blend of Tarski's ideas (about monoids of equidecomposability types), Dobbertin's ideas (about V-measures), and a more ring-theoretical/universal algebraic approach to Boolean inverse semigroups, thus giving rise to an algebraic theory of equidecomposability types semigroups.

The purpose of the present work is to consolidate that framework. We are dealing with *refinement monoids*, that is, commutative monoids satisfying the refinement axiom, stating that any equation of the form $a_0 + a_1 = b_0 + b_1$ has a common refinement (cf. Sect. 1.5). A congruence relation Γ on a refinement monoid is a *V-congruence* (cf. Sect. 2.6) if for any relation of the form $a_0 + a_1 \mathrel{\Gamma} b$, there is a decomposition $b = b_0 + b_1$ with each $a_i \mathrel{\Gamma} b_i$.

The main focus of attention of the present work is the study of commutative monoids of the form M/Γ, where M and Γ are both well understood.

Two fundamental examples of that situation are the following:

(1) Every ring $\mathcal{B}$ of subsets of a set Ω, closed under the action of a group G, gives rise to a commutative monoid $\mathbb{Z}^+\langle\mathcal{B}\rangle$, which turns out to be the positive cone of an Abelian lattice-ordered group (cf. Example 2.2.7), and to a V-congruence $\simeq_G$ on $\mathbb{Z}^+\langle\mathcal{B}\rangle$ (cf. Sect. 2.8). The monoid of equidecomposability types $\mathbb{Z}^+\langle\mathcal{B}\rangle /\!\!/ G$, alluded to in Sect. 1.2.1, is then nothing else than the quotient monoid $\mathbb{Z}^+\langle\mathcal{B}\rangle/{\simeq_G}$.
(2) Every Boolean inverse semigroup S gives rise to a conical refinement monoid, called the *type monoid* of S and denoted by $\operatorname{Typ} S$ (cf. Definition 4.1.3), see Wallis [116], Kudryavtseva et al. [71]. If B denotes the generalized Boolean algebra of all idempotents of S, then $\operatorname{Typ} S = \mathbb{Z}^+\langle B\rangle/\mathscr{D}^+$, where $\mathscr{D}^+$ denotes the monoid congruence on $\mathbb{Z}^+\langle B\rangle$ generated by the restriction to B of Green's relation $\mathscr{D}$.

It will turn out that the two classes of monoids, described in (1) and (2) above, are identical. This will be stated formally in Proposition 4.8.5. These monoids are all *conical refinement monoids* (this is an important difference with the monoids, constructed from *separated graphs*, introduced in Ara and Goodearl [9]). The converse fails: by a series of counterexamples of cardinality $\aleph_2$, originating in Wehrung [124], not every conical refinement monoid is isomorphic to some $\mathbb{Z}^+\langle B\rangle /\!\!/ G$ (equivalently, $\operatorname{Typ} S$).

1.2.5 Levels of Non-commutativity

In a scale measuring levels of non-commutativity of various existing mathematical theories, the position of Boolean inverse semigroups is quite modest. Such semigroups are not necessarily commutative, so they definitely stand above Boolean algebras. On the other hand, all idempotents of an inverse semigroup commute, thus inverse semigroups stand below rings in the above-mentioned hierarchy.

Nonetheless, the idea that many idempotents commute is implicit in various works on the representation problems mentioned in Sect. 1.2.3. This idea makes Boolean inverse semigroups a potentially fruitful paradigm, for studying even less commutative ring-theoretical questions. More specifically: we are given a ring-theoretical question (e.g., the representation problem of countable conical refinement monoids as $\mathrm{V}(R)$ for a regular ring R); could one first get a hint, by solving a related question for Boolean inverse semigroups?

The present work partly aims at answering the above question in the positive.

1.2.6 A New Link with Universal Algebra: Biases

The environments, of such objects as semigroups (inverse or not) or Boolean algebras, are traditionally quite friendly to universal algebra and model theory. While the axioms defining Boolean inverse semigroups are originally stated in the language of inverse semigroups (a binary operation for the multiplication, a unary operation for the inversion, and a constant for the zero) enriched by a symbol for the binary orthogonal join $\oplus$, which is a *partial* operation, we find in Sect. 3.2 an alternate axiomatization of Boolean inverse semigroups, obtained via the introduction of two new (full) binary operations, that we call the *skew difference* $\oslash$ and the *skew join* $\triangledown$. We thus obtain a *variety of algebras* (in the sense of universal algebra), that is, the class of all structures satisfying a given set of identities. We call those structures *biases*. Hence, Boolean inverse semigroups are definitionally equivalent to biases. This enables us to apply directly tools of universal algebra, notably free algebras, to Boolean inverse semigroups.

An unexpected byproduct of that study is the following: *The class of all Boolean inverse semigroups is congruence-permutable* (cf. Theorem 3.4.11). This means that any two bias congruences of any Boolean inverse semigroup permute. Hence *the lattice of all bias congruences of a given Boolean inverse semigroup is modular.*[6] As should be expected by readers more familiar with universal algebra, congruence-permutability is achieved via a so-called *Mal'cev term*. This term is written out in (3.4.5). It involves the inverse semigroup operations, together with the two new operations $\oslash$ and $\triangledown$.

[6]Recall that a lattice $(L, \vee, \wedge)$ is *modular* if $x \wedge (y \vee z) = (x \wedge y) \vee z$ whenever $x, y, z \in L$ with $z \leq x$.

The congruence-permutability result, for Boolean inverse semigroups (biases), says that those structures are much closer to rings than to semigroups.

1.2.7 Why Should Our Inverse Semigroups Be Boolean?

A few experiments suggest that the introduction of the type monoid that we give in Definition 4.1.3 makes the most sense for those inverse semigroups that are, in addition, Boolean.

Nevertheless, the real reason, why the present work insists on Booleanity, lies in the link with ring theory. Given a ring R, endowed with an involutory anti-automorphism $x \mapsto x^*$ (then we say that R is an *involutary ring*), we are often dealing with multiplicative subsemigroups S of R, which are also inverse semigroups, satisfying that $x^* = x^{-1}$ for any $x \in S$. Then we say that S is an *inverse semigroup in* R (Definition 6.1.4). It turns out that in such a case, S is always contained in a larger *Boolean* inverse subsemigroup $\overline{S}$ in R (cf. Theorem 6.1.7). This is illustrated on Fig. 1.1. Hence, the embedding problem of an inverse semigroup S into an involutary ring R is essentially the same when considering only *Boolean* inverse semigroups.

For a Boolean inverse semigroup S and a unital ring K, the *additive enveloping* K*-algebra* $K\langle S\rangle$ can be defined as the universal K-algebra, containing a copy of S centralizing K, such that finite orthogonal joins in S are turned to finite sums in $K\langle S\rangle$.

We prove in Theorem 6.5.2 that for any unital involutary ring K, any (not necessarily Boolean) inverse semigroup S, and any equation system Σ consisting of (formal) equations of the form $\bigoplus_{i=1}^{m} x_i = \bigoplus_{j=1}^{n} y_j$, the involutary K-algebra $K(S, \Sigma)$ defined by generators S, centralizing K, and satisfying Σ, has the form $K\langle S^{\Sigma}\rangle$ for a suitable *Boolean* inverse semigroup S^{Σ} (independent of K). This applies, in particular, to the *Leavitt path algebras* $\mathsf{L}_K(E)$, or, more generally, to the algebras $\mathsf{L}_K^{\mathrm{ab}}(E, C)$ introduced in Ara and Exel [7]. Analogues of those results also hold for C*-algebras (see, in particular, Theorem 6.4.11).

Unlike the methods chosen in works such as Exel et al. [43], Renault [97], Duncan and Paterson [36], Paterson [93], Steinberg [104, 105], we will handle the additive enveloping algebras $K\langle S\rangle$ without much mention of topology, choosing instead to emphasize the use of the *refinement property*. In particular, we describe $K\langle S\rangle$ as a quotient, defined in terms of finite subrings of the idempotents, of the contracted semigroup algebra $K[S]_0$. As a consequence, two elements of $K[S]_0$

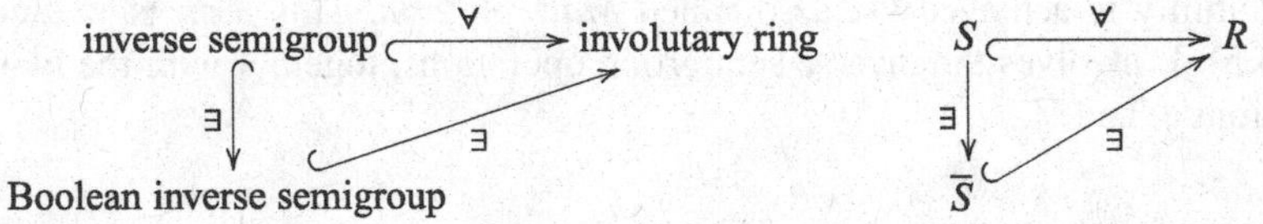

Fig. 1.1 Embedding an inverse semigroup into an involutary ring

represent the same element of $K\langle S\rangle$ iff they are "locally" equal in $K[S]_0$ (cf. Lemma 6.4.6).

1.2.8 Short Descriptions of the Chapters

Chapter 1 is devoted to introducing the present work, as well for motivation as for notation and terminology. We also survey some known basic results, mainly about Boolean rings and refinement monoids, that will be needed later.

Chapter 2 serves partly as a survey, partly as a gentle introduction to some new results, the latter about tensor products of refinement monoids (Sects. 2.5 and 2.6) and the quotient $M/\!\!/G$ of a refinement monoid M by the monoid congruence generated by the action of a group G (Sects. 2.8–2.10). Based on ideas originating in Tarski [109], we give the relevant definition of a partial commutative monoid. For those structures, the relevant morphisms are called *V-homomorphisms* and the relevant relations *V-relations*. Every partial commutative monoid P has an enveloping (full) commutative monoid $\mathrm{U_{mon}}(P)$, which behaves well with respect to V-homomorphisms and V-relations. The behavior of the quotient monoid $M/\!\!/G$ is discussed in detail, with many positive and negative results.

Boolean inverse semigroups come into play in Chap. 3. In Sect. 3.2, we establish the above-mentioned identity between Boolean inverse semigroups and the variety of all biases. In further sections, we prove that homomorphisms, congruences, and ideals of biases are identical to additive semigroup homomorphisms, additive congruences, and additive ideals, respectively. We establish further results for known structures such as generalized rook matrices and Exel's regular representation, and we introduce the crossed product of a Boolean inverse semigroup by a group action.

Type monoids of Boolean inverse semigroups come into play in Chap. 4. We establish that the type monoid functor preserves finite direct products and directed colimits, and that it turns any crossed product to a quotient of the form $M/\!\!/G$ (cf. Sect. 2.8). We also establish that the o-ideals of a type monoid of a Boolean inverse semigroup S are in one-to-one correspondence with the additive ideals of S, and that the type monoid functor behaves well with respect to quotients. Moreover, we prove the identity between equidecomposability types monoids $\mathbb{Z}^+\langle B\rangle/\!\!/G$ and type monoids of Boolean inverse semigroups $\mathrm{Typ}\,S$, and we prove that every countable conical refinement monoid has this form (Theorem 4.8.9).

In Chap. 5 we investigate type monoids of special classes of Boolean inverse semigroups. We show, in Sect. 5.1, how arguments, very similar to those already known for rings and operator algebras, make it possible to represent the positive cone of every dimension group of cardinality at most $\aleph_1$ as the type monoid of a directed colimit of finite products of finite symmetric inverse semigroups (Theorem 5.1.10). In Sect. 5.2, we introduce a different argument, which enables us to represent the positive cone of any Abelian lattice-ordered group (Theorem 5.2.7), in a functorial way (Theorem 5.2.18). In Sect. 5.3, we introduce an inverse semigroup version of supramenability, which we call *fork-nilpotence* and which implies strong separativity of the type monoid. In Sect. 5.4, we investigate (and, mostly, survey) the effect of various completeness assumptions on the type monoid.

It is in Chap. 6 that we start relating Boolean inverse semigroups and involutary (semi)rings. Any Boolean inverse semigroup S is a partial refinement monoid, which, by virtue of the results of Sect. 2.1, gives rise to the enveloping monoid $\mathrm{U}_{\mathrm{mon}}(S)$. This commutative monoid turns out to be cancellative, and it is in fact the positive cone of a dimension group. This dimension group is, in turn, endowed with a natural structure of involutary ring, which we denote by $\mathbb{Z}\langle S\rangle$. Tensoring with any unital ring K, we obtain the above-mentioned additive enveloping K-algebra $K\langle S\rangle$. Leavitt path algebras, and in fact much more general types of algebras, are a particular case of the $K\langle S\rangle$ construction (Theorem 6.5.2). The additive enveloping K-algebra functor $S \mapsto K\langle S\rangle$ does not turn additive semigroup embeddings to ring embeddings, but it does so in a number of significant cases. Further constructions arising from either involutary semirings or involutary rings are also discussed, such as the Boolean unitization construction (Sect. 6.6) and the tensor product (of Boolean inverse semigroups) construction (Sects. 6.8 and 6.9).

More detailed summaries will be given at the beginning of each chapter.

We illustrate this work with a large number of examples and counterexamples, often showing the optimality of our results' assumptions, occasionally solving known open problems.

1.3 Basic Concepts

1.3.1 Sets, Functions, Relations

We denote disjoint unions by writing $\sqcup$ instead of $\cup$. We set $[n] = \{1, 2, \dots, n\}$, for any nonnegative integer n.

We denote by $\operatorname{dom} f$ (resp., $\operatorname{rng} f$) the domain (resp., the range) of a function f. Furthermore, we denote by $f[X]$ the image under f of $X \cap \operatorname{dom} f$, and by $f^{-1}[X]$ the inverse image under f of $X \cap \operatorname{rng} f$, for any set X. We also denote by $f\restriction_X$ the restriction of f to $X \cap \operatorname{dom} f$. The *kernel* of f is

$$\operatorname{Ker} f = \{(x, y) \in (\operatorname{dom} f) \times (\operatorname{dom} f) \mid f(x) = f(y)\} .$$

We denote by $\operatorname{Pow} \Omega$ the powerset of any set Ω. A subset $\mathcal{B}$ of $\operatorname{Pow} \Omega$ is a *ring of subsets* of Ω if $\mathcal{B}$ is closed under finite union and set-theoretical difference.

For a binary relation Γ on a set Ω, the statement $(x, y) \in \Gamma$, for $x, y \in \Omega$, will often be abbreviated $x \mathrel{\Gamma} y$. If, in addition, Γ is an equivalence relation, we will sometimes write this statement $x \equiv_\Gamma y$ or $x \equiv y \pmod{\Gamma}$.

The composition of two binary relations Γ_0 and Γ_1 is denoted by

$$\Gamma_0 \circ \Gamma_1 = \big\{(x, z) \mid (\exists y)\big((x, y) \in \Gamma_0 \text{ and } (y, z) \in \Gamma_1\big)\big\} ;$$

We also set $\Gamma^{-1} = \{(y,x) \mid (x,y) \in \Gamma\}$. For an equivalence relation Γ on a set Ω, we will usually denote by x/Γ the equivalence class of x with respect to Γ.

Following a convention in use notably in Goodearl [49], a binary relation $\lhd$ on a set E and elements $a_1, \dots, a_m, b_1, \dots, b_n$ of E, the conjunction of all relations $a_i \lhd b_j$, for $1 \le i \le m$ and $1 \le j \le n$, will often be written in the form

$$\begin{matrix} a_1 & & b_1 \\ a_2 & & b_2 \\ \vdots & \lhd & \vdots \\ a_m & & b_n \end{matrix}$$

Likewise, for $X, Y \subseteq E$, the notation $X \lhd Y$ means that $x \lhd y$ for all $(x,y) \in X \times Y$. We write $a \lhd X$ (resp., $X \lhd a$) instead of $\{a\} \lhd X$ (resp., $X \lhd \{a\}$).

The *length* of a finite sequence $x = (x_0, \cdots, x_{n-1})$ is the integer $\operatorname{len}(x) = n$. We set $X^{<n} = \{s \in X^{<\omega} \mid \operatorname{len}(s) < n\}$, for any set X and any $n \in \mathbb{Z}^+$. We denote by $p \frown q$ the concatenation of finite sequences p and q. We say that p is a *prefix of* q if $q = p \frown r$ for some finite sequence r.

1.3.2 Partially Ordered Sets (Posets)

For subsets X and Y in a partially preordered set $(P, \le)$, we will write

$$X \downarrow Y = \{x \in X \mid (\exists y \in Y)(x \le y)\},$$
$$X \uparrow Y = \{x \in X \mid (\exists y \in Y)(x \ge y)\},$$

and we will say that X is a *lower subset* of P (resp., an *upper subset* of P) if $X = P \downarrow X$ (resp., $X = P \uparrow X$). We will also write $\downarrow X$ (resp., $\uparrow X$) instead of $P \downarrow X$ (resp., $P \uparrow X$) in case P is understood, and $X \downarrow a$ (resp., $X \uparrow a$) instead of $X \downarrow \{a\}$ (resp., $X \uparrow \{a\}$) for $a \in P$. The least element (resp., largest element) of P will usually be denoted by 0_P (resp., 1_P) if it exists. An *atom* of P is a minimal element of $P \setminus \{0_P\}$. We denote by $\operatorname{At} P$ the set of all atoms of P.

For posets P and Q, a map $f: P \to Q$ is *isotone* (resp., *antitone*) if $x \le y$ implies that $f(x) \le f(y)$ (resp., $f(y) \le f(x)$) for all $x, y \in P$.

A poset P is

- *σ-complete* if every nonempty countable subset of P has a join and a meet,
- *conditionally σ-complete* if every nonempty countable upper bounded subset of P has a join and every nonempty countable lower bounded subset of P has a meet.

1.3.3 Equidecomposability, Supramenability

For an action $\alpha\colon G \times \Omega \to \Omega$ of a group G on a set Ω, we will often denote by $\alpha_g\colon \Omega \to \Omega$, $x \mapsto gx$ the left translation by an element $g \in G$. We say that two subsets X and Y of Ω are *α-equidecomposable*, in symbol $X \simeq_\alpha Y$, if there are finite partitions $X = \bigsqcup_{i=1}^n X_i$ and $Y = \bigsqcup_{i=1}^n Y_i$, together with $g_1, \ldots, g_n \in G$, such that each $Y_i = g_i X_i$. We say that a subset X of Ω is *α-paradoxical* (cf. Wagon [115, Definition 1.1]) if there are disjoint subsets X_0 and X_1 of Ω such that each $X_i \simeq_\alpha X$. Equivalently (cf. Wagon [115, Corollary 3.6]), there is a partition $X = X_0 \sqcup X_1$ such that each $X_i \simeq_\alpha X$.

In case the action of G is understood, we will often say "G-equidecomposable", "G-paradoxical", and write $X \simeq_G Y$ instead of "α-equidecomposable", "α-paradoxical", and $X \simeq_\alpha Y$, respectively.

It will often be the case that the pieces X_i and Y_i will be kept inside a given Boolean ring $\mathcal{B}$ of subsets of Ω, closed under the action of G, in which case we will say that X and Y are *G-equidecomposable with pieces from* $\mathcal{B}$. For a commutative monoid M, a map $\mu\colon \mathcal{B} \to M$ is a *premeasure* if $\mu(\varnothing) = 0$ and $\mu(X \cup Y) = \mu(X) + \mu(Y)$ whenever X and Y are disjoint elements of $\mathcal{B}$. We say that μ is *G-invariant* if $\mu(gX) = \mu(X)$ for all $g \in G$ and all $X \in \mathcal{B}$.

The group G is *supramenable* (cf. Wagon [115, Definition 12.1]) if for every nonempty $A \subseteq G$, there is a G-invariant premeasure $\mu\colon \operatorname{Pow} G \to [0, \infty]$ such that $\mu(A) = 1$. This is equivalent to saying that no nonempty subset A of G is paradoxical with respect to the natural left action of G on itself (cf. Wagon [115, Chap. 12]).

For a subset S in a group G and a positive integer n, we denote by $S^{(n)}$ the set of all products of n elements of S, and we define $\gamma_S(n)$ as the cardinality of $S^{(n)}$. Observing that $\gamma_S(n) \geq 1$ and $\gamma_S(mn) \leq \gamma_S(m)\gamma_S(n)$, it follows from Fekete's Lemma (cf. Fekete [44, p. 233]) that the real number $\lambda(S) = \lim_{n\to\infty} \gamma_S(n)^{1/n}$ exists, and $1 \leq \lambda(S) \leq \operatorname{card} S$. Observing that $\lambda(S) \leq \lambda(S^{(m)}) \leq \lambda(S)^m$ for every positive integer m, it follows that if $\lambda(S) = 1$ for some finite generating subset S of G, then $\lambda(S) = 1$ for every finite generating subset S of G.

The group G is *exponentially bounded* (cf. Rosenblatt [100], see also Wagon [115, Chap. 12]) if $\lambda(S) = 1$ for every nonempty finite subset S of G. Equivalently, for any nonempty finite subset S of G and any real number $b > 1$, $\gamma_S(n) < b^n$ for all large enough n.

Every nilpotent group is exponentially bounded and every exponentially bounded group is supramenable. There are exponentially bounded groups that are not nilpotent, and not even with polynomial growth (cf. Grigorchuk [55]), but it is still unknown whether every supramenable group is exponentially bounded (cf. Wagon [115, Problem 12]). If the free semigroup with two generators embeds into a group G, then G is not supramenable. The converse fails (cf. Ol'shanskiĭ [90]).

1.3.4 Nonstable K-Theory of Rings

All our rings will be associative, but not necessarily commutative or unital.

Two idempotent elements a, b in a ring R are *Murray–von Neumann equivalent*, in symbol $a \sim b$ (or $a \sim_R b$ in case R needs to be specified), if there are $x, y \in R$ such that $a = yx$ and $b = xy$. In that case, x and y may be taken in such a way that $x = xyx$ and $y = yxy$.

We denote by $\mathrm{M}_n(R)$ the ring of all $n \times n$ matrices over R, and we embed $\mathrm{M}_n(R)$ into $\mathrm{M}_{n+1}(R)$ via $x \mapsto \begin{pmatrix} x & 0 \\ 0 & 0 \end{pmatrix}$. Then we denote by $\mathrm{M}_\infty(R)$ the union of all $\mathrm{M}_n(R)$, for n a natural number.

We denote by $\mathrm{V}(R)$ the set of all Murray–von Neumann equivalence classes of idempotent elements of $\mathrm{M}_\infty(R)$ (cf. Goodearl [51, § 4] for the unital case, Ara [5, § 3] for the general case). We denote by $[a]$, or $[a]_R$ in case R needs to be specified, the Murray–von Neumann equivalence class of a matrix a. Those equivalence classes can be added, by setting

$$[a] + [b] = [a + b]\,, \text{ whenever } a \text{ and } b \text{ are idempotent elements of } \mathrm{M}_\infty(R) \text{ with } ab = ba = 0\,.$$

This addition endows $\mathrm{V}(R)$ with a structure of a commutative monoid, which encodes the *nonstable K-theory* of R.

1.4 Distributive Lattices and Boolean Rings

In this section we shall recall a few basic facts about the topological representation of distributive[7] lattices, Boolean algebras, and Boolean rings. The material of this section originates in Birkhoff [23], Stone [106] (notably Theorem 67 of that paper), and Stone [107] (notably Theorem 4 of that paper). It can be found in Grätzer [53, § 2.5].

A *filter* (resp., *ideal*) of a distributive lattice D is a nonempty upper subset (resp., lower subset) of $(D, \leq)$ closed under nonempty finite meets (resp., joins). A filter $\mathfrak{p}$ of D, with $\mathfrak{p} \neq D$ (we say that $\mathfrak{p}$ is a *proper filter*), is *prime* if $x \vee y \in \mathfrak{p}$ implies that either $x \in \mathfrak{p}$ or $y \in \mathfrak{p}$, for all $x, y \in D$.

Denoting by Ω the set of all prime filters of D, we set $\Omega(a) = \{\mathfrak{p} \in \Omega \mid a \in \mathfrak{p}\}$, for each $a \in D$.

Proposition 1.4.1 *Let D be a distributive lattice with zero. Then the assignment $a \mapsto \Omega(a)$ defines a zero-preserving lattice embedding from D into the powerset algebra of Ω.*

---

[7] A lattice $(L, \vee, \wedge)$ is *distributive* if $x \wedge (y \vee z) = (x \wedge y) \vee (x \wedge z)$ for all $x, y, z \in L$.

It is interesting to describe the range of the map $a \mapsto \Omega(a)$ in topological terms. Since $\Omega(a) \cap \Omega(b) = \Omega(a \wedge b)$ whenever $a, b \in D$, the subsets $\Omega(a)$, where $a \in D$, form the basis of a topology on Ω. The set Ω of all prime filters of D, endowed with that topology, is often called the *prime spectrum* of D.

Theorem 1.4.2 *Let D be a distributive lattice with zero. Then the prime spectrum Ω of D is a locally compact topological space. In fact, the subsets $\Omega(a)$, where $a \in D$, are exactly the compact open subsets of Ω.*

A (not necessarily unital) ring B is *Boolean* if $x^2 = x$ for all $x \in B$. It follows that B is commutative and $2x = 0$ for all $x \in B$. A *generalized Boolean algebra* is defined as a distributive lattice with zero, endowed with a binary operation $\smallsetminus$, called the *difference operation*, satisfying the identities

$$0 = \mathsf{y} \wedge (\mathsf{x} \smallsetminus \mathsf{y}) \quad \text{and} \quad \mathsf{x} = (\mathsf{x} \wedge \mathsf{y}) \vee (\mathsf{x} \smallsetminus \mathsf{y}),$$

Generalized Boolean algebras can be identified with Boolean rings (cf. Stone [106, Theorem 4]). The meet, the join, and the difference are then given by the rules

$$\mathsf{x} \wedge \mathsf{y} = \mathsf{x} \cdot \mathsf{y}, \quad \mathsf{x} \vee \mathsf{y} = \mathsf{x} + \mathsf{y} + \mathsf{x} \cdot \mathsf{y}, \quad \mathsf{x} \smallsetminus \mathsf{y} = \mathsf{x} + \mathsf{x} \cdot \mathsf{y}, \tag{1.4.1}$$

and conversely, those operations define in turn the addition and the multiplication, via the formulas

$$\mathsf{x} \cdot \mathsf{y} = \mathsf{x} \wedge \mathsf{y}, \quad \mathsf{x} + \mathsf{y} = (\mathsf{x} \vee \mathsf{y}) \smallsetminus (\mathsf{x} \wedge \mathsf{y}). \tag{1.4.2}$$

While Boolean rings can be defined by the identities defining rings, together with the single idempotent identity $\mathsf{x}^2 = \mathsf{x}$, generalized Boolean algebras can be defined by the following set of identities:

$$\begin{aligned}
\mathsf{x} \vee (\mathsf{y} \vee \mathsf{z}) &= (\mathsf{x} \vee \mathsf{y}) \vee \mathsf{z} \\
\mathsf{x} \vee \mathsf{y} &= \mathsf{y} \vee \mathsf{x} \\
\mathsf{x} \vee \mathsf{x} &= \mathsf{x} \\
\mathsf{x} \vee 0 &= \mathsf{x} \\
\mathsf{x} \wedge (\mathsf{y} \wedge \mathsf{z}) &= (\mathsf{x} \wedge \mathsf{y}) \wedge \mathsf{z} \\
\mathsf{x} \wedge \mathsf{y} &= \mathsf{y} \wedge \mathsf{x} \\
\mathsf{x} \wedge \mathsf{x} &= \mathsf{x} \\
\mathsf{x} \wedge (\mathsf{x} \vee \mathsf{y}) &= \mathsf{x} \\
\mathsf{x} \vee (\mathsf{x} \wedge \mathsf{y}) &= \mathsf{x} \\
\mathsf{x} \wedge (\mathsf{y} \vee \mathsf{z}) &= (\mathsf{x} \wedge \mathsf{y}) \vee (\mathsf{x} \wedge \mathsf{z}) \\
\mathsf{x} &= (\mathsf{x} \smallsetminus \mathsf{y}) \vee (\mathsf{x} \wedge \mathsf{y}) \\
0 &= \mathsf{y} \wedge (\mathsf{x} \smallsetminus \mathsf{y}).
\end{aligned} \tag{1.4.3}$$

The *natural ordering* on a Boolean ring is defined by $\mathsf{x} \leq \mathsf{y} \Leftrightarrow \mathsf{xy} = \mathsf{x}$. A *Boolean algebra* is a generalized Boolean algebra with a largest element. Boolean algebras are, via the correspondence described above, definitionally equivalent to unital Boolean rings.

Remark 1.4.3 We shall mention here a point about terminology. In many references, generalized Boolean algebras are simply called Boolean algebras, which means that Boolean algebras are not assumed to be unital. This occurs, in particular, in the abundant already existing literature about Boolean inverse semigroups, a topic that we will handle from Chap. 3 on. In order to allay that confusion, we shall often use the ring terminology, thus dealing with Boolean rings (generalized Boolean algebras) and unital Boolean rings (Boolean algebras).

An *ultrafilter* of a distributive lattice is a maximal proper filter. The following result relates prime filters of a Boolean ring B with ultrafilters of the principal ideals of B.

Lemma 1.4.4 (Folklore) *Let $\mathfrak{p}$ be a filter of a Boolean ring B and let $a \in \mathfrak{p}$. Then $\mathfrak{p}$ is prime iff $\mathfrak{p} \downarrow a$ is an ultrafilter of $B \downarrow a$.*

The following result specializes Theorem 1.4.2 to Boolean rings.

Theorem 1.4.5 *The prime spectrum of any Boolean ring is a locally compact, Hausdorff, and zero-dimensional[8] topological space.*

1.5 Commutative Monoids, Refinement Monoids

A *partially preordered commutative monoid* is a structure $(M, +, 0, \leq)$, where $(M, +, 0)$ is a commutative monoid and $\leq$ is a preordering on M (i.e., a reflexive, transitive binary relation) which is compatible with the addition (i.e., $x \leq y$ implies that $x + z \leq y + z$, for all $x, y, z \in M$). We say that M is

- *cancellative* if $x + z = y + z$ (resp., $x + z \leq y + z$) implies that $x = y$ (resp., $x \leq y$), for all $x, y, z \in M$.
- *m-power cancellative* if $mx = my$ implies that $x = y$, for all $x, y \in M$.
- *m-unperforated*, where m is a positive integer, if $mx \leq my$ implies that $x \leq y$, for all $x, y \in M$.
- *directed* if the preordered set $(M, \leq)$ is (upward) directed, that is, for all $x, y \in M$ there exists $z \in M$ such that $\begin{matrix} x \\ y \end{matrix} \leq z$.

We also say that M is *power cancellative* (resp., *unperforated*) if it is m-power cancellative (resp., m-unperforated) for every positive integer m.

[8]A topological space is *zero-dimensional* if it has a basis of clopen (i.e., simultaneously closed and open) sets.

The *positive cone* (resp., *strict positive cone*) of M is $M^{+} = \{x \in M \mid 0 \leq x\}$ (resp., $M^{++} = \{x \in M^{+} \mid x \nleq 0\}$). We denote by $\mathbb{Z}$ ($\mathbb{Q}$, $\mathbb{R}$, $\mathbb{C}$, respectively) the set of all integers (rationals, reals, complex numbers, respectively), endowed with the structure appropriate to the context, and we set $\mathbb{N} = \mathbb{Z}^{++}$.

In any partially preordered commutative monoid,

$$a \propto b \text{ if there is } n \in \mathbb{N} \text{ such that } a \leq nb\,; \tag{1.5.1}$$

$$a \asymp b \text{ if } a \propto b \text{ and } b \propto a\,. \tag{1.5.2}$$

An element e in a partially ordered commutative monoid M is an *order-unit* if $0 \leq e$ and $x \propto e$ for every $x \in M$. We say that M is *simple*[9] if $M^{+} \neq \{0\}$ and every element of $M^{+} \setminus \{0\}$ is an order-unit of M.

In any commutative monoid M,

$$x \leq^{+} y \text{ if there exists } z \text{ such that } y = x + z\,; \tag{1.5.3}$$

$$x <^{+} y \text{ if there exists } z \neq 0 \text{ such that } y = x + z\,. \tag{1.5.4}$$

The binary relation $\leq^{+}$ is a preordering on M, compatible with the addition on M. It is usually called the *algebraic preordering* on M. The structure $(M, +, 0, \leq^{+})$ is a partially preordered commutative monoid. The binary relation $<^{+}$ is transitive iff M is *conical*, that is, $x + y = 0$ implies that $x = y = 0$, for all $x, y \in M$. For example, the monoid $\mathrm{V}(R)$ (cf. Sect. 1.3.4) is conical for any ring R.

A nonempty subset I of M is an *o-ideal* of M if $x + y \in I$ iff $\{x, y\} \subseteq I$, for all $x, y \in M$. We denote by $M|e$ the o-ideal of M generated by e: that is, $M|e = \{x \in M \mid x \propto e\}$ (cf. (1.5.1)). Observe that M is simple iff it has exactly two o-ideals, namely $\{0\}$ and M.

For a submonoid N of a commutative monoid M, the binary relation $\approx_N$ defined by

$$x \approx_N y \text{ if } (\exists u, v \in N)(x + u = y + v)$$

is a monoid congruence of M (cf. Wehrung [121, Lemma 2.8]). We write M/N instead of $M/{\approx_N}$.

A *pointed commutative monoid* is a pair (M, u), where M is a commutative monoid and $u \in M$. We say that u is *directly finite* in M, or, alternatively, that (M, u) is directly finite, if $x + u = u$ implies that $x = 0$, for all $x \in M$. We say that M is *stably finite* if every element of M is directly finite.

The monoid M satisfies the *refinement property*, or, equivalently, M is a *refinement monoid*, if for all positive integers m and n and all elements $a_1, \ldots, a_m, b_1, \ldots,$

---

[9] This definition is tailored to accommodate both cases of simple commutative monoids and simple partially ordered Abelian groups.

b_n of M, if $\sum_{i=1}^{m} a_i = \sum_{j=1}^{n} b_n$, then there are elements $c_{i,j} \in M$, for $1 \leq i \leq m$ and $1 \leq j \leq n$, such that

$$a_i = \sum_{j=1}^{n} c_{i,j} \text{ whenever } 1 \leq i \leq m\,, \text{ and } b_j = \sum_{i=1}^{m} c_{i,j} \text{ whenever } 1 \leq j \leq n\,. \tag{1.5.5}$$

It is well known that it is sufficient to verify that property for $m = n = 2$. The relations (1.5.5) are often recorded in the format of a *refinement matrix*,

	b_1	b_2	$\ldots$	b_n
a_1	$c_{1,1}$	$c_{1,2}$	$\ldots$	$c_{1,n}$
$a_2$1	$c_{2,1}$	$c_{2,2}$	$\ldots$	$c_{2,n}$
$\vdots$	$\vdots$	$\vdots$	$\ddots$	$\vdots$
a_m	$c_{m,1}$	$c_{m,2}$	$\ldots$	$c_{m,n}$

or sometimes

	$b_j(1 \leq j \leq n)$
$a_i(1 \leq i \leq m)$	$c_{i,j}$

Every refinement monoid satisfies the *Riesz decomposition property*, that is, whenever $a \leq^{+} b_1 + b_2$, there are $a_1 \leq^{+} b_1$ and $a_2 \leq^{+} b_2$ such that $a = a_1 + a_2$. A classical example of a commutative monoid with Riesz decomposition but without refinement is $M = \{0, 1, \infty\}$, with $1 + 1 = 1 + \infty = \infty$.

The class of all rings R such that $\mathrm{V}(R)$ (cf. Sect. 1.3.4) satisfies refinement includes the so-called *exchange rings* (cf. Ara [5, Proposition 1.5]), thus it also includes the smaller class of all *regular rings*. The result for unital regular rings is stated in Goodearl [50, Theorem 2.8].

Refinement monoids were formally introduced, independently, in Dobbertin [32] and Grillet [56], and probably in other places as well. Their origin can be traced back to Tarski [109].

Definition 1.5.1 For a semigroup S, we denote by $S^{\sqcup 0}$ the semigroup obtained from S by adding a new zero element 0 (i.e., $x \cdot 0 = 0 \cdot x = 0$ for all $x \in S \cup \{0\}$).

A commutative monoid M is *regular* if $2x \leq^{+} x$ for every $x \in M$.

Whenever G is an Abelian group, both commutative monoids G and $G^{\sqcup 0}$ are refinement monoids. Moreover, the commutative monoids of the form $G^{\sqcup 0}$ are exactly the conical simple regular ones.

Example 1.5.2 A $(\vee, 0)$-*semilattice* is a commutative monoid in which every element is idempotent, endowed with its algebraic preordering, which can then be defined by $x \leq^{+} y$ iff $x + y = y$, and which is then a partial ordering. The binary addition $+$ is then the binary join with respect to the partial ordering $\leq^{+}$, so we usually denote it by $\vee$.

A $(\vee, 0)$-semilattice S is *distributive* if the ideal lattice of S is distributive (cf. Grätzer [53, Sect. II.5.1]). Equivalently, for all $a, b, c \in S$ such that $c \leq^{+} a \vee b$, there are $x \leq^+ a$ and $y \leq^+ b$ in S such that $z = x \vee y$. It is well known (and easy to verify) that *a* $(\vee, 0)$*-semilattice is a refinement monoid iff it is distributive*. In particular, *a lattice with zero is a refinement monoid under join iff it is distributive*.

Example 1.5.3 For any partially ordered Abelian group G, the positive cone G^+ of G is a refinement monoid iff G is an *interpolation group*, that is, whenever $a_i \leq b_j$ in G for all $i, j \in \{0, 1\}$, there exists $x \in G$ such that $a_i \leq x \leq b_j$ for all $i, j \in \{0, 1\}$ (cf. Goodearl [49, Proposition 2.1]). Directed interpolation groups that are *unperforated* (i.e., $mx \geq 0$ implies $x \geq 0$, whenever $n \in \mathbb{N}$) are called *dimension groups* (cf. Goodearl [49, Chap. 3]). Every Abelian lattice-ordered group is a dimension group. A partially ordered Abelian group is *simplicial* it is it isomorphic to $\mathbb{Z}^n$, endowed with its componentwise ordering, for some $n \in \mathbb{Z}^+$. A *simplicial monoid* is the positive cone of a simplicial group. It is well known (cf. Grillet [57] or Effros et al. [38]) that the dimension groups are exactly the directed colimits of simplicial groups.

A partially ordered Abelian group G is *discrete* if $G^+ = \{0\}$.

For elements a and b in a partially ordered Abelian group G, we define

$$a \ll b \text{ if } b - a \text{ is an order-unit of } G\,, \tag{1.5.6}$$

$$a \underline{\ll} b \text{ if either } a = b \text{ or } a \ll b\,. \tag{1.5.7}$$

The binary relation $\underline{\ll}$ is a partial ordering on G, and it endows G with a structure of a partially ordered Abelian group. We will denote $G^s = (G, \underline{\ll})$. This partially ordered Abelian group is either discrete or simple. In fact, $(G, \leq)$ is either discrete or simple iff $G = G^s$, that is, the orderings $\leq$ and $\underline{\ll}$ are identical. Hence, the simple partially ordered Abelian groups are exactly the non-discrete ones of the form $(G, \underline{\ll})$.

Lemma 1.5.4 *Let M be a refinement monoid. Then the set of all order-units of M is downward directed.*

Proof Let a and b be order-units of M. Since $a \asymp b$ and M is a refinement monoid, it follows from Wehrung [122, Corollary 3.2] that there are finite sets I and J, together with elements a_i $(i \in I)$ and b_j $(j \in J)$ of M such that $a = \sum_{i \in I} a_i$, $b = \sum_{j \in J} b_j$, and $\{a_i \mid i \in I\} = \{b_j \mid j \in J\}$. Fix a repetition-free enumeration $(c_k \mid k < n)$ of that set, and set $c = \sum_{k<n} c_k$. Then $c \leq^+ \begin{matrix} a \\ b \end{matrix}$. Since every a_i is equal to some c_k, we get $a \propto c$, so c is an order-unit of M. □

Proposition 1.5.5 *The following statements hold, for any interpolation group G:*

(1) *G^s is an interpolation group iff either $G \cong \mathbb{Z}$ or every order-unit of G is the sum of two order-units of G.*
(2) *If G is a dimension group and every order-unit of G is the sum of two order-units of G, then G^s is a dimension group.*

Proof (1) First observe that $\mathbb{Z}^{s} = \mathbb{Z}$ is a dimension group. Now suppose that every order-unit of G is the sum of two order-units of G and let $a_0, a_1, b_0, b_1 \in (G^{s})^+$ such that $a_0 + a_1 = b_0 + b_1$. We must find a refinement for that equation. Suppose first that one of the a_i or b_j is zero. We may assume that $a_0 = 0$. A refinement is then provided by the matrix

	b_0	b_1
$a_0 = 0$	0	0
a_1	b_0	b_1

Suppose now that all a_i, b_j are order-units. By applying Lemma 1.5.4 to the refinement monoid G^+, we get an order-unit e such that $e \leq a_i$ and $e \leq b_i$ for all $i \in \{0, 1\}$. By assumption, $e = e_0 + e_1$ for order-units e_0 and e_1 of G. Since G^+ is a refinement monoid, there is a refinement matrix of the form

	$b_0 - e$	$b_1 - e$
$a_0 - e$	$c_{0,0}$	$c_{0,1}$
$a_1 - e$	$c_{1,0}$	$c_{1,1}$

in G^+ .

Therefore, we get the following refinement matrix, whose entries are all order-units:

	b_0	b_1
a_0	$c_{0,0} + e_0$	$c_{0,1} + e_1$
a_1	$c_{1,0} + e_1$	$c_{1,1} + e_0$

in $(G^{s})^+$.

Hence G^{s} is an interpolation group.

Suppose, conversely, that G^{s} is an interpolation group and $G \ncong \mathbb{Z}$. We must prove that any order-unit e of G is a sum of two order-units. Since G is noncyclic and by Goodearl [49, Lemma 14.5], there are $a, b \in G^{++}$ such that $e = a + b$. Suppose, towards a contradiction, that e is not a sum of two-order-units of G. Since $(G^{s})^+$ has refinement and e, $e+a$, and $e+b$ are all order-units with $(e+a)+(e+b) = e+e+e$, there is a refinement matrix of the form

	e	e	e
$e + a$	a_0	a_1	a_2
$e + b$	b_0	b_1	b_2

in $(G^{s})^+$.

For each $i < 3$, since $e = a_i + b_i$ with $0 \ll \frac{a_i}{b_i}$ and e is not the sum of two order-units, we obtain that either $a_i = 0$ or $b_i = 0$. It follows that there are distinct indices i and j such that either $a_i = a_j = 0$ or $b_i = b_j = 0$. We may thus assume that $a_0 = a_1 = 0$. It follows that $b_0 = b_1 = e$, so $e + b = b_0 + b_1 + b_2 = 2e + b_2$, and so $b = e + b_2 \geq e$. Since $b \leq e$, it follows that $b = e$, thus $a = 0$, a contradiction.

(2) By (1), it suffices to prove that G^s is unperforated. This follows trivially from the unperforation of G. □

1.6 Weak Comparability and Strict Unperforation

Weak comparability is a monoid-theoretical concept, introduced in Ara and Pardo [11]. Although it seems rather weak at first sight, it turns out that under certain conditions, it implies cancellativity.

Definition 1.6.1 For a commutative monoid M, we set

$$\mathrm{comp}(a : b) = \left\{k \in \mathbb{N} \mid (\forall x \in M)(kx \leq^+ b \Rightarrow x \leq^+ a\right\}, \quad \text{for all } a, b \in M .$$

(*Observe that* $\mathrm{comp}(a : b)$ *is either empty or of the form* $\mathbb{N} \uparrow k$ *for some* $k \in \mathbb{N}$.) The *weak comparability set* of M is defined as

$$I = \{e \in M \mid (\forall y \in M \setminus \{0\})(\mathrm{comp}(y : e) \neq \varnothing)\} .$$

An element $e \in M$ has *finite index* if there is $k \in \mathbb{Z}^+$ such that $(k+1)x \leq^+ e \Rightarrow x \leq^+ 0$ for every $x \in M$. Such an integer k is called an *index* of e.

The notation $\mathrm{comp}(a : b)$ is designed in such a way that the assignment $(a, b) \mapsto \mathrm{comp}(a : b)$ is isotone in a and antitone in b.

Lemma 1.6.2 *Let M be a commutative monoid. If M is not conical, then the weak comparability set of M consists exactly of those elements of M with finite index.*

Proof There are nonzero $a, b \in M$ such that $0 = a + b$. Let e be an element in the weak comparability set of M. From $a \neq 0$ it follows that $\mathrm{comp}(a : e)$ is nonempty; pick k in that set. Every $x \in M$ such that $kx \leq^+ e$ satisfies $x \leq^+ a$, thus $x \leq^+ 0$. Hence e has finite index. The converse statement, that every element with finite index belongs to the weak comparability set, is trivial. □

Lemma 1.6.3 *Let M be a refinement monoid. Then the set F of all elements of M with finite index is an o-ideal of M.*

Proof It is trivial that F is a lower subset of M. Now let $a, b \in F$ and let $k - 1$ be a common index of both a and b. Let $x \in M$ such that $2kx \leq^+ a + b$. By Wehrung [119, Lemma 2.3], there is a decomposition $x = u + v$ such that $ku \leq^+ a$ and $kv \leq^+ b$. It follows that $u, v \leq^+ 0$, so $x \leq^+ 0$. Therefore, $2k - 1$ is an index of $a + b$. □

Lemma 1.6.4 *Let M be a refinement monoid. Then the weak comparability set C of M is an o-ideal of M.*

Proof It is trivial that C is a lower subset of M, so it suffices to prove that $a+b \in C$ whenever $a, b \in C$. Let $y \in M \setminus \{0\}$, we must prove that the set $\mathrm{comp}(y : a+b)$ is nonempty. We separate cases.

Case 1. $2t \leq^+ y$ for some $t \in M \setminus \{0\}$. Pick $k \in \mathrm{comp}(t : a) \cap \mathrm{comp}(t : b)$ and let $x \in M$ such that $2kx \leq^+ a+b$. By Wehrung [119, Lemma 2.3], there is a decomposition $x = u + v$ such that $ku \leq^+ a$ and $kv \leq^+ b$. From $k \in \mathrm{comp}(t : a)$ and $k \in \mathrm{comp}(t : b)$ it follows that $u \leq^+ t$ and $v \leq^+ t$. Hence, $x \leq^+ 2t \leq^+ y$. Therefore, $2k \in \mathrm{comp}(y : a+b)$.

Case 2. $2t \leq^+ y$ implies that $t = 0$, for all $t \in M$. Let k be an element of $\mathrm{comp}(y : a) \cap \mathrm{comp}(y : b)$ and let $x \in M$ such that $4kx \leq^+ a+b$. By Wehrung [119, Lemma 2.3], there is a decomposition $x = u + v$ such that $2ku \leq^+ a$ and $2kv \leq^+ b$. From $k \in \mathrm{comp}(y : a)$ and $k \in \mathrm{comp}(y : b)$ it follows that $2u \leq^+ y$ and $2v \leq^+ y$. By assumption, it follows that $u = v = 0$, whence $x = 0$. Therefore, $4k \in \mathrm{comp}(y : a+b)$.

In any case, $\mathrm{comp}(y : a+b)$ is nonempty. □

By virtue of Lemma 1.6.3, the weak comparability set of a refinement monoid will often be called its *weak comparability ideal.*

Accordingly, we set the following definition.

Definition 1.6.5 A commutative monoid M satisfies *weak comparability* if its weak comparability set is M itself.

The definition of weak comparability of a pointed monoid (M, u) introduced in Ara and Pardo [11] is equivalent to saying that u belongs to the weak comparability set of M. Hence, by virtue of Lemma 1.6.4, if M is a simple refinement monoid and $u \in M \setminus \{0\}$, then M satisfies weak comparability (in the sense of Definition 1.6.5) iff (M, u) satisfies weak comparability (in the sense of Ara and Pardo [11]).

Simple commutative monoids are particularly interesting in the conical, stably finite case. The other cases are taken care of by the following easy description.

Proposition 1.6.6 *The following statements hold, for any simple commutative monoid M:*

(1) *If M is not conical, then M is an Abelian group.*
(2) *If M is not stably finite, then M is regular. In particular, 0 is the only directly finite element of M.*

Proof (1) Let $a, b \in M$ such that $0 = a+b$. Since M is simple, every $x \in M$ satisfies $x \propto a$, thus $x \leq^+ 0$, that is, x has an additive inverse.

(2) Let $a, b \in M$ with $a + b = b$ and $a \neq 0$. Since M is simple, there is $m \in \mathbb{N}$ such that $b \leq^+ ma$. Hence $2b \leq^+ ma + b = b$, so there exists $h \in M$ such that $b = 2b + h$. It follows that the element $e = b + h$ is idempotent. Since $a + b = b$ and $b \leq^+ e$, we get $a + e = e$. In particular, $e \neq 0$, thus M is not a group. By (1), it follows that M is conical. Since M is simple, $x \asymp e$ for all $x \in M \setminus \{0\}$, thus, since $e = 2e$, we get $2x \leq^+ x$. □

The following definition is stated, in the language of nonstable K-theory of rings, in Blackadar [24]. Our statement involves the binary relation $<^+$ introduced in (1.5.4).

Definition 1.6.7 Let M be a commutative monoid and let m be a positive integer. We say that M is *strictly m-unperforated* if $mx <^+ my$ implies that $x <^+ y$, for all $x, y \in M$.

Proposition 1.6.8 *Let M be a simple conical refinement monoid and let m be an integer with $m \geq 2$. Then M satisfies weak comparability iff it is strictly m-unperforated.*

Proof If M is not stably finite, then, since M is conical and by Proposition 1.6.6, $M = G^{\sqcup 0}$ for some Abelian group G. It follows easily that M has weak comparability and is strictly unperforated.

Suppose from now on that M is both conical and stably finite. If M satisfies weak comparability, then, by Ara and Pardo [11, Corollary 1.8], it is cancellative, so it is the positive cone of some interpolation group. By Ara et al. [12, Theorem 4.2], it follows that M is strictly unperforated.

Suppose, conversely, that M is strictly m-unperforated. We argue as in the proof of Ara et al. [12, Theorem 4.2]. Given $e, y \in M \setminus \{0\}$, we must prove that $\mathrm{comp}(y : e) \neq \varnothing$. Since M is simple, there is $k \in \mathbb{N}$ such that $e \leq^+ (m^k - 1)y$. Let $x \in M$ such that $m^k x \leq^+ e$. Then $m^k x \leq^+ (m^k - 1)y <^+ m^k y$, thus, since M is conical, $m^k x <^+ m^k y$. Since M is strictly m-unperforated, it follows that $x <^+ y$. Therefore, $m^k \in \mathrm{comp}(y : e)$, so e belongs to the weak comparability ideal of M. □

Chapter 2
Partial Commutative Monoids

Many constructions of commutative monoids start with a set P endowed with a partial addition $\oplus$. The partial structure $(P, \oplus)$ is then extended to a full commutative monoid, which works then as the "enveloping monoid of P". Although this process has been mostly studied in case P satisfies the refinement axiom (this originates in Tarski [109]), the initial part of the work does not require that axiom.

Section 2.1 deals mainly with the extension process of a partial commutative monoid P to its enveloping (full) commutative monoid, denoted by $\mathrm{U}_{\mathrm{mon}}(P)$. This subject is pursued in Sect. 2.2, which deals with the special case of refinement monoids. Section 2.3 states some material about so-called *multiple-free* partial refinement monoids. Section 2.4 establishes some material about the important concept of *V-relation*.

We claim no originality for most results of Sects. 2.1–2.4, which are often known in some form. However, in the few cases where well-defined bibliographical sources could be found, those were not necessarily easily applicable to our context, so we felt that precise formulations were required.

Section 2.5 introduces some material about tensor products of commutative monoids, extending some of the work of Wehrung [121]. Section 2.6 deals with tensor products of V-relations. The material about tensor products of commutative monoids will be applied to Boolean inverse semigroups in Chap. 6.

Section 2.7 gives a few sufficient conditions, for certain cancellativity properties of partial commutative monoids, to be transferrable from a given conical partial refinement monoid P to its enveloping monoid $\mathrm{U}_{\mathrm{mon}}(P)$.

Section 2.8 introduces, for a group G acting by automorphisms on a commutative monoid M, the range $M /\!\!/ G$ of the universal G-invariant measure on M. Section 2.9 initiates the study of the cancellativity properties that can be transferred from M to $M /\!\!/ G$. Section 2.10 illustrates the difficulties of such "cancellativity transfer" results, notably with a class of counterexamples.

F. Wehrung, *Refinement Monoids, Equidecomposability Types, and Boolean Inverse Semigroups*, Lecture Notes in Mathematics 2188,
DOI 10.1007/978-3-319-61599-8_2

2.1 The Enveloping Monoid of a Partial Commutative Monoid

In this section we shall describe how to universally embed a partial commutative monoid (as introduced shortly) $(P, \oplus, 0)$, into a full commutative monoid $\mathrm{U}_{\mathrm{mon}}(P, \oplus, 0)$, in such a way that the partial monoid is a lower subset of the full monoid. None of the results of this section requires any refinement assumption on P.

Definition 2.1.1 A *partial commutative monoid* is a structure $(P, \oplus, 0)$, where P is a set, $0 \in P$, and $\oplus$ is a partial binary operation on P satisfying the following properties, for all $x, y, z \in P$:

(PC1) *Associativity*: $x \oplus (y \oplus z)$ is defined iff $(x \oplus y) \oplus z$ is defined, and then the two values are equal.
(PC2) *Commutativity*: $x \oplus y$ is defined iff $y \oplus x$ is defined, and then the two values are equal.
(PC3) *Zero element*: $x \oplus 0$ is defined with value x.

The *algebraic preordering* on P is defined by

$$x \leq^{\oplus} y \quad \text{if} \quad (\exists z)(y = x \oplus z)\,, \qquad \text{for all } x, y \in P\,. \tag{2.1.1}$$

Recall from Sect. 1.5 that if P is a full monoid (as opposed to a partial monoid), then we emphasize this point by writing $\leq^{+}$ instead of $\leq^{\oplus}$.

Observe that if we assume commutativity, then associativity amounts to verifying that $u = (x \oplus y) \oplus z$ implies $u = x \oplus (y \oplus z)$.

It follows immediately from (PC1) and (PC3) that the binary relation $\leq^{\oplus}$ is indeed a preordering.

Definition 2.1.2 Let M be a partial commutative monoid.

(1) Whenever $X \subseteq M$, we define $X^{\oplus}$ as the set of all elements of M of the form $\bigoplus_{i<n} x_i$, where n is a nonnegative integer and all $x_i \in P$. We say that X is
  - $\oplus$-*closed*, or a *partial submonoid of* M, if $X = X^{\oplus}$,
  - an *o-ideal* of M if it is both a partial submonoid of M and a lower subset of P with respect to the algebraic preordering of P (*this extends the definition given in Sect.* 1.5 *for full monoids*),
  - a *generating subset* of M if $M = X^{\oplus}$.

(2) For partial commutative monoids P and Q, a map $f: P \to Q$ is
  - *conical* if $f^{-1}\{0_Q\} = \{0_P\}$;
  - a *homomorphism* (of partial monoids) if $f(0_P) = 0_Q$ and $x = x_0 \oplus x_1$ implies that $f(x) = f(x_0) \oplus f(x_1)$, for all $x, x_0, x_1 \in P$;
  - a *V-homomorphism* if it is a homomorphism and for all $x \in P$ and all $y_0, y_1 \in Q$, if $f(x) = y_0 \oplus y_1$, then there are $x_0, x_1 \in P$ such that $x = x_0 \oplus x_1$ and each $f(x_i) = y_i$;
  - a *V-embedding* if it is a one-to-one V-homomorphism.

(3) A *lower interval* of M is a nonempty lower subset P of $(M, \leq^{\oplus})$, endowed with the partial addition defined by

$$z = x \oplus_P y \quad \text{if} \quad z = x \oplus_M y\,, \qquad \text{for all } x, y, z \in P\,. \tag{2.1.2}$$

(*It is then straightforward to verify that* $(P, \oplus_P, 0)$ *is also a partial commutative monoid.*) Equivalently, $P \subseteq M$ and the inclusion map from P into M is a V-embedding.

Here and at many other places, the "V" in "V-homomorphism", "V-embedding", and so on, stands as the initial letter of "Vaught", having in mind his thesis [114].

We emphasize that while lower subsets are defined for partially preordered sets, lower intervals are defined for partial commutative monoids.

From now on, until the end of this section, let $(P, \oplus, 0)$ be a partial commutative monoid.

Definition 2.1.3 The *one point completion* of P consists of the set $P^{\sqcup\infty} = P \cup \{\infty\}$ (for a new element ∞), endowed with the binary operation $+$ defined by

$$x + \infty = \infty + x = \infty\,, \qquad \text{for all } x \in P^{\sqcup\infty}\,,$$

$$x + y = \begin{cases} z\,, & \text{if } z = x \oplus y \text{ in } P \\ \infty\,, & \text{otherwise} \end{cases}\,, \qquad \text{for all } x, y \in P\,.$$

The proof of the following result is straightforward.

Proposition 2.1.4 *The one point completion* $P^{\sqcup\infty}$ *is a commutative monoid, for every partial commutative monoid* P*. Furthermore,* P *is a lower interval of* $P^{\sqcup\infty}$.

Proposition 2.1.4 makes it possible to translate problems about partial monoids to problems about full monoids. In particular, dealing with finite sums in P becomes a triviality. Formally, finite sums can be defined as follows.

Definition 2.1.5 For a finite set I, a family $(x_i \mid i \in I)$ of elements in a partial commutative monoid P, and $x \in P$, let $x = \bigoplus_{i \in I} x_i$ hold if $x = \sum_{i \in I} x_i$ within the one point completion $P^{\sqcup\infty}$.

We say that $\bigoplus_{i \in I} x_i$ is *defined* if there exists $x \in P$, called the *value* of the finite sum, such that $x = \bigoplus_{i \in I} x_i$.

Note that by definition, the value of the finite sum $\bigoplus_{i \in I} x_i$, if it exists, is the unique value of the sum $\sum_{i \in I} x_i$ in the full monoid $P^{\sqcup\infty}$. For example, $x = \bigoplus_{i \in \{0,1\}} x_i$ iff $x = x_0 \oplus x_1$. For the sake of readability, we shall often write finite sums as $x \oplus y$, $\bigoplus_{i \in I} x_i$ in partial monoids, and $x + y$, $\sum_{i \in I} x_i$ in full monoids.

As a further immediate application of Proposition 2.1.4, we get the following.

Lemma 2.1.6 *The following statements hold, for every element* x *and every finite family* $(x_i \mid i \in I)$ *of elements in a partial commutative monoid* P.

(1) *If* $\bigoplus_{i\in I} x_i$ *is defined, then* $\bigoplus_{i\in J} x_i$ *is defined for every* $J \subseteq I$*, and the inequality* $\bigoplus_{i\in J} x_i \leq^{\oplus} \bigoplus_{i\in I} x_i$ *holds.*
(2) *Let* $I = \bigsqcup_{j\in J} I_j$*. Then* $x = \bigoplus_{i\in I} x_i$ *iff* $y_j = \bigoplus_{i\in I_j} x_i$ *is defined for every* $j \in J$ *and* $x = \bigoplus_{j\in J} y_j$*.*
(3) *Let* J *be a set and let* $\sigma: J \to I$ *be a bijection. Then* $\bigoplus_{i\in I} x_i$ *is defined iff* $\bigoplus_{j\in J} x_{\sigma(j)}$ *is defined, and then the two values are equal.*

Denote by $\mathrm{F_{mon}}(X)$ the free commutative monoid on the set X, for any set X. It can be realized as the additive monoid of all maps from X to $\mathbb{Z}^+$ with finite support. Identifying every element $x \in X$ with the characteristic function $\dot{x}$ of $\{x\}$, we obtain that the elements of $\mathrm{F_{mon}}(X)$ are the finite sums of elements of X.

We shall realize the enveloping monoid of the partial commutative monoid P as the quotient of $\mathrm{F_{mon}}(P)$ by a certain monoid congruence. We define binary relations $\stackrel{\circ}{=}$, $\to$, and $\sim$ on $\mathrm{F_{mon}}(P)$ as follows. For $u, v \in \mathrm{F_{mon}}(P)$,

$$u \stackrel{\circ}{=} v \text{ if } u\restriction_{P\setminus\{0\}} = v\restriction_{P\setminus\{0\}}\,,$$

$$u \to v \text{ if there are } w \in \mathrm{F_{mon}}(P) \text{ and } x, y, z \in P \text{ such that}$$

$$z = x \oplus y\,,\quad u \stackrel{\circ}{=} w + \dot{z}\,,\quad \text{and}\quad v \stackrel{\circ}{=} w + \dot{x} + \dot{y}\,,$$

$$u \sim v \text{ if either } u \to v \text{ or } v \to u\,.$$

Furthermore, we denote by $\equiv$ the transitive closure of $\sim$. From $u \stackrel{\circ}{=} u+\dot{0} \stackrel{\circ}{=} u+2\cdot\dot{0}$ it follows that $\sim$ is reflexive (take $x = y = 0$ in the definition of $\to$); in fact, $\sim$ contains $\stackrel{\circ}{=}$. Since $\sim$ is trivially symmetric, it follows that $\equiv$ is an equivalence relation on $\mathrm{F_{mon}}(P)$. Moreover, $\to$ is compatible with the addition on $\mathrm{F_{mon}}(P)$ (i.e., $u \to v$ implies $u + w \to v + w$), hence so are $\sim$ and $\equiv$. In particular, $\equiv$ is a monoid congruence of $\mathrm{F_{mon}}(P)$. We denote by $[u]$ the $\equiv$-equivalence class of an element $u \in \mathrm{F_{mon}}(P)$ and we set $\varepsilon_P(x) = [\dot{x}]$, for all $x \in P$. The quotient monoid $\mathrm{U_{mon}}(P) = \mathrm{F_{mon}}(P)/{\equiv}$ is a commutative monoid, and ε_P is a homomorphism of partial monoids from P to $\mathrm{U_{mon}}(P)$ (indeed, $z = x \oplus y$ implies that $\dot{z} \to \dot{x} + \dot{y}$, thus $\varepsilon_P(z) = \varepsilon_P(x) + \varepsilon_P(y)$). Since every element of $\mathrm{F_{mon}}(P)$ is a sum of elements of the form $\dot{x}$, the range of ε_P generates $\mathrm{U_{mon}}(P)$ as a monoid.

Proposition 2.1.7 *The monoid* $\mathrm{U_{mon}}(P)$*, endowed with the homomorphism* $\varepsilon_P: P \to \mathrm{U_{mon}}(P)$ *of partial monoids, is the free commutative monoid on the partial commutative monoid* P*.*

Proof We must prove that for every commutative monoid N and every homomorphism $f: P \to N$ of partial monoids, there exists a unique monoid homomorphism $\overline{f}: \mathrm{U_{mon}}(P) \to N$ such that $f = \overline{f} \circ \varepsilon_P$. The uniqueness follows from the fact that the range of ε_P generates $\mathrm{U_{mon}}(P)$. For the existence, let $\varphi: \mathrm{F_{mon}}(P) \to N$ the unique monoid homomorphism such that $\varphi(\dot{x}) = f(x)$ for each $x \in P$. It is straightforward to verify that $u \equiv v$ implies that $\varphi(u) = \varphi(v)$, for all $u, v \in \mathrm{F_{mon}}(P)$. Define $\overline{f}([u]) = \varphi(u)$. □

By virtue of Proposition 2.1.7, we shall call $\mathrm{U_{mon}}(P)$ the *enveloping monoid* of P.

The description of the enveloping monoid $\mathrm{U_{mon}}(P)$ via $\stackrel{\circ}{=}$, $\rightarrow$, and $\sim$, given by Proposition 2.1.7, will be applied in Example 2.7.15.

The following result shows that the map ε_P identifies P with a lower interval of the enveloping monoid $\mathrm{U_{mon}}(P)$. We say that P is *conical* if $0 = x \oplus y$ implies $x = 0$, for all $x, y \in P$.

Proposition 2.1.8 *The homomorphism ε_P is a V-embedding from P into $\mathrm{U_{mon}}(P)$. Furthermore, if P is conical, then so is $\mathrm{U_{mon}}(P)$.*

Proof By Proposition 2.1.7, there is a unique monoid homomorphism $\psi\colon \mathrm{U_{mon}}(P) \to P^{\sqcup\infty}$ such that $\psi \circ \varepsilon_P$ is the inclusion map from P into $P^{\sqcup\infty}$. In particular, ε_P is one-to-one.

Claim $\varepsilon_P[P] = \psi^{-1}[P]$.

Proof of Claim. For each $x \in P$, $\psi(\varepsilon_P(x)) = x \in P$. Conversely, let $\boldsymbol{x} \in \psi^{-1}[P]$. Write $\boldsymbol{x} = \sum_{i<n} \varepsilon_P(x_i)$, where $n \in \mathbb{Z}^+$ and each $x_i \in P$, and set $x = \psi(\boldsymbol{x})$. Since $x = \sum_{i<n} x_i$ (within $P^{\sqcup\infty}$) belongs to P, we get $x = \bigoplus_{i<n} x_i$ (within P). Hence, $\boldsymbol{x} = \varepsilon_P(\bigoplus_{i<n} x_i) = \varepsilon_P(x)$ belongs to $\varepsilon_P[P]$. □ Claim.

Now let $z \in P$ and let $\boldsymbol{x}, \boldsymbol{y} \in \mathrm{U_{mon}}(P)$ such that $\varepsilon_P(z) = \boldsymbol{x} + \boldsymbol{y}$. Setting $x = \psi(\boldsymbol{x})$ and $y = \psi(\boldsymbol{y})$, it follows that $z = x + y$ (within $P^{\sqcup\infty}$), thus, since P is a lower subset of $P^{\sqcup\infty}$, x and y both belong to P. By the Claim above, $\boldsymbol{x} = \varepsilon_P(x')$ and $\boldsymbol{y} = \varepsilon_P(y')$ for some $x', y' \in P$. Moreover, $x' = (\psi \circ \varepsilon_P)(x') = \psi(\boldsymbol{x}) = x$, and, similarly, $y' = y$, thus $\boldsymbol{x} = \varepsilon_P(x)$ and $\boldsymbol{y} = \varepsilon_P(y)$. This completes the proof that ε_P is a V-embedding.

By the above, any $\boldsymbol{x}, \boldsymbol{y} \in \mathrm{U_{mon}}(P)$ such that $\boldsymbol{x} + \boldsymbol{y} \doteq 0$ have the form $\boldsymbol{x} = \varepsilon_P(x)$ and $\boldsymbol{y} = \varepsilon_P(y)$, where $x, y \in P$ with $x + y = 0$. Hence, if P is conical, then so is $\mathrm{U_{mon}}(P)$. □

2.2 Partial Refinement Monoids

The construction of the enveloping monoid $\mathrm{U_{mon}}(P)$ assumes a special significance in case P satisfies the extension, introduced in Definition 2.2.1, of the refinement axiom originally defined for full commutative monoids in Sect. 1.5. In particular, this leads to another perspective on the enveloping monoid $\mathbb{Z}^+\langle B\rangle$ of a generalized Boolean algebra B (cf. Example 2.2.7).

Definition 2.2.1 A partial commutative monoid $(P, \oplus, 0)$ is a *partial refinement monoid* if it satisfies the *refinement property*, that is, for all $a_0, a_1, b_0, b_1 \in P$ with $a_0 \oplus a_1 = b_0 \oplus b_1$, there are elements $c_{i,j} \in P$, for $i, j \in \{0, 1\}$, such that $a_i = c_{i,0} \oplus c_{i,1}$ and $b_i = c_{0,i} \oplus c_{1,i}$ for every $i \in \{0, 1\}$. If P is a full monoid, then we say that P is a *refinement monoid*.

It is an easy exercise to prove, by induction, that the statement of the definition of a partial refinement monoid extends to finite sums of elements.

Proposition 2.2.2 *Let $(P, \oplus, 0)$ be a partial refinement monoid, let m, n be positive integers (resp., if P is conical, nonnegative integers), and let $a_0, \ldots, a_{m-1}, b_0, \ldots, b_{n-1}$ be elements of P such that $\bigoplus_{i<m} a_i = \bigoplus_{j<n} b_j$. Then there are elements $c_{i,j} \in P$, for $i < m$ and $j < n$, such that $a_i = \bigoplus_{j<n} c_{i,j}$ for all $i < m$ and $b_j = \bigoplus_{i<m} c_{i,j}$ for all $j < n$.*

We keep for partial refinement monoids the refinement matrix notation introduced in Sect. 1.5.

Although Proposition 2.1.4 is generally helpful in proofs of statements like Proposition 2.2.2, this help cannot be pushed too far. For example, for the partial commutative monoid $P = \{0, 1\}$, with $1 \oplus 1$ undefined, the one point completion $P^{\sqcup\infty} = \{0, 1, \infty\}$ does not satisfy refinement (e.g., there is no refinement for the equation $1 + 1 = 1 + \infty$). As we shall see shortly, this problem does not occur with the enveloping monoid construction $\mathrm{U}_{\mathrm{mon}}(P)$.

For a partial commutative monoid P, it follows from Proposition 2.1.8 that the canonical homomorphism $\varepsilon_P \colon P \to \mathrm{U}_{\mathrm{mon}}(P)$ identifies P with a lower interval of $\mathrm{U}_{\mathrm{mon}}(P)$. Since $\mathrm{U}_{\mathrm{mon}}(P)$ is generated by the range of ε_P, it follows that the elements of $\mathrm{U}_{\mathrm{mon}}(P)$ are exactly the finite sums of elements of P. Furthermore, it follows from Proposition 2.1.7 that for finite sequences $(a_i \mid i < m)$ and $(b_j \mid j < n)$ of elements of P,

$$\sum_{i<m} a_i = \sum_{j<n} b_j \text{ in } \mathrm{U}_{\mathrm{mon}}(P) \iff \sum_{i<m} \dot{a}_i \equiv \sum_{j<n} \dot{b}_j \text{ in } \mathrm{F}_{\mathrm{mon}}(P),$$

where $\equiv$ is the monoid congruence of $\mathrm{F}_{\mathrm{mon}}(P)$ introduced in Sect. 2.1. Our next result gives a convenient description of $\equiv$ in case P is a partial refinement monoid. The techniques underlying the proof of that result originate in Tarski [109], and they are nowadays well understood. They are pursued, in particular, in Wehrung [122, Chap. 4].

Theorem 2.2.3 *Let $(P, \oplus, 0)$ be a partial refinement monoid and let $(a_i \mid i < m)$ and $(b_j \mid j < n)$ be finite sequences of elements of P, with $m, n > 0$ (or just $m, n \geq 0$ in case P is conical). Then $\sum_{i<m} a_i = \sum_{j<n} b_j$ in $\mathrm{U}_{\mathrm{mon}}(P)$ iff there are elements $c_{i,j} \in P$, for $i < m$ and $j < n$, such that $a_i = \bigoplus_{j<n} c_{i,j}$ for each $i < m$ and $b_j = \bigoplus_{i<m} c_{i,j}$ for each $j < n$. Furthermore, $\mathrm{U}_{\mathrm{mon}}(P)$ is a refinement monoid.*

Proof Denote by S the set of all nonempty finite sequences of elements of P. For $\vec{a} = (a_i \mid i < p)$ and $\vec{b} = (b_j \mid j < q)$ in S, say that $\vec{a} \approx \vec{b}$ if there are $x_{i,j} \in P$, for $i < p$ and $j < q$, such that $a_i = \bigoplus_{j<q} x_{i,j}$ for all $i < p$ and $b_j = \bigoplus_{i<p} x_{i,j}$ for all $j < q$. Now let $\vec{c} = (c_k \mid k < r)$ in S and suppose that $\vec{a} \approx \vec{b}$, via elements $x_{i,j} \in P$, and $\vec{b} \approx \vec{c}$, via elements $y_{j,k} \in P$. For each $j < q$, since $b_j = \bigoplus_{i<p} x_{i,j} = \bigoplus_{k<r} y_{j,k}$ and by Proposition 2.2.2, there are $v_{i,j,k} \in P$, for $i < p$ and $k < r$, such that $x_{i,j} = \bigoplus_{k<r} v_{i,j,k}$ for each $i < p$ and $y_{j,k} = \bigoplus_{i<p} v_{i,j,k}$ for each $k < r$. Therefore, we obtain that $a_i = \bigoplus_{k<r} z_{i,k}$ for each $i < p$ and $c_k = \bigoplus_{i<p} z_{i,k}$ for each $k < r$, where $z_{i,k} = \bigoplus_{j<q} v_{i,j,k}$, and so $\vec{a} \approx \vec{c}$. It is obvious that $\approx$ is reflexive and symmetric, thus it is an equivalence relation on S. Further, it is obvious that $\approx$ is compatible with

concatenation of finite sequences, and that $\vec{a} \approx \vec{b}$ whenever $\vec{a}$ and $\vec{b}$ are obtained from one another by a permutation. Therefore, $\approx$ is a semigroup congruence of S, with commutative quotient M. Actually, $\vec{a} \approx \vec{b}$ whenever $\vec{b}$ is obtained from $\vec{a}$ by concatenating the one-element sequence (0), hence the $\approx$-equivalence class of (0) is the neutral element of M, which is thus a commutative monoid.

The map $f: P \to M$ that sends every $x \in P$ to the $\approx$-equivalence class of (x) is a homomorphism of partial monoids, thus, by Proposition 2.1.7, f extends to a unique monoid homomorphism $g: \mathrm{U_{mon}}(P) \to M$. Now let $\vec{a} = (a_i \mid i < m)$ and $\vec{b} = (b_j \mid j < n)$ in S. It is trivial that $\vec{a} \approx \vec{b}$ implies that $\sum_{i<m} a_i = \sum_{j<n} b_j$ in $\mathrm{U_{mon}}(P)$. Conversely, if $\sum_{i<m} a_i = \sum_{j<n} b_j$ in $\mathrm{U_{mon}}(P)$, then, applying the homomorphism g, we obtain that $\sum_{i<m} f(a_i) = \sum_{j<n} f(b_j)$ in M, that is, by definition of the equality in M, $\vec{a} \approx \vec{b}$. This establishes the characterization of the equality between elements of $\mathrm{U_{mon}}(P)$.

Now let $a_{(0)}, a_{(1)}, b_{(0)}, b_{(1)} \in \mathrm{U_{mon}}(P)$ such that $a_{(0)} + a_{(1)} = b_{(0)} + b_{(1)}$. There are nonempty finite sets I_0, I_1, J_0, J_1 such that $I_0 \cap I_1 = J_0 \cap J_1 = \varnothing$, together with elements $a_i \in P$, for $i \in I_0 \cup I_1$, and $b_j \in P$, for $j \in J_0 \cup J_1$, such that $a_{(p)} = \sum_{i \in I_p} a_i$ for each $p \in \{0, 1\}$ and $b_{(q)} = \sum_{j \in J_q} b_j$ for each $q \in \{0, 1\}$. The equality $a_{(0)} + a_{(1)} = b_{(0)} + b_{(1)}$ means that $\sum_{i \in I_0 \cup I_1} a_i = \sum_{j \in J_0 \cup J_1} b_j$, hence, by the paragraph above, there are elements $c_{i,j} \in P$, for $i \in I_0 \cup I_1$ and $j \in J_0 \cup J_1$, such that $a_i = \bigoplus_j c_{i,j}$ for each i and $b_j = \bigoplus_i c_{i,j}$ for each j. Setting $c_{(p,q)} = \sum_{(i,j) \in I_p \times J_q} c_{i,j}$, we obtain that $a_{(i)} = c_{(i,0)} + c_{(i,1)}$ and $b_{(j)} = c_{(0,j)} + c_{(1,j)}$, thus completing the proof that $\mathrm{U_{mon}}(P)$ is a refinement monoid. $\square$

Easy examples show that a partial commutative monoid P may generate more than one full commutative monoid. However, if we require the full monoid be a refinement monoid (which requires P be a partial refinement monoid), we are led to the following uniqueness result.

Proposition 2.2.4 *Let P be a partial refinement monoid and let M be a refinement monoid. Let $f: P \to M$ be a homomorphism of partial monoids, and denote by $\tilde{f}: \mathrm{U_{mon}}(P) \to M$ the unique extension of f to a monoid homomorphism. If f is a V-homomorphism (resp., a V-embedding), then so is $\tilde{f}$. Furthermore, if f is a V-embedding and the range of f generates M, then $\tilde{f}$ is an isomorphism.*

Proof Let $\tilde{f}(c) = a + b$, where $c \in \mathrm{U_{mon}}(P)$ and $a, b \in M$. Write $c = \sum_{i<n} c_i$, where each $c_i \in P$. By applying refinement to the equation $\sum_{i<n} f(c_i) = a + b$, we obtain a refinement matrix

	a	b
$f(c_i)$	a_i	b_i

within M.

Let $i < n$. Since f is a V-homomorphism and $f(c_i) = a_i + b_i$, there is a decomposition $c_i = x_i \oplus y_i$ such that $f(x_i) = a_i$ and $f(y_i) = b_i$. Setting $x = \sum_{i<n} x_i$ and $y = \sum_{i<n} y_i$, it follows that $f(x) = a$, $f(y) = b$, and $c = x + y$. Therefore, $\tilde{f}$ is a V-homomorphism.

Suppose from now on that f is a V-embedding, and let $a, b \in \mathrm{U_{mon}}(P)$ such that $f(a) = f(b)$. Write $a = \sum_{i<m} a_i$ and $b = \sum_{j<n} b_j$, where $m, n \in \mathbb{N}$ and all a_i, b_j belong to P. Since $\sum_{i<m} f(a_i) = \sum_{j<n} f(b_j)$ and M is a refinement monoid, there are elements $c_{i,j} \in M$, for $i < m$ and $j < n$, such that each $f(a_i) = \sum_{j<n} c_{i,j}$ and each $f(b_j) = \sum_{i<m} c_{i,j}$. Since f is a V-homomorphism, there are decompositions $a_i = \bigoplus_{j<n} a_{i,j}$ and $b = \bigoplus_{i<m} b_{i,j}$ in P such that each $c_{i,j} = f(a_{i,j}) = f(b_{i,j})$. Since f is one-to-one, $a_{i,j} = b_{i,j}$ for all i, j, so $a = \sum_{i,j} a_{i,j} = b$, thus completing the proof that $\tilde{f}$ is one-to-one.

The final statement of Proposition 2.2.4 follows trivially. □

Part of Proposition 2.2.4 can be paraphrased by stating that *for every partial refinement monoid P, the full monoid* $\mathrm{U_{mon}}(P)$ *is the unique refinement monoid in which P is a generating lower interval.*

Because of the universal property defining $\mathrm{U_{mon}}(P)$, every homomorphism $f: P \to Q$ of partial commutative monoids extends, with respect to the canonical embeddings $\varepsilon_P: P \hookrightarrow \mathrm{U_{mon}}(P)$ and $\varepsilon_Q: Q \hookrightarrow \mathrm{U_{mon}}(Q)$, to a unique monoid homomorphism $\mathrm{U_{mon}}(f): \mathrm{U_{mon}}(P) \to \mathrm{U_{mon}}(Q)$ (i.e., $\mathrm{U_{mon}}(f) \circ \varepsilon_P = \varepsilon_Q \circ f$). If P and Q both satisfy refinement and f is a V-homomorphism, more can be said.

Proposition 2.2.5 *Let P and Q be partial refinement monoids and let* $f: P \to Q$ *be a V-homomorphism (resp., V-embedding). Then* $\mathrm{U_{mon}}(f)$ *is a V-homomorphism (resp., V-embedding) from* $\mathrm{U_{mon}}(P)$ *into* $\mathrm{U_{mon}}(Q)$*. Furthermore, if f is a V-embedding and the range of f generates Q, then* $\mathrm{U_{mon}}(f)$ *is an isomorphism.*

Proof Since f and ε_Q are both V-homomorphisms, so is the map $\varepsilon_Q \circ f = \mathrm{U_{mon}}(f) \circ \varepsilon_P$. Since $\mathrm{U_{mon}}(Q)$ is a refinement monoid (cf. Theorem 2.2.3) and by Proposition 2.2.4, $\mathrm{U_{mon}}(f) \circ \varepsilon_P$ extends to a unique V-homomorphism from $\mathrm{U_{mon}}(P)$ into $\mathrm{U_{mon}}(Q)$, which is necessarily equal to $\mathrm{U_{mon}}(f)$. If f is a V-embedding, then so is $\varepsilon_Q \circ f$, thus, again by Proposition 2.2.4, $\mathrm{U_{mon}}(f)$ is a V-embedding.

Suppose, finally, that the range of f generates Q. Since $\varepsilon_Q[Q]$ generates $\mathrm{U_{mon}}(Q)$, the range of $\varepsilon_Q \circ f$ generates $\mathrm{U_{mon}}(Q)$. It follows that $\mathrm{U_{mon}}(f)$ it surjective. In particular, if f is a V-embedding, then, by the above, so is $\mathrm{U_{mon}}(f)$, and thus $\mathrm{U_{mon}}(f)$ is an isomorphism. □

The following easy example shows that "V-embedding" cannot be weakened to "embedding" in the statement of Proposition 2.2.4.

Example 2.2.6 The commutative monoid $M = (\mathbb{Z}^+)^4$ is a conical refinement monoid. Setting $a_0 = (1,1,0,0)$, $a_1 = (0,0,1,1)$, $b_0 = (1,0,1,0)$, and $b_1 = (0,1,0,1)$, the subset $P = \{0, a_0, a_1, b_0, b_1\}$ is a lower subset of M. The corresponding lower interval is a conical partial refinement monoid, with no non-trivial sums (i.e., $x \oplus y$ is defined iff $0 \in \{x, y\}$). By the universal property of $\mathrm{U_{mon}}(P)$ (cf. Proposition 2.1.7), the inclusion map from P into M extends to a unique homomorphism $f: \mathrm{U_{mon}}(P) \to M$ of commutative monoids. Since P has no non-trivial sums, $a_0 +_{\mathrm{U_{mon}}(P)} a_1 \neq b_0 +_{\mathrm{U_{mon}}(P)} b_1$ (e.g., use Theorem 2.2.3). Nevertheless, $a_0 +_M a_1 = b_0 +_M b_1$, that is, $f(a_0 +_{\mathrm{U_{mon}}(P)} a_1) = f(b_0 +_{\mathrm{U_{mon}}(P)} b_1)$. Therefore, the map f is not one-to-one.

For an analogue of Example 2.2.6 for the so-called Boolean inverse semigroups (cf. Definition 3.1.6), see Example 6.4.1.

The following class of examples, based on Boolean rings, will be crucial.

Example 2.2.7 Let B be a Boolean ring (cf. Sect. 1.4). Instead of endowing B with its join operation $\vee$ (cf. Example 1.5.2), let us endow it with its *disjoint sum* operation, given by

$$z = x \oplus y \quad \text{if} \quad z = x \vee y \text{ and } x \wedge y = 0\,, \qquad \text{for all } x, y, z \in B\,.$$

If $\bigoplus_{i<m} a_i = \bigoplus_{j<n} b_j$ in B, then we get a refinement $\left(c_{i,j} \mid i < m \text{ and } j < n\right)$ by setting $c_{i,j} = a_i \wedge b_j$. Hence, $(B, \oplus, 0)$ *is a conical partial refinement monoid.*

By Theorem 2.2.1, the enveloping monoid $\mathbb{Z}^+\langle B\rangle = \mathrm{U}_{\mathrm{mon}}(B, \oplus, 0)$ is a conical refinement monoid. We shall now outline a convenient description of that monoid, given in Wehrung [122, Sect. 2.3]. Denoting by Ω the Stone space of B (cf. Sect. 1.4), B can be identified with the algebra of all compact open subsets of Ω. Then $\mathbb{Z}^+\langle B\rangle$ can be represented as the additive monoid of all maps $x\colon \Omega \to \mathbb{Z}^+$ with finite range such that $x^{-1}\{n\} \in B$ for every $n > 0$. The canonical map from B into $\mathbb{Z}^+\langle B\rangle$ assigns to every element of B its characteristic function.

While the description of $\mathbb{Z}^+\langle B\rangle$ stated above depends on the given representation of B as an algebra of subsets, it is easy to make this representation intrinsic. Indeed, $\mathbb{Z}^+\langle B\rangle$ can be defined as the commutative monoid freely generated by generators $\mathbf{1}_a$ (thought of as the "characteristic function" of a), for $a \in B$, subjected to the relations $\mathbf{1}_0 = 0$ and $\mathbf{1}_{a\vee b} + \mathbf{1}_{a\wedge b} = \mathbf{1}_a + \mathbf{1}_b$ for $a, b \in B$. It is the positive cone of a lattice-ordered group (cf. Example 1.5.3), which we shall naturally denote by $\mathbb{Z}\langle B\rangle$. If B is represented as an algebra of subsets of a set Ω, then $\mathbb{Z}\langle B\rangle$ is the additive group of all maps $x\colon \Omega \to \mathbb{Z}$ with finite range such that $x^{-1}\{n\} \in B$ for every $n \neq 0$ (we say that x is *B-measurable*), ordered componentwise.

Example 2.2.8 An *MV-algebra* (cf. Chang [27], Cignoli et al. [29]) can be defined as a cancellative conical partial refinement monoid with largest element e, lattice-ordered under $\leq^{\oplus}$, in which $x \oplus 1$ is defined iff $x = 0$, for every element x. Mundici proves, in Theorems 2.5 and 3.8 of [86], that MV-algebras are exactly the lower intervals, of positive cones of Abelian lattice-ordered groups, with a largest element.

Hence every MV-algebra A is isomorphic to the interval $[0, u]$, for some element u in the positive cone of an Abelian lattice-ordered group G. It follows from Proposition 2.2.4 that if we add the condition that u be an order-unit, then G is unique: namely, G is the universal group of the commutative monoid $\mathrm{U}_{\mathrm{mon}}(A)$.

2.3 Disjunctive Addition in a Partial Conical Refinement Monoid

For technical reasons, we will need to put some emphasis on partial commutative monoids in which all finite sums are meet-orthogonal, in the sense given by the following definition.

Definition 2.3.1 Two elements a and b in a partial conical commutative monoid P are *meet-orthogonal*, in notation $a \wedge b = 0$, if 0 is the only lower bound of $\{a, b\}$ with respect to $\leq^{\oplus}$. Observe that $(P, \leq^{\oplus})$ need not be a meet-semilattice. For $a, b, c \in P$, let $c = a \boxplus b$ hold if $c = a \oplus b$ and $a \wedge b = 0$. After Tarski [109, Definition 8.16], we shall call $\boxplus$ the *disjunctive addition of* P.

Proposition 2.3.2 *For all elements a, b, c in a conical partial refinement monoid P, if $a \wedge c = b \wedge c = 0$ and $a \oplus b$ is defined, then $(a \oplus b) \wedge c = 0$. Furthermore, $(P, \boxplus, 0)$ is a conical partial refinement monoid.*

Proof Suppose that $a \wedge c = b \wedge c = 0$ and that $a \oplus b$ is defined, and let $x \leq^{\oplus} \genfrac{}{}{0pt}{}{a \oplus b}{c}$ in P. By Riesz decomposition, there are $u \leq^{\oplus} a$ and $v \leq^{\oplus} b$ such that $x = u \oplus v$. From $u \leq^{\oplus} \genfrac{}{}{0pt}{}{a}{c}$ and $a \wedge c = 0$ it follows that $u = 0$. Likewise, $v = 0$, so $x = 0$, thus proving that $(a \oplus b) \wedge c = 0$.

Now let $a, b, c \in P$ such that the element $e = (a \boxplus b) \boxplus c$ is defined in P. In particular, $e = (a \oplus b) \oplus c$, thus $e = a \oplus (b \oplus c)$. From $(a \oplus b) \wedge c = 0$ it follows that $b \wedge c = 0$, thus $b \oplus c = b \boxplus c$. Furthermore, $a \wedge b = a \wedge c = 0$, thus, since $b \oplus c = b \boxplus c$ and by the paragraph above, $a \wedge (b \boxplus c) = 0$, so $e = a \boxplus (b \boxplus c)$. Therefore, $(P, \boxplus, 0)$ is a partial commutative monoid. It is trivially conical. Let $a_0 \boxplus a_1 = b_0 \boxplus b_1$ in P. Since $a_0 \oplus a_1 = b_0 \oplus b_1$ and since P is a partial refinement monoid, there is a refinement matrix of the form

	b_0	b_1
a_0	$c_{0,0}$	$c_{0,1}$
a_1	$c_{1,0}$	$c_{1,1}$

within $(P, \oplus)$.

Since $b_0 \wedge b_1 = 0$, it follows from $c_{0,0} \leq^{\oplus} b_0$ and $c_{0,1} \leq^{\oplus} b_1$ that $c_{0,0} \wedge c_{0,1} = 0$; whence $a_0 = c_{0,0} \boxplus c_{0,1}$. Likewise, $a_1 = c_{1,0} \boxplus c_{1,1}$, $b_0 = c_{0,0} \boxplus c_{1,0}$, and $b_1 = c_{1,0} \boxplus c_{1,1}$, thus completing the proof that $(P, \boxplus, 0)$ is a partial refinement monoid. □

Definition 2.3.3 We shall call $(P, \boxplus, 0)$ the *multiple-free part* of $(P, \oplus, 0)$. Furthermore, we shall say that a partial conical commutative monoid P is *multiple-free*[1] if $c = a \oplus b$ implies that $a \wedge b = 0$, for all $a, b, c \in P$. This means that the partial operations $\oplus$ and $\boxplus$ are identical.

The definition of the index of an element in a commutative monoid (cf. Definition 1.6.1) also applies to partial commutative monoids:

Definition 2.3.4 Let P be a partial commutative monoid. The *index* of an element $a \in P$ is defined as the largest $n \in \mathbb{Z}^{+}$ for which there is $x \in P$ such that $nx \leq^{\oplus} a$ and

[1]The terminology "multiple-free" is borrowed from Tarski [109].

$x \not\leq^{\oplus} 0$, if it exists; and ∞, otherwise. We say that a is *cancellable* if $a \oplus x = a \oplus y$ implies that $x = y$, for all $x, y \in P$.

The proof of the following lemma is a straightforward exercise.

Lemma 2.3.5 *A partial conical commutative monoid is multiple-free iff every element of P has index at most* 1 *in P.*

The main connection between index and cancellativity is given by the following lemma.

Lemma 2.3.6 *Let P be a conical partial refinement monoid. Then every element of P with finite index is cancellable.*

Proof Let $a \in P$ with finite index n, and let $x, y \in P$ such that $a \oplus x = a \oplus y$. Since the proof of Wehrung [118, Lemma 1.11] obviously extends from full refinement monoids to partial refinement monoids, there are $u, v, z \in P$ such that $x = u \oplus z$, $y = v \oplus z$, and $\begin{matrix}(n+1)u \\ (n+1)v\end{matrix} \leq^{\oplus} a$. Since a has index n and P is conical, it follows that $u = v = 0$, so $x = z = y$. □

As the following example shows, the assumption of conicality cannot be dropped from the statement of Lemma 2.3.6.

Example 2.3.7 A refinement monoid M, with nonzero elements $a, b \in M$ such that $a \leq^{+} 0$, b has index 1, $a + b = b$, and every element of M has finite index.

Proof Let M be the commutative monoid defined by the generators a, a', b subjected to the relations $a + a' = 0$ and $a + b = b$. Then M can be realized as $\mathbb{Z} \sqcup \{mb \mid m \in \mathbb{N}\}$, with $a = 1$ and $a' = -1$, and $n + mb = mb$ whenever $n \in \mathbb{Z}$ and $m \in \mathbb{N}$. It is straightforward to verify that M is a refinement monoid. Set $\iota(n) = 0$ whenever $n \in \mathbb{Z}$, and $\iota(mb) = m$ for each $m \in \mathbb{N}$. Then ι is a homomorphism from M onto $\mathbb{Z}^{+}$. In fact, $\iota(x)$ is the index of x, for any $x \in M$. In particular, the index of b is 1. □

We will need later the following lemma about elements of index at most 1.

Lemma 2.3.8 *Let a and b be elements in a conical partial refinement monoid P, with $a \boxplus b$ defined. If a and b are both of index at most* 1, *then so is $a \boxplus b$.*

Proof Let $c \in P$ such that $2c \leq^{\oplus} a \oplus b$. Since P has refinement, there are $a' \leq^{\oplus} a$ and $b' \leq^{\oplus} b$ such that $2c = a' \oplus b'$. By Wehrung [118, Lemma 1.9] (whose proof remains valid in partial refinement monoids), there are $u, v, w \in P$ such that $a' = 2u \oplus w$, $b' = 2v \oplus w$, and $c = u \oplus v \oplus w$. Since a' and b' both have index at most 1 and P is conical, $u = v = 0$. From $a' \wedge b' = 0$ it follows that $w = 0$. Therefore, $c = 0$. □

Lemma 2.3.9 *The following statements hold, for any pairwise meet-orthogonal elements a, b, c in a conical partial refinement monoid P:*

(1) *If $a \oplus c$ and $b \oplus c$ both exist, then the lower bounds of $\{a \oplus c, b \oplus c\}$ are exactly the lower bounds of c.*

(2) *If $a \oplus c$ and $b \oplus c$ both exist, then $\{a \oplus c, b \oplus c\}$ is bounded above iff $a \oplus b \oplus c$ exists, and then the upper bounds of $\{a \oplus c, b \oplus c\}$ are exactly the upper bounds of $a \oplus b \oplus c$.*

Proof We prove only the non-trivial containments.

(1) Let $x \in P$ such that $x \leq^{\oplus} \begin{matrix} a \oplus c \\ b \oplus c \end{matrix}$. By Riesz decomposition in P, there are decompositions $x = u \oplus x_0 = v \oplus x_1$ in P, with $u \leq^{\oplus} a$, $v \leq^{\oplus} b$, and $\begin{matrix} x_0 \\ x_1 \end{matrix} \leq^{\oplus} c$. By refinement and since a, b, c are pairwise meet-orthogonal, there is a refinement matrix of the following form:

	v	x_1
u	0	0
x_0	0	y

within $(P, \oplus)$.

Therefore, $x = y \leq^{\oplus} c$.

(2) First observe that if $a \oplus b \oplus c$ is defined, then it is greater than or equal to $\{a \oplus c, b \oplus c\}$. Now let $x \in P$ such that $\begin{matrix} a \oplus c \\ b \oplus c \end{matrix} \leq^{\oplus} x$. There are $x_0, x_1 \in P$ such that $x = a \oplus c \oplus x_0 = b \oplus x \oplus x_1$. Since a, b, c are pairwise meet-orthogonal and by Proposition 2.2.2, there is a refinement matrix of the form

	b	c	x_1
a	0	0	a
c	0	c'	y_1
x_0	b	y_0	z

within $(P, \oplus)$.

In particular, $b \leq^{\oplus} x_0$, thus, since $x = a \oplus c \oplus x_0$, the element $a \oplus b \oplus c$ is defined and below x. □

The following result is, essentially, a reformulation of Tarski [109, Theorem 15.16]. We include a proof for convenience.

Proposition 2.3.10 *The algebraic preordering $\leq^{\oplus}$ on any multiple-free conical partial commutative monoid P is antisymmetric. Furthermore, if P has refinement and $\leq^{\oplus}$ is upward directed, then it endows P with a structure of a generalized Boolean algebra, in such a way that for any $a, b \in P$, $a \oplus b$ exists iff $a \wedge b = 0$, and in this case $a \oplus b$ is the join of $\{a, b\}$ in P.*

Proof Let $a, b \in P$ such that $a \leq^{\oplus} b \leq^{\oplus} a$. There are $x, y \in P$ such that $b = a \oplus x$ and $a = b \oplus y$. It follows that $a = a \oplus x \oplus y$, thus $a = a \oplus 2x \oplus 2y$. Since P

is both conical and multiple-free, $x = y = 0$, thus $a = b$. This proves that $\leq^{\oplus}$ is antisymmetric.

Suppose from now on that $(P, \leq^{\oplus})$ is upward directed.

Claim 1 For any $a, b \in P$, $a \oplus b$ exists iff $a \wedge b = 0$, and then $a \oplus b$ is the join of $\{a, b\}$ in P.

Proof of Claim. Since P is multiple-free, if $a \oplus b$ exists, then $a \wedge b = 0$. Suppose, conversely, that $a \wedge b = 0$. Since $(P, \leq^{\oplus})$ is upward directed, $\{a, b\}$ is bounded above P. By Lemma 2.3.9, $a \oplus b$ is defined and it is the join of $\{a, b\}$ in P. □ Claim 1.

Claim 2 $(P, \leq^{\oplus})$ is a lattice with zero.

Proof of Claim. Let $a, b \in P$. Since P is upward directed, there are $e \in P$ and $a', b' \in P$ such that $e = a \oplus a' = b \oplus b'$. By our assumption, there is a refinement matrix of the following form:

	b	b'
a	$c_{0,0}$	$c_{0,1}$
a'	$c_{1,0}$	$c_{1,1}$

within $(P, \oplus)$.

Since P is multiple-free, the $c_{i,j}$ are pairwise meet-orthogonal. Since $\leq^{\oplus}$ is antisymmetric and by Lemma 2.3.9, it follows that $a \vee b = c_{0,0} \oplus c_{0,1} \oplus c_{1,0}$ and $a \wedge b = c_{0,0}$. □ Claim 2.

For all $a \leq^{\oplus} b$ in P, there exists, by the definition of $\leq^{\oplus}$, an element $c \in P$ such that $b = a \oplus c$. By Claim 1, this means that $0 = a \wedge c$ and $b = a \vee c$. In other words, the lattice $(P, \leq^{\oplus})$ is sectionally complemented. Hence, in order to conclude the proof, it suffices to prove that this lattice is distributive. We must prove that $a \wedge (b \vee c) \leq (a \wedge b) \vee (a \wedge c)$, for all $a, b, c \in P$. (The converse inequality is trivial.) Since $b \wedge c \leq^{\oplus} b$, there is $b' \in P$ such that $b = (b \wedge c) \oplus b'$. Observe that $b' \wedge c = b' \wedge (b \wedge c) = 0$ while $b' \oplus c = b \vee c$, so we must prove the inequality $a \wedge (b' \oplus c) \leq (a \wedge b') \oplus (a \wedge c)$. Since P has refinement, the left hand side x of that inequality can be written $x = v \oplus w$, for some $v \leq^{\oplus} b'$ and $w \leq^{\oplus} c$. Since $v \leq^{\oplus} a \wedge b'$ and $w \leq^{\oplus} a \wedge c$, the desired inequality follows. □

2.4 V-Relations on Partial Commutative Monoids

Vaught relations originate in Vaught's thesis [114], and have been much studied since, notably by Dobbertin. Essentially, a Vaught relation is a binary relation Γ such that any relation of the form $x \, \Gamma \sum_{i<n} y_i$ occurs "for a good reason". The following more precise definition extends Dobbertin [32, Definition 1.2] from full commutative monoids to partial commutative monoids.

Definition 2.4.1 Let P and Q be partial commutative monoids. A binary relation $\Gamma \subseteq P \times Q$ is

- *left conical* if $0_P \mathbin{\Gamma} y$ implies that $y = 0_Q$, for all $y \in Q$;
- a *left V-relation* if whenever n is a positive integer, $x = \bigoplus_{i<n} x_i$ in P, $y \in Q$, and $x \mathbin{\Gamma} y$, there are $y_0, \dots, y_{n-1} \in Q$ such that $y = \bigoplus_{i<n} y_i$ and $x_i \mathbin{\Gamma} y_i$ for each $i < n$ (*it is sufficient to verify this for* $n = 2$);
- *right conical* if Γ^{-1} is left conical;
- *conical* if it is simultaneously left and right conical;
- a *right V-relation* if Γ^{-1} is a left V-relation;
- a *V-relation* if it is simultaneously a left V-relation and a right V-relation;
- a *V-equivalence* if it is both an equivalence relation and a V-relation;
- *additive* if it is a partial submonoid of $P \times Q$; that is, whenever n is a nonnegative integer, $x = \bigoplus_{i<n} x_i$ in P, $y = \bigoplus_{i<n} y_i$ in Q, and $x_i \mathbin{\Gamma} y_i$ for each $i < n$, then $x \mathbin{\Gamma} y$ (*it is sufficient to verify this for* $n = 0$ *and* $n = 2$).

Observe that the partial submonoid $\Gamma^{\oplus}$ (cf. Definition 2.1.2) is the least additive relation on $P \times Q$ containing Γ. We will sometimes refer to that relation as the *additive closure of* Γ *in* $P \times Q$. It is the set of all pairs $(x, y) \in P \times Q$ for which there are decompositions $x = \bigoplus_{i<n} x_i$, $y = \bigoplus_{i<n} y_i$, and $x_i \mathbin{\Gamma} y_i$ for each $i < n$.

Note that the V-homomorphisms, considered for full commutative monoids in Dobbertin [32], are identical to our conical V-homomorphisms.

The proof of the following two lemmas are straightforward exercises.

Lemma 2.4.2

(1) *Any composition of V-relations is a V-relation.*
(2) *Any union of V-relations is a V-relation.*
(3) *Let P be a partial commutative monoid and let $\Gamma \subseteq P \times P$. If Γ is a V-relation, then so is the equivalence closure of Γ (i.e., the least equivalence relation containing Γ).*
(4) *The set of all V-equivalences on P is a complete lattice under set inclusion.*

The join and the meet, of a collection $(\Gamma_i \mid i \in I)$ of V-equivalences, in the poset of all V-equivalences, can be evaluated as follows: $\bigvee_{i \in I} \Gamma_i$ is the equivalence closure of $\bigcup_{i \in I} \Gamma_i$, and $\bigwedge_{i \in I} \Gamma_i$ is the join of all V-equivalences contained in all Γ_i (there is always one such equivalence, namely the identity).

Lemma 2.4.3 *The following statements hold, for all partial refinement monoids P and Q and any $\Gamma \subseteq P \times Q$.*

(1) *If Γ is a V-relation, then so is the additive closure of Γ in $P \times Q$.*
(2) *If $P = Q$ and Γ is a V-equivalence, then so is the additive closure $\Gamma^{\oplus}$ of Γ in $P \times P$.*
(3) *The set of all additive V-equivalences on P is a complete lattice under set inclusion.*

The join and the meet, of a collection $(\Gamma_i \mid i \in I)$ of additive V-equivalences, in the lattice of all additive V-equivalences on P, can be evaluated as follows: $\bigvee_{i \in I} \Gamma_i$

is the equivalence generated by $\left(\bigcup_{i\in I}\Gamma_i\right)^{\oplus}$, and $\bigwedge_{i\in I}\Gamma_i$ is the join of all additive V-equivalences contained in all Γ_i.

The following lemma says that every additive V-equivalence, on a partial commutative monoid, yields a structure of a partial commutative monoid on the quotient.

Lemma 2.4.4 *Let P be a partial commutative monoid and let $\Gamma \subseteq P \times P$ be an additive V-equivalence. Then the quotient P/Γ can be endowed with a structure of a partial commutative monoid, with addition defined by*

$$\boldsymbol{z} = \boldsymbol{x} \oplus \boldsymbol{y} \quad \textit{if} \quad (\exists x \in \boldsymbol{x})(\exists y \in \boldsymbol{y})(\exists z \in \boldsymbol{z})(z = x \oplus y)\,. \tag{2.4.1}$$

Furthermore, the following statements hold:

(1) *The canonical projection $\pi\colon P \twoheadrightarrow P/\Gamma$ is a V-homomorphism.*
(2) *If P and Γ are both conical, then so is P/Γ.*
(3) *If P satisfies refinement, then so does P/Γ.*

Proof We first prove that (2.4.1) defines, indeed, a partial operation on P/Γ. We must prove that if the statements intended to mean that $\boldsymbol{u} = \boldsymbol{x} \oplus \boldsymbol{y}$ and $\boldsymbol{v} = \boldsymbol{x} \oplus \boldsymbol{y}$ both hold in P/Γ, then $\boldsymbol{u} = \boldsymbol{v}$. By assumption, there are $u \in \boldsymbol{u}$, $v \in \boldsymbol{v}$, $x_0, x_1 \in \boldsymbol{x}$, and $y_0, y_1 \in \boldsymbol{y}$ such that $u = x_0 \oplus y_0$ and $v = x_1 \oplus y_1$. Since $x_0 \mathrel{\Gamma} x_1$, $y_0 \mathrel{\Gamma} y_1$, and since Γ is additive, it follows that $u \mathrel{\Gamma} v$; whence $\boldsymbol{u} = \boldsymbol{v}$.

Let $\boldsymbol{x}, \boldsymbol{y} \in P/\Gamma$ and let $z \in P$ such that $\pi(z) = \boldsymbol{x} \oplus \boldsymbol{y}$. Since $\boldsymbol{x} \oplus \boldsymbol{y}$ is defined, there are $x' \in \boldsymbol{x}$, $y' \in \boldsymbol{y}$, and $z' \in \boldsymbol{z}$ such that $z' = x' \oplus y'$. Since $z \mathrel{\Gamma} z' = x' \oplus y'$ and Γ is a V-relation, there is a decomposition $z = x \oplus y$ in P such that $x \mathrel{\Gamma} x'$ and $y \mathrel{\Gamma} y'$. Observe that $\boldsymbol{x} = \pi(x)$ and $\boldsymbol{y} = \pi(y)$. Once we will have proved that P/Γ is a partial commutative monoid, this will ensure that π is a V-homomorphism.

Let $\boldsymbol{x}, \boldsymbol{y}, \boldsymbol{z}, \boldsymbol{u} \in P/\Gamma$ and suppose that $\boldsymbol{u} = (\boldsymbol{x} \oplus \boldsymbol{y}) \oplus \boldsymbol{z}$. There are $u \in \boldsymbol{u}$, $z' \in \boldsymbol{x} \oplus \boldsymbol{y}$, and $z \in \boldsymbol{z}$ such that $u = z' \oplus z$. By the paragraph above, there are $x \in \boldsymbol{x}$ and $y \in \boldsymbol{y}$ such that $z' = x \oplus y$. Hence $u = (x \oplus y) \oplus z$. Since P is a partial commutative monoid, it follows that $u = x \oplus (y \oplus z)$. Hence $\boldsymbol{u} = \boldsymbol{x} \oplus (\boldsymbol{y} \oplus \boldsymbol{z})$. Therefore, P/Γ is a partial commutative monoid. Item (1) follows.

The verification of (2) is straightforward.

Towards (3), suppose that P is a refinement monoid and let $\boldsymbol{u} = \boldsymbol{x}_0 \oplus \boldsymbol{x}_1 = \boldsymbol{y}_0 \oplus \boldsymbol{y}_1$ in P/Γ. Pick $u \in \boldsymbol{u}$. By (1) above, there are $x_i \in \boldsymbol{x}_i$ and $y_i \in \boldsymbol{y}_i$, for $i \in \{0, 1\}$, such that $u = x_0 \oplus x_1 = y_0 \oplus y_1$. Since P satisfies refinement, there are $z_{i,j} \in P$, for $i, j \in \{0, 1\}$, such that each $x_i = z_{i,0} \oplus z_{i,1}$ and each $y_i = z_{0,i} \oplus z_{1,i}$. Setting $\boldsymbol{z}_{i,j} = z_{i,j}/\Gamma$, we obtain that each $\boldsymbol{x}_i = \boldsymbol{z}_{i,0} \oplus \boldsymbol{z}_{i,1}$ and each $\boldsymbol{y}_i = \boldsymbol{z}_{0,i} \oplus \boldsymbol{z}_{1,i}$. □

The following result is a version of the First Isomorphism Theorem for V-homomorphisms.

Lemma 2.4.5 *Let P and Q be partial commutative monoids. Then every V-homomorphism $\varphi\colon P \to Q$ induces a V-embedding from $P/\operatorname{Ker}\varphi$, endowed with its canonical structure of partial commutative monoid (cf. Lemma 2.4.4), into Q.*

Proof Set $\boldsymbol{\theta} = \operatorname{Ker}\varphi$. The canonical map $\psi\colon P/\boldsymbol{\theta} \hookrightarrow Q$ is obviously a one-to-one homomorphism of partial commutative monoids. Let $\boldsymbol{x} \in P/\boldsymbol{\theta}$ and let $y_0, y_1 \in Q$

such that $\psi(\boldsymbol{x}) = y_0 \oplus y_1$. Picking $x \in \boldsymbol{x}$, this means that $\varphi(x) = y_0 \oplus y_1$, thus, since φ is a V-homomorphism, there is a decomposition $x = x_0 \oplus x_1$ in P such that each $\varphi(x_i) = y_i$. Setting $\boldsymbol{x}_i = x_i/\boldsymbol{\theta}$, it follows that $\boldsymbol{x} = \boldsymbol{x}_0 \oplus \boldsymbol{x}_1$ and each $\psi(\boldsymbol{x}_i) = y_i$. Therefore, ψ is a V-embedding. □

The following result says that additive V-equivalences on a conical partial refinement monoid P are essentially the same as the additive V-equivalences on the enveloping monoid of P, and that this identification extends to the quotient monoids.

Theorem 2.4.6 *Let P be a partial refinement monoid and let Γ be an additive V-relation on P. Then there exists a unique additive V-relation $\mathrm{U_{mon}}(\Gamma)$ on $\mathrm{U_{mon}}(P)$ such that $\Gamma = \mathrm{U_{mon}}(\Gamma) \cap (P \times P)$. Furthermore, if Γ is a V-equivalence on P, then the following statements hold:*

(1) *Γ and $\mathrm{U_{mon}}(\Gamma)$ are both V-equivalences on $\mathrm{U_{mon}}(P)$.*
(2) *Denote by $\pi\colon P \twoheadrightarrow P/\Gamma$ and $\overline{\pi}\colon \mathrm{U_{mon}}(P) \twoheadrightarrow \mathrm{U_{mon}}(P)/\mathrm{U_{mon}}(\Gamma)$ the canonical projections and by $\varepsilon_P\colon P \hookrightarrow \mathrm{U_{mon}}(P)$ the canonical V-embedding. Then there is a unique map $\eta\colon P/\Gamma \to \mathrm{U_{mon}}(P)/\mathrm{U_{mon}}(\Gamma)$ such that $\eta \circ \pi = \overline{\pi} \circ \varepsilon_P$, and this map is a V-embedding with generating range.*
(3) *P/Γ and $\mathrm{U_{mon}}(P)/\mathrm{U_{mon}}(\Gamma)$ both satisfy refinement. Furthermore the V-embedding η induces an isomorphism $\mathrm{U_{mon}}(P)/\mathrm{U_{mon}}(\Gamma) \cong \mathrm{U_{mon}}(P/\Gamma)$.*

The conclusion of Theorem 2.4.6(2) is illustrated on Fig. 2.1.

Proof In order to ease notation, we set $M = \mathrm{U_{mon}}(P)$. We may assume that ε_P is the inclusion map from P into M.

The first part of (1) is obvious: since P is a lower interval of M and Γ is a V-relation on P, it is also a V-relation on M. Furthermore, by Lemma 2.4.3, the additive closure $\mathrm{U_{mon}}(\Gamma) = \Gamma^+$ of Γ in M is an additive V-relation on M. (*Consistently with an earlier convention, we emphasize the fact that the addition on M is defined everywhere, by writing Γ^+ instead of $\Gamma^\oplus$.*) Still by Lemma 2.4.3, if Γ is a V-equivalence (on P), then so is Γ^+ (on M). It is obvious that $\Gamma \subseteq \Gamma^\oplus \cap (P \times P)$. Conversely, let $(x, y) \in \Gamma^+ \cap (P \times P)$. By the definition of Γ^+, there are decompositions $x = \sum_{i<n} x_i$ and $y = \sum_{i<n} y_i$ (both within M) such that each $x_i \mathrel{\Gamma} y_i$. Since P is a lower interval of M, both relations $x = \bigoplus_{i<n} x_i$ and $y = \bigoplus_{i<n} y_i$ hold in P. Since Γ is additive in P, it follows that $x \mathrel{\Gamma} y$. Therefore, $\Gamma = \Gamma^+ \cap (P \times P)$.

Now let $\widetilde{\Gamma}$ be an additive V-relation on M such that $\Gamma = \widetilde{\Gamma} \cap (P \times P)$. From $\Gamma \subseteq \widetilde{\Gamma}$ and by the additivity of $\widetilde{\Gamma}$ it follows that $\Gamma^+ \subseteq \widetilde{\Gamma}$. Conversely, let $(x, y) \in \widetilde{\Gamma}$. Since P generates M, there exists a decomposition $x = \bigoplus_{i \in I} x_i$ where I is finite and all $x_i \in P$. Since $\widetilde{\Gamma}$ is a V-relation, there exists a decomposition $y = \bigoplus_{i \in I} y_i$ where all $x_i \widetilde{\Gamma} y_i$. Since P generates M, each y_i can be written as a sum $y_i = \bigoplus_{j \in J_i} y_{i,j}$,

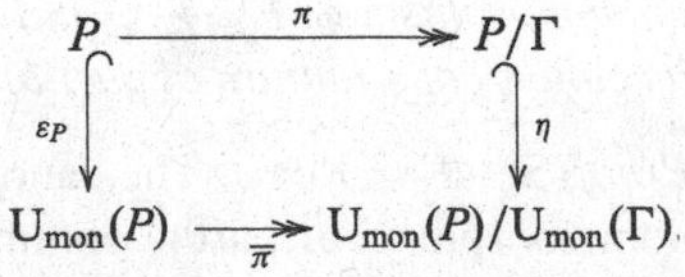

Fig. 2.1 The canonical lower embedding $\eta\colon P/\Gamma \hookrightarrow \mathrm{U_{mon}}(P)/\mathrm{U_{mon}}(\Gamma)$

with J_i finite and all $y_{i,j} \in P$. Since $\widetilde{\Gamma}$ is a V-relation, each x_i can be written as a sum $x_i = \bigoplus_{j \in J_i} x_{i,j}$ where each $x_{i,j} \widetilde{\Gamma} y_{i,j}$. From $x_i \in P$ it follows that $x_{i,j} \in P$. Since $\widetilde{\Gamma} \cap (P \times P) = \Gamma$, it follows that $x_{i,j} \; \Gamma \; y_{i,j}$, for all possible values of i and j. Since $x = \bigoplus_{i \in I, \, j \in J_i} x_{i,j}$ and $y = \bigoplus_{i \in I, \, j \in J_i} y_{i,j}$, it follows that $x \; \Gamma^+ \; y$. Therefore, $\Gamma^+ = \widetilde{\Gamma}$, thus proving the uniqueness statement on Γ^+.

Suppose from now on that Γ is an additive V-equivalence on P. The homomorphism $\varphi \colon P \to M/\Gamma^+, x \mapsto x/\Gamma^+$ is the composite of the V-embedding $P \hookrightarrow M$ and the V-homomorphism $M \twoheadrightarrow M/\Gamma^+$, thus it is a V-homomorphism. The statement $\Gamma = \Gamma^+ \cap (P \times P)$, proved above, means exactly that Γ is the kernel of φ. By Lemma 2.4.5, φ induces a V-embedding $\eta \colon P/\Gamma \hookrightarrow M/\Gamma^+$. It is trivial that the range of η generates M/Γ^+.

Item (3) follows from Lemma 2.4.4. The conclusion that $M/\Gamma^+ \cong \mathrm{U_{mon}}(P/\Gamma)$ follows then from Proposition 2.2.4. □

Observe that in Theorem 2.4.6, for every additive V-equivalence Γ on P, the binary relation $\mathrm{U_{mon}}(\Gamma)$ is the unique additive V-equivalence on $\mathrm{U_{mon}}(P)$ such that $\Gamma = \mathrm{U_{mon}}(\Gamma) \cap (P \times P)$. Hence, *the assignment $\Gamma \mapsto \mathrm{U_{mon}}(\Gamma)$ defines an isomorphism from the lattice of all additive V-equivalences on P onto the lattice of all additive V-equivalences on $\mathrm{U_{mon}}(P)$.*

Proposition 2.4.7 *Let P be a lower interval in a partial refinement monoid Q and let Γ be an additive V-equivalence on Q. Then the relation $\Gamma_P = \Gamma \cap (P \times P)$ is an additive V-equivalence on P, and the assignment $\varphi \colon x/\Gamma_P \mapsto x/\Gamma$ defines a V-embedding from P/Γ_P into Q/Γ.*

Proof It is obvious that Γ_P is an additive V-equivalence on P. The map $P \to Q/\Gamma$, $x \mapsto x/\Gamma$ is the composite of the V-embedding $P \hookrightarrow Q$ and the V-homomorphism $Q \twoheadrightarrow Q/\Gamma$, thus it is a V-homomorphism. By Lemma 2.4.5, the induced map $\varphi \colon x/\Gamma_P \mapsto x/\Gamma$ is a V-embedding. □

2.5 Tensor Products of Commutative Monoids

Tensor products of commutative monoids are defined the same way as tensor products of modules, via an obvious reformulation of the definition of a bimorphism. Following the terminology in use in Wehrung [121], we say that for commutative monoids M, N, P, a map $f \colon M \times N \to P$ is a *bimorphism* if $f(x, \_)$ is a monoid homomorphism from N to P for each $x \in M$, and $f(\_, y)$ is a monoid homomorphism from M to P for each $y \in N$. We say that the bimorphism f is *universal* if for every commutative monoid Z and every bimorphism $g \colon M \times N \to Z$, there is a unique monoid homomorphism $\overline{g} \colon P \to Z$ such that $g(x, y) = \overline{g}(f(x, y))$ for all $(x, y) \in M \times N$. For all commutative monoids M and N, there are a commutative monoid P and a universal bimorphism from $M \times N$ to P. Then P is unique up to isomorphism, and we denote it by $M \otimes N$. Further, the universal bimorphism is denoted by $(x, y) \mapsto x \otimes y$ (cf. Wehrung [121, § 1]). The monoid $M \otimes N$ is called the *tensor product* of M and N. This monoid is, in [121], denoted by $M \otimes^{\mathrm{cm}} N$. We will call *pure tensors* the elements of the form $x \otimes y$, where $(x, y) \in M \times N$.

Lemma 2.5.1 *Let M and N be conical commutative monoids. Then $M \otimes N$ is also conical. Furthermore, $x \otimes y \neq 0$ whenever $x \in M \setminus \{0\}$ and $y \in N \setminus \{0\}$.*

Proof Denoting by $\mathbf{2} = \{0, 1\}$ the two-element semilattice, let the map $\nu_M : M \twoheadrightarrow \mathbf{2}$ be defined by $\nu(x) = 0$ iff $x = 0$. Define ν_N similarly. Since M and N are both conical, ν_M and ν_N are both monoid homomorphisms. It follows that the assignment $M \times N \to \mathbf{2}$, $(x, y) \mapsto \nu_M(x)\nu_N(y)$ is a monoid bimorphism. Hence, there is a unique monoid homomorphism $\nu : M \otimes N \to \mathbf{2}$ such that $\nu(x \otimes y) = \nu_M(x)\nu_N(y)$ for all $(x, y) \in M \times N$. If $\sum_{i<n}(x_i \otimes y_i) = 0$, then $\nu\bigl(\sum_{i<n}(x_i \otimes y_i)\bigr) = 0$, thus each $\nu_M(x_i)\nu_N(y_i) = 0$, that is, either $x_i = 0$ or $y_i = 0$. This implies that $x_i \otimes y_i = 0$. □

The author's paper [121] introduces a few tools designed for the study of tensor products of refinement monoids. In particular, the tensor product of two refinement monoids (resp., of two conical refinement monoids) is a refinement monoid (resp., a conical refinement monoid), see Theorems 2.7 and 2.9 in [121].

We shall fix, until the end of this section, conical refinement monoids M and N. We will use modifications[2] of the objects denoted by $\mathcal{C}$, $\to^0$, $\to$, $\to^*$, and $\equiv$ in [121, § 2].

Denote by $\mathrm{F}_{\mathrm{mon}}(M \times N)$ the free commutative monoid on the set $M \times N$. For $(x, y) \in M \times N$, we shall denote by $x \bullet y$ the corresponding element of $\mathrm{F}_{\mathrm{mon}}(M \times N)$. This notation change is introduced in order not to confuse the sum $(x_0, y_0) + (x_1, y_1)$ in $M \times N$ (which has value $(x_0 + x_1, y_0 + y_1)$) and the sum $x_0 \bullet y_0 + x_1 \bullet y_1$ (which is the sum of two elements of the canonical basis of $\mathrm{F}_{\mathrm{mon}}(M \times N)$).

We define binary relations $\to$, $\to^*$, and $\leftrightarrows$ on $\mathrm{F}_{\mathrm{mon}}(M \times N)$ as follows:

- $u \to v$ if there are decompositions $u = \sum_{k \in K}(a_k \bullet b_k)$ in $\mathrm{F}_{\mathrm{mon}}(M \times N)$, each $a_k = \sum_{i \in I_k} a_{k,i}$ in M, each $b_k = \sum_{j \in J_k} b_{k,j}$ in N, with I and all I_k, J_k finite (*possibly empty*), such that, setting

$$\overline{K} = \{(k, i, j) \mid k \in K\,,\ i \in I_k\,,\ \text{and}\ j \in J_k\}$$

  the equation $v = \sum_{(k,i,j) \in \overline{K}}(a_{k,i} \bullet b_{k,j})$ holds.
- The binary relation $\to^*$ is the transitive closure of $\to$.
- For any $u, v \in \mathrm{F}_{\mathrm{mon}}(M \times N)$, $u \leftrightarrows v$ if there is $w \in \mathrm{F}_{\mathrm{mon}}(M \times N)$ such that $u \to^* w$ and $v \to^* w$.

The following lemma is the analogue, for our newly defined relations $\to$ and $\to^*$, of [121, Lemma 2.1]. We omit the proof, which is straightforward (*observe that the proof of* (1) *in that lemma arises from the possibility to use empty sums within the*

[2]Although those modifications might not be, strictly speaking, necessary, I feel that they provide a slightly better presentation than the one of my earlier paper [121], while at the same time gently introducing the reader to those concepts about tensor products of commutative monoids necessary to follow all parts of the present work.

definition of $\rightarrow$). Recall that additive relations and V-relations are both introduced in Definition 2.4.1.

Lemma 2.5.2 *The following statements hold:*

(1) $x \bullet 0 \rightarrow 0$ *for all* $x \in M$, *and* $0 \bullet y \rightarrow 0$ *for all* $y \in N$.
(2) *Both relations* $\rightarrow$ *and* $\rightarrow^*$ *are reflexive, additive left V-relations.*

The following lemma is an analogue of [121, Lemma 2.3]. The proof is similar, only noticeably easier, because we do not need to worry about which elements are nonzero.

Lemma 2.5.3 *The relation* $\rightarrow$ *is confluent, that is, for all elements* $u, v, v' \in \mathrm{F}_{\mathrm{mon}}(M \times N)$, *if* $u \rightarrow v$ *and* $u \rightarrow v'$, *then there is* $w \in \mathrm{F}_{\mathrm{mon}}(M \times N)$ *such that* $v \rightarrow w$ *and* $v' \rightarrow w$. *Further, the relation* $\rightarrow^*$ *is confluent.*

Proof Since $\rightarrow^*$ is the transitive closure of $\rightarrow$, it suffices to prove that $\rightarrow$ is confluent (this is often expressed as "local confluence implies confluence"). Further, since $\rightarrow$ is an additive left V-relation, it suffices to consider the case where $u = a \bullet b$ for some $(a, b) \in M \times N$. In that case, there are decompositions $a = \sum_{i \in I} a_i = \sum_{i' \in I'} a'_{i'}$ in M and $b = \sum_{j \in J} b_j = \sum_{j' \in J'} b_{j'}$ in N, with I, I', J, J' finite (possibly empty), such that $v = \sum_{(i,j) \in I \times J} (a_i \bullet b_j)$ and $v' = \sum_{(i',j') \in I' \times J'} (a_{i'} \bullet b_{j'})$. Since M and N are conical refinement monoids, there are refinement matrices as follows:

$$\begin{array}{c|c|} \multicolumn{1}{c}{} & \multicolumn{1}{|c|}{a'_{i'} (i' \in I')} \\ \hline a_i (i \in I) & a_{i,i'} \\ \hline \end{array} \quad \text{in } M, \quad \text{and} \quad \begin{array}{c|c|} \multicolumn{1}{c}{} & \multicolumn{1}{|c|}{b'_{j'} (j' \in J')} \\ \hline b_j (j \in J) & b_{j,j'} \\ \hline \end{array} \quad \text{in } N.$$

Setting $w = \sum_{((i,i'),(j,j')) \in (I \times I') \times (J \times J')} (a_{i,i'} \bullet b_{j,j'})$, it follows easily that $v \rightarrow w$ and $v' \rightarrow w$. □

As an easy consequence of Lemmas 2.5.2 and 2.5.3, we obtain the following analogue of [121, Lemma 2.4].

Lemma 2.5.4 *The binary relation* $\leftrightarrows$ *is a monoid congruence on* $\mathrm{F}_{\mathrm{mon}}(M \times N)$.

Proposition 2.5.5 *The tensor product of* M *and* N *is* $\mathrm{F}_{\mathrm{mon}}(M \times N)/{\leftrightarrows}$, *with* $x \otimes y$ *defined as* $x \bullet y/{\leftrightarrows}$, *for each* $(x, y) \in M \times N$.

Proof Setting $P = \mathrm{F}_{\mathrm{mon}}(M \times N)/{\leftrightarrows}$ and, temporarily, $x \odot y = x \bullet y/{\leftrightarrows}$ whenever $(x, y) \in M \times N$, it suffices to prove that $\odot$ is a universal bimorphism.

First observe that $x \bullet 0 \rightarrow 0$ for each $x \in M$ (cf. Lemma 2.5.2); whence $x \odot 0 = 0$. Likewise $0 \odot y = 0$ whenever $y \in N$. For all $x_0, x_1 \in M$ and all $y \in N$, $(x_0 + x_1) \bullet y \rightarrow x_0 \bullet y + x_1 \bullet y$, thus $(x_0 + x_1) \odot y = (x_0 \odot y) + (x_1 \odot y)$. Likewise, $x \odot (y_0 + y_1) = (x \odot y_0) + (x \odot y_1)$, for all $x \in M$ and all $y_0, y_1 \in N$. Therefore, $\odot$ is a bimorphism. By the universal property of the tensor product, there is a unique monoid homomorphism $\varphi\colon M \otimes N \rightarrow P$ such that $\varphi(x \otimes y) = x \odot y$ for each $(x, y) \in M \times N$.

Conversely, the unique monoid homomorphism $\Phi\colon \mathrm{F}_{\mathrm{mon}}(M \times N) \to M \otimes N$, sending each $x \bullet y$ to $x \otimes y$, satisfies the implication $(u \to v) \Rightarrow \Phi(u) = \Phi(v)$, for all $u, v \in \mathrm{F}_{\mathrm{mon}}(M \times N)$. Hence, Φ is constant on all $\leftrightarrows$-equivalence classes, so it factors, through $\leftrightarrows$, to a monoid homomorphism $\psi\colon P \to M \otimes N$. By construction, φ and ψ are mutually inverse isomorphisms. □

A similar construction is used to prove, in Wehrung [121, Theorem 2.7], that $M \otimes N$ is a conical refinement monoid.

The following lemma will be used in Sect. 6.8.

Lemma 2.5.6 *Let M and N be conical refinement monoids and let $(a, b) \in M \times N$. If a has index at most 1 in M and b has index at most 1 in N, then $a \otimes b$ has index at most 1 in $M \otimes N$.*

Proof Denote by $M_{(1)}$ (resp., $N_{(1)}$) the set of all elements of M (resp., N) with index at most 1. Denote by $\mathcal{I}$ the set of elements of $\mathrm{F}_{\mathrm{mon}}(M \times N)$ of the form $\sum_{i<n}(a_i \bullet b_i)$, where $n \in \mathbb{Z}^+$ and the $a_i \otimes b_i$, for $i < n$, are pairwise meet-orthogonal within $M \otimes N$.

Claim Whenever $u \in \mathcal{I}$ and $v \in \mathrm{F}_{\mathrm{mon}}(M \times N)$, if $u \to^* v$, then $v \in \mathcal{I}$.

Proof of Claim. Since $\to^*$ is the transitive closure of $\to$, it suffices to consider the case where $u \to v$. Write $u = \sum_{k \in K}(a_k \bullet b_k)$, for a finite set K and $(a_k, b_k) \in M_{(1)} \times N_{(1)}$, with the $a_k \otimes b_k$ pairwise meet-orthogonal within $M \otimes N$. By the definition of $\to$, for each $k \in K$, there are decompositions $a_k = \sum_{i \in I_k} a_{k,i}$ in M and $b_k = \sum_{j \in J_k} b_{k,j}$ in N, such that $v = \sum_{k \in K} \sum_{(i,j) \in I_k \times J_k}(a_{k,i} \bullet b_{k,j})$. Let $k, k' \in K$, $(i, j) \in I_k \times J_k$, and $(i', j') \in I_{k'} \times J_{k'}$, with $(k, i, j) \neq (k', i', j')$; we must prove that $a_{k,i} \otimes b_{k,j}$ and $a_{k',i'} \otimes b_{k',j'}$ are meet-orthogonal within $M \otimes N$. Suppose first that $k \neq k'$. Since $a_{k,i} \otimes b_{k,j} \leq^+ a_k \otimes b_k$, $a_{k',i'} \otimes b_{k',j'} \leq^+ a_{k'} \otimes b_{k'}$, and $a_k \otimes b_k$ and $a_{k'} \otimes b_{k'}$ are meet-orthogonal within $M \otimes N$, the desired conclusion follows. Suppose now that $k = k'$. Then $(i, j) \neq (i', j')$, say $i \neq i'$. Since a_k has index at most 1 in M and $a_{k,i} + a_{k,i'} \leq^+ a_k$, it follows that $a_{k,i}$ and $a_{k,i'}$ are orthogonal within M. Therefore, the desired conclusion follows from Wehrung [121, Corollary 2.11]. Since each $(a_{k,i}, b_{k,j}) \in M_{(1)} \times N_{(1)}$, it follows that $v \in \mathcal{I}$. □ Claim.

Now let a and b be elements of $M_{(1)}$ and $N_{(1)}$, respectively, and let $\boldsymbol{u} \in M \otimes N$ such that $2\boldsymbol{u} \leq a \otimes b$. Write $\boldsymbol{u} = u/{\equiv}$, with $u \in \mathrm{F}_{\mathrm{mon}}(M \times N)$. By Proposition 2.5.5, there are $v, w \in \mathrm{F}_{\mathrm{mon}}(M \times N)$ such that $a \bullet b \to^* w$ and $2u + v \to^* w$. By Lemma 2.5.2, there is a decomposition $w = u_0 + u_1 + v'$ in $\mathrm{F}_{\mathrm{mon}}(M \times N)$ such that $u \to^* u_0$, $u \to^* u_1$, and $v \to^* v'$. By Lemma 2.5.3, there is $u' \in \mathrm{F}_{\mathrm{mon}}(M \times N)$ such that $u_0 \to^* u'$ and $u_1 \to^* u'$. By Lemma 2.5.2 again, it follows that $w \to^* 2u' + v'$. Therefore, $a \bullet b \to^* 2u' + v'$, thus, since $a \bullet b \in \mathcal{I}$ and by the Claim above, $2u' + v' \in \mathcal{I}$. Writing $u' = \sum_{i<n}(x_i \bullet y_i)$, with each $x_i \in M \setminus \{0\}$ and $y_i \in N \setminus \{0\}$, it follows that $x_i \otimes y_i$ is meet-orthogonal from itself, for each $i < n$; a contradiction unless $n = 0$. It follows that $u' = 0$, thus $\boldsymbol{u} = u'/{\equiv} = 0$. □

2.6 Tensor Products of Conical V-Homomorphisms and V-Equivalences

While the tensor product of o-ideals in commutative monoids behaves, in its basic aspects, like the tensor product of ideals in modules, we will need to focus attention on a matter far more specific to refinement monoids, with no obvious module-theoretical analogue. For conical refinement monoids M and N, instead of tensoring o-ideals of M and N, we will need to tensor monoid congruences of M and N.

A *V-congruence* of a commutative monoid M is defined as a V-equivalence on M (cf. Definition 2.4.1) which is also a monoid congruence of M.

Obviously, the kernel $\operatorname{Ker} f$ of any V-homomorphism f, with domain M, is a V-congruence of M. Conversely, for every V-congruence $\boldsymbol{\alpha}$ of M, the canonical projection $\alpha: M \twoheadrightarrow M/\boldsymbol{\alpha}$ is a V-homomorphism with kernel $\boldsymbol{\alpha}$. Furthermore, the congruence $\boldsymbol{\alpha}$ is conical (as a binary relation) iff the homomorphism α is conical.

For commutative monoids M, M', N, N', a standard application of the universal property of the tensor product yields that any pair of monoid homomorphisms $f: M \to M'$ and $g: N \to N'$ gives rise to a unique monoid homomorphism $f \otimes g: M \otimes N \to M' \otimes N'$ such that $(f \otimes g)(x \otimes y) = f(x) \otimes g(y)$ for all $(x, y) \in M \times N$.

In that context, for each $u = \sum_{i \in I}(x_i \bullet y_i)$ in $\mathrm{F_{mon}}(M \times N)$, we set

$$(f \bullet g)(u) = \sum_{i \in I} \big(f(x_i) \bullet g(y_i)\big)\,.$$

Observe that $f \bullet g$ is a monoid homomorphism from $\mathrm{F_{mon}}(M \times N)$ to $\mathrm{F_{mon}}(M' \times N')$. Furthermore, it is obvious that $f \bullet g$ is a conical V-homomorphism.

Lemma 2.6.1 *Let M, M', N, N' be conical refinement monoids, let $f: M \to M'$ and $g: N \to N'$ be conical V-homomorphisms. Then for all elements $u \in \mathrm{F_{mon}}(M \times N)$ and $v \in \mathrm{F_{mon}}(M' \times N')$, if $(f \bullet g)(u) \to v$, then there exists $u' \in \mathrm{F_{mon}}(M \times N)$ such that $u \to u'$ and $(f \bullet g)(u') = v$. A similar statement holds for $\to^*$.*

Proof It suffices to prove the statement about $\to$. Furthermore, by Lemma 2.5.2, it suffices to consider the case where $u = x \bullet y$, where $(x, y) \in M \times N$. By the definition of $\to$, there are decompositions $f(x) = \sum_{i \in I} x_i$, in M', and $g(y) = \sum_{j \in J} y_j$, in N', with I and J finite, such that $v = \sum_{(i,j) \in I \times J}(x_i \bullet y_j)$. Since f and g are both conical V-homomorphisms (*the conicality assumption on f and g is included in order to take care of the case where either I or J is empty*), there are decompositions $x = \sum_{i \in I} x'_i$ and $y = \sum_{j \in J} y'_j$ such that each $f(x'_i) = x_i$ and each $g(y'_j) = y_j$. The element $u' = \sum_{(i,j) \in I \times J}(x'_i \bullet y'_j)$ is as required. □

Theorem 2.6.2 *Let M, M', N, N' be conical refinement monoids, let $f: M \to M'$ and $g: N \to N'$ be conical V-homomorphisms. Then $f \otimes g$ is a conical V-homomorphism from $M \otimes N$ to $M' \otimes N'$.*

Proof For each $x \in M \setminus \{0\}$ and $y \in N \setminus \{0\}$, it follows from the conicality of both f and g that $f(x) \neq 0$ and $g(y) \neq 0$. It follows (cf. Lemma 2.5.1) that $f(x) \otimes g(y) \neq 0$. Since $M' \otimes N'$ is conical, $f \otimes g$ is conical.

Now we prove that $f \otimes g$ is a V-homomorphism. To ease the notation, we use the same symbols to denote the binary relations $\to^*$ and $\leftrightarrows$, on $\mathrm{F_{mon}}(M \times N)$ and $\mathrm{F_{mon}}(M' \times N')$. Let $\boldsymbol{u} \in M \otimes N$ and $\boldsymbol{v}_0, \boldsymbol{v}_1 \in M' \otimes N'$ such that $(f \otimes g)(\boldsymbol{u}) = \boldsymbol{v}_0 + \boldsymbol{v}_1$. Pick $u \in \mathrm{F_{mon}}(M \times N)$ and $v_0, v_1 \in \mathrm{F_{mon}}(M' \times N')$ such that $\boldsymbol{u} = u/{\leftrightarrows}$ and $\boldsymbol{v}_i = v_i/{\leftrightarrows}$ whenever $i \in \{0, 1\}$. Since $(f \bullet g)(u) \leftrightarrows v_0 + v_1$, there is $w \in \mathrm{F_{mon}}(M' \otimes N')$ such that $(f \bullet g)(u) \to^* w$ and $v_0 + v_1 \to^* w$. The latter relation, together with Lemma 2.5.2, implies that there is a decomposition $w = w_0 + w_1$ such that each $v_i \to^* w_i$. Since $(f \bullet g)(u) \to^* w$ and by Lemma 2.6.1, there is $u' \in \mathrm{F_{mon}}(M \times N)$ such that $u \to^* u'$ and $(f \bullet g)(u') = w$. Since $w = w_0 + w_1$ and $f \bullet g$ is a V-homomorphism, there is a decomposition $u' = u_0 + u_1$ in $\mathrm{F_{mon}}(M \times N)$ such that each $(f \bullet g)(u_i) = w_i$. Setting $\boldsymbol{u}_i = u_i/{\leftrightarrows}$ whenever $i \in \{0, 1\}$, we get $\boldsymbol{u} = (u/{\leftrightarrows}) = (u'/{\leftrightarrows}) = \boldsymbol{u}_0 + \boldsymbol{u}_1$. Further, $(f \otimes g)(\boldsymbol{u}_i) = \big((f \bullet g)(u_i)/{\leftrightarrows}\big) = (w_i/{\leftrightarrows}) = (v_i/{\leftrightarrows}) = \boldsymbol{v}_i$, for each $i \in \{0, 1\}$. □

Notation 2.6.3 Let M and N be commutative monoids and let $\boldsymbol{\alpha}$ (resp., $\boldsymbol{\beta}$) be a congruence of M (resp., N). We denote by $\boldsymbol{\alpha} \otimes \boldsymbol{\beta}$ the kernel of the canonical homomorphism $\alpha \otimes \beta \colon M \otimes N \twoheadrightarrow (M/\boldsymbol{\alpha}) \otimes (N/\boldsymbol{\beta})$.

We obtain easily, from Theorem 2.6.2, the following corollary.

Corollary 2.6.4 *Let M and N be conical refinement monoids and let $\boldsymbol{\alpha}$ (resp., $\boldsymbol{\beta}$) be a conical V-congruence on M (resp., N). Then $\boldsymbol{\alpha} \otimes \boldsymbol{\beta}$ is a conical V-congruence on $M \otimes N$.*

2.7 Refinement-Spreading Properties in Partial Refinement Monoids

In a number of cases, various cancellativity properties of a refinement monoid M can be directly verified on a generating lower interval of M. In this section, we shall record some properties for which this transfer principle works, and some for which it does not.

Definition 2.7.1 A property (Φ), formulated in the language of partial commutative monoids, is *refinement-spreading* if for every partial refinement monoid P satisfying (Φ), the enveloping monoid $\mathrm{U_{mon}}(P)$ also satisfies (Φ).

For a partial refinement monoid P, it follows from Proposition 2.2.4 that $\mathrm{U_{mon}}(P)$ is the unique refinement monoid in which P is a generating lower interval. Hence, *the property (Φ) is refinement-spreading iff for every generating lower interval P in a refinement monoid M, P satisfies (Φ) implies that M satisfies (Φ).*

2.7.1 Conicality

A trivial prototype of a refinement-spreading property is given by conicality. Indeed, whenever P is a lower interval in a commutative monoid M, then all $x, y \in M$ such that $x + y = 0$ already belong to P. Hence we get:

Proposition 2.7.2 *Conicality is refinement-spreading.*

2.7.2 Cancellativity

A partial commutative monoid P is *cancellative* if $x \oplus z = y \oplus z$ implies that $x = y$, for all $x, y, z \in P$. It is trivial that every lower interval in a cancellative partial commutative monoid is cancellative.

Lemma 2.7.3 *Let M be a refinement monoid and let I and J be the subsets of M defined as*

$$I = \{e \in M \mid (\forall a, b, c \in M)(a + c = b + c = e \Rightarrow a = b\} \,,$$
$$J = \{c \in M \mid (\forall a, b \in M)(a + c = b + c \Rightarrow a = b)\} \,.$$

Then $I = J$ and this set is an o-ideal of M.

Proof It is obvious that $J \subseteq I$ and that J is an o-ideal of M. Now let $c \in I$, and let $a, b \in M$ such that $a + c = b + c$. Since M is a refinement monoid, there is a refinement matrix of the following form:

$$\begin{array}{c|c|c|}\cline{2-3} & b & c \\ \hline \multicolumn{1}{|c|}{a} & d & a' \\ \hline \multicolumn{1}{|c|}{c} & b' & c' \\ \hline \end{array} \tag{2.7.1}$$

Since $c = a' + c' = b' + c'$ and $c \in I$, we get $a' = b'$, thus $a = d + a' = d + b' = b$, thus completing the proof that $c \in J$. □

The following corollary is immediate.

Corollary 2.7.4 *Cancellativity is refinement-spreading.*

The result of Corollary 2.7.4 is partly contained in Dvurečenskij and Pulmannová [37, Theorem 1.7.12] (the latter result is formulated in the language of *effect algebras*).

As witnessed by Example 2.3.7, the weaker form of cancellativity $x + z = y + z \Rightarrow x \leq^{+} y$ does not imply cancellativity, even for refinement monoids.

2.7.3 Strong Separativity

A partial commutative monoid P is *strongly separative* if $x \oplus z = y \oplus 2z$ implies that $x = y \oplus z$, for all $x, y, z \in P$. If P is a full commutative monoid, then this is easily seen to be equivalent to saying that $x + z = 2z$ implies $x = z$, for all $x, z \in P$ (cf. Brookfield [25, Proposition 8.12]).

Lemma 2.7.5 *Let M be a refinement monoid and let $a, b, c \in M$. If $a + c = b + 2c$, then there are $d, \overline{a}, \overline{b}, \overline{c} \in M$ such that $a = d + \overline{a}$, $b + c = d + \overline{b} + \overline{c}$, and $c = \overline{a} + \overline{c} = \overline{b} + 2\overline{c}$.*

Proof Since $a + c = b + 2c$, there exists a refinement matrix of the form

	b	c	c
a	a'	c_1	c_2
c	b'	c_3	c_4

(2.7.2)

Since $c = c_1 + c_3 = c_2 + c_4$, there exists a refinement matrix of the form

	c_2	c_4
c_1	d_1	d_2
c_3	d_3	$\overline{c}$

(2.7.3)

Set $\overline{a} = d_1 + d_2 + d_3$, $\overline{b} = b' + d_2 + d_3$, and $d = a' + d_1$. Now $c = c_1 + c_3 = b' + c_3 + c_4$ yields, using (2.7.2) and (2.7.3), that $c = d_1 + d_2 + d_3 + \overline{c} = \overline{a} + \overline{c}$ and $c = b' + d_2 + d_3 + 2\overline{c} = \overline{b} + 2\overline{c}$. Furthermore, $a = a' + c_1 + c_2 = a' + 2d_1 + d_2 + d_3 = d + \overline{a}$ and $b + c = a' + b' + d_1 + d_2 + d_3 + \overline{c} = d + \overline{b} + \overline{c}$. □

Lemma 2.7.6 *Let M be a refinement monoid and let I and J be the subsets of M defined as*

$$I = \{e \in M \mid (\forall a, b, c \in M)(a + c = b + 2c \leq^+ e \Rightarrow b + c \leq^+ a)\},$$
$$J = \{e \in M \mid (\forall a, b, c \in M)((a + c = b + 2c \text{ and } c \leq^+ e) \Rightarrow a = b + c)\}.$$

Then $I = J$ and this set is an o-ideal of M.

Proof It is trivial that $J \subseteq I$ and that I and J are both lower subsets of M. Now let $a, b, c \in M$ and $e \in I$ such that $a + c = b + 2c$ and $c \leq^+ e$. Let $d, \overline{a}, \overline{b}, \overline{c}$ be elements satisfying the conclusion of Lemma 2.7.5. Since $\overline{a} + \overline{c} = \overline{b} + 2\overline{c} = c \leq^+ e$ and $e \in I$, there exists $h \in M$ such that

$$\overline{a} = \overline{b} + \overline{c} + h.$$

In particular, $c + h = \overline{b} + 2\overline{c} + h = \overline{a} + \overline{c} = c$. Therefore,

$$\begin{aligned} a &= d + \overline{a} \\ &= d + \overline{b} + \overline{c} + h \\ &= b + c + h \\ &= b + c\,. \end{aligned}$$

Therefore, $e \in J$, so we have proved that $I = J$.

Let $e_0, e_1 \in J$ and let $a, b, c \in M$ with $a + c = b + 2c$ and $c \leq^+ e_0 + e_1$. By Riesz decomposition, $c = c_0 + c_1$ for some $c_0 \leq^+ e_0$ and $c_1 \leq^+ e_1$. Since $a + c_0 + c_1 = b + 2c_0 + 2c_1$ and $c_1 \in J$, we get that $a + c_0 = b + 2c_0 + c_1$. Since $c_0 \in J$, it follows that $a = b + c_0 + c_1$, that is, $a = b + c$; whence $e_0 + e_1 \in J$. Therefore, J is a submonoid of M. □

Hence we get the following.

Corollary 2.7.7 *Let P be a generating lower interval in a refinement monoid M. Then M is strongly separative iff $x + z = y + 2z \in P$ implies that $y + z \leq^+ x$, for all $x, y, z \in P$. In particular, strong separativity is refinement-spreading.*

Proof It is trivial that the strong separativity of M implies the given condition. Suppose, conversely, that this condition holds. By keeping the notation of Lemma 2.7.6, this means that $P \subseteq I$. By that lemma, it follows that $P \subseteq J$, hence, since P generates M and J is an o-ideal, we get $J = M$, which means that M is strongly separative. □

2.7.4 Order-Separativity

A partial commutative monoid P is *order-separative* if $x \oplus z = y \oplus 2z$ implies that $x \leq^{\oplus} y \oplus z$, for all $x, y, z \in P$. This is easily seen to be equivalent to: whenever $x \oplus z = y \oplus z$ and $z \leq^{\oplus} y$, then $x \leq^{\oplus} y$. If P is a full commutative monoid, then this is also equivalent to saying that $x + z = 2z$ implies $x \leq^+ z$, for all $x, z \in P$.

Lemma 2.7.8 *Let M be a refinement monoid and let I and J be the subsets of M defined as*

$$I = \{e \in M \mid (\forall a, b, c \in M)(a + c = b + 2c \leq^+ e \Rightarrow a \leq^+ b + c)\}\,,$$

$$J = \{e \in M \mid (\forall a, b, c \in M)\big((a + c = b + 2c \text{ and } c \leq^+ e) \Rightarrow a \leq^+ b + c\big)\}\,.$$

Then $I = J$ and this set is an o-ideal of M.

Proof It is trivial that $J \subseteq I$ and that I and J are both lower subsets of M. Now let $a, b, c \in M$ and $e \in I$ such that $a+c = b+2c$ and $c \leq^+ e$. Let $d, \bar{a}, \bar{b}, \bar{c}$ be elements of M satisfying the conclusion of Lemma 2.7.5. Since $\bar{a} + \bar{c} = \bar{b} + 2\bar{c} = c \leq^+ e$ and $e \in I$, we get $\bar{a} \leq^+ \bar{b} + \bar{c}$, hence $a = d + \bar{a} \leq^+ d + \bar{b} + \bar{c} = b + c$, thus completing the proof that $e \in J$.

Let $e_0, e_1 \in J$ and let $a, b, c \in M$ with $a + c = b + 2c$ and $c \leq^+ e_0 + e_1$. By Riesz decomposition, $c = c_0 + c_1$ for some $c_0 \leq^+ e_0$ and $c_1 \leq^+ e_1$. Since $a + c_0 + c_1 = b + 2c_0 + 2c_1$ and $c_1 \in J$, we get that $a + c_0 \leq^+ b + 2c_0 + c_1$, so $a + c_0 + h = b + 2c_0 + c_1$ for some $h \in M$. Since $c_0 \in J$, it follows that $a + h \leq^+ b + c_0 + c_1$, that is, $a + h \leq^+ b + c$, so $a \leq^+ b + c$; whence $e_0 + e_1 \in J$. Therefore, J is a submonoid of M. □

In a similar manner as for Corollary 2.7.7, we get the following.

Corollary 2.7.9 *Let P be a generating lower interval in a refinement monoid M. Then M is order-separative iff $x + z = y + 2z$ implies that $x \leq^+ y + z$, for all $x, y, z \in P$. In particular, order-separativity is refinement-spreading.*

2.7.5 Separativity

A *partial refinement monoid P* is *separative* if $x \oplus 2z = y \oplus 2z$ implies that $x \oplus z = y \oplus z$, for all $x, y, z \in P$. This is easily seen to be equivalent to saying that whenever $x \oplus z = y \oplus z$, $z \leq^{\oplus} x$, and $z \leq^{\oplus} y$, then $x = y$.

We do not state the definition above for arbitrary, non-refinement partial commutative monoids. The reason for this is that even for full commutative monoids, the classical concept of separativity due to Hewitt and Zuckerman [59, § 4] (viz., $2x = 2y = x + y \Rightarrow x = y$) is stronger than the concept of separativity defined above (e.g., the commutative monoid freely generated by a, b subjected to the relations $2a = 2b = a + b$ satisfies the former definition of separativity, but not the latter). Nevertheless, it is well known (cf. Ara et al. [13, Lemma 2.1]) that the two concepts are equivalent for (full) refinement monoids, which will be sufficient for our purposes.

Lemma 2.7.10 *Let M be a refinement monoid and let I and J be the subsets of M defined as*

$$I = \{e \in M \mid (\forall a, b, c \in M)((a + c = b + c \leq^+ e \text{ and } c \leq^+ a, b) \Rightarrow a \leq^+ b))\},$$

$$J = \{e \in M \mid (\forall a, b, c \in M)((a + c = b + c \text{ and } c \leq^+ a, b, e) \Rightarrow a = b)\}.$$

Then $I = J$ and this set is an o-ideal of M.

Proof It is trivial that $J \subseteq I$ and that I and J are both lower subsets of M. Now let $e \in I$ and $a, b, c \in M$ such that $a + c = b + c$ and $c \leq^+ a, b, e$. By Ara et al. [13, Lemma 2.7], there is a refinement matrix of the form

$$\begin{array}{c|c|c} & b & c \\ \hline a & d & a' \\ \hline b & b' & c' \end{array} \tag{2.7.4}$$

with $c' \leq^+ a', b'$. Since $c = a' + c' = b' + c' \leq^+ e$, it follows from the assumption $e \in I$ that $a' \leq^+ b'$, so $b' = a' + h$ for some $h \in M$. It follows that

$$a + h = d + a' + h = d + b' = b\,,$$
$$c + h = a' + c' + h = b' + c' = c\,.$$

Further, from $c + h = c$ and $c \leq^+ a$ it follows that $a + h = a$; whence $a = b$, thus completing the proof that $e \in J$. Therefore, $I = J$.

Let $e_0, e_1 \in J$ and let $a, b, c \in M$ with $a + c = b + c$ and $c \leq^+ a, b, e_0 + e_1$. By Riesz decomposition, $c = c_0 + c_1$ for some $c_0 \leq^+ e_0$ and $c_1 \leq^+ e_1$. Since $a + c_0 + c_1 = b + c_0 + c_1$ and $c_1 \leq^+ a + c_0, b + c_0, e_1$, it follows from $e_1 \in J$ that $a + c_0 = b + c_0$. A similar argument, involving the assumption that $e_0 \in J$, yields $a = b$. Hence $e_0 + e_1 \in J$, so J is an o-ideal of M. □

In a similar manner as Corollary 2.7.7, we obtain the following corollary, contained in Chen [28, Theorem 1].

Corollary 2.7.11 *Let P be a generating lower interval in a refinement monoid M. Then M is separative iff $x + z = y + z \in P$ and $z \leq^+ x, y$ implies that $x \leq^+ y$, for all $x, y, z \in P$. In particular, separativity is refinement-spreading.*

Observe the obvious implications

cancellative ⇒ strongly separative ⇒ order-separative ⇒ separative ,

holding for all partial refinement monoids. Recall that none of those implications can be reversed, even for conical refinement monoids.

For arbitrary commutative monoids, order-separativity (i.e., $a + b = 2b$ implies $a \leq^+ b$) does not imply separativity (i.e., $2a = a + b = 2b$ implies $a = b$). However, the implication holds, non-trivially, for refinement monoids: this fact is contained in Corollary 2.7.11.

The following section shows that not every natural-looking cancellation property of commutative monoids is refinement-spreading.

2.7.6 Antisymmetry and Stable Finiteness

A partial commutative monoid P is

- *antisymmetric* if its algebraic preordering $\leq^{\oplus}$ is antisymmetric (i.e., if $\leq^{\oplus}$ is a partial ordering);
- *stably finite* if $y = x \oplus y$ implies that $x = 0$, for all $x, y \in P$ (see Sect. 1.5 for the case of full monoids).

It is clear that every stably finite commutative monoid is antisymmetric. The converse already fails for the two-element semilattice.

Denote by A the commutative monoid with underlying set $\{0, 1, 2, 3\}$, with addition defined by its zero element 0 together with the relations

$$1 + 1 = 2\,,\ 1 + 2 = 3\,,\ \text{and}\ 1 + 3 = 2\,.$$

The table of A is represented in Table 2.1. Observe that A is conical. Setting $a = 1$, the monoid A satisfies the relations $4a = 2a$ and $3a \neq 2a$. In particular, $3a \leq^{+} 2a \leq^{+} 3a$ and $3a \neq 2a$, thus A is not antisymmetric. Furthermore, a is a directly finite order-unit of A (i.e., $x + a = a$ implies $x = 0$, for any $x \in A$).

By Proposition 1.5 and Theorem 1.8 in Wehrung [123], A has a *unitary* embedding into a conical refinement monoid M. "Unitary" means that $\{u, u + x\} \subseteq A$ implies that $x \in A$, for all $u, x \in M$. In particular, a is directly finite in M, because $a + x = a$ implies $x \in A$ (we are using $a \in A$ and the unitarity), thus, by the direct finiteness of a in A, we get $x = 0$. We may replace M by $M \downarrow A$, which ensures that a is an order-unit of M. Nevertheless, M is not antisymmetric, because A is not. In particular, we get the following.

Proposition 2.7.12 *Neither antisymmetry nor stable finiteness is refinement-spreading.*

Those observations bear an intriguing connection with a well known open problem in the nonstable K-theory of (von Neumann) regular rings.

A unital ring R is directly finite if $xy = 1$ implies $yx = 1$, for all $x, y \in R$. We say that R is stably finite if all matrix rings over R are directly finite. These concepts can be read on the conical commutative monoid $\mathrm{V}(R)$ (cf. Sect. 1.3.4) and its partial submonoid $P = \{[x] \mid x \in R \text{ idempotent}\}$, where $[x]$ denotes the Murray–von Neumann equivalence class of the 1×1 matrix with entry x (cf. Goodearl [50, § 5]). Namely, R is stably finite iff $\mathrm{V}(R)$ is stably finite, and directly finite iff the

Table 2.1 A non-antisymmetric commutative monoid with directly finite order-unit

+	0	1	2	3
0	0	1	2	3
1	1	2	3	2
2	2	3	2	3
3	3	2	3	2

partial commutative monoid P is stably finite (i.e., $[x] + [1] = [1]$ iff $x = 0$, for every idempotent $x \in R$).

It is not known whether every directly finite unital regular ring is stably finite, see Goodearl [50, Problem 1]. This means that it is not known whether the stable finiteness of the partial refinement monoid $\{[x] \mid x^2 = x \in R\}$ always carries over to the full refinement monoid $\mathrm{V}(R)$.

In the following series of observations, we show that without the refinement property, even cancellativity may not spread from partial monoids to full monoids.

Example 2.7.13 A finite, conical, cancellative partial commutative monoid P which cannot be embedded into any cancellative full commutative monoid.

Proof If a partial commutative monoid P can be embedded into a cancellative full commutative monoid M, then it satisfies the following implication:

$$\left[a_i \oplus b_i = c_i \oplus d_i \ (\forall i \in \{1,2\}) \ \& \ b_1 \oplus b_2 = d_1 \oplus d_2 \ \& \ a = a_1 \oplus a_2 \ \& \ c = c_1 \oplus c_2\right] \Rightarrow a = c\,, \tag{2.7.5}$$

for all $a, c, a_1, a_2, b_1, b_2, c_1, c_2, d_1, d_2 \in P$. Indeed, arguing within M and adding together the first two equations in (2.7.5), we get $a + (b_1 + b_2) = c + (d_1 + d_2)$, thus, since $b_1 + b_2 = d_1 + d_2$ and M is cancellative, we get $a = c$.

A computer search, based on the `Mace4` component of McCune's `Prover9 - Mace4` software (cf. McCune [81]), yields the cancellative partial commutative monoid P, with universe $\{0, 1, 2, 3, 4, 5, 6, 7, 8\}$, whose table is represented in Table 2.2. The empty entries of the table correspond to the places where $\oplus$ is undefined. This partial commutative monoid is an *effect algebra*, that is, a cancellative partial commutative monoid with a largest element e such that $x \oplus e$ is defined iff $x = 0$, for every element x (here, $e = 8$); see, for example, Foulis and Bennett [47] for more information about effect algebras.

Take $a_1 = a_2 = d_1 = 2$, $b_1 = b_2 = c_1 = 4$, $c_2 = d_2 = 7$. Then $a_1 \oplus b_1 = c_1 \oplus d_1 = a_2 \oplus b_2 = c_2 \oplus d_2 = 6$, $b_1 \oplus b_2 = d_1 \oplus d_2 = 5$, $a_1 \oplus a_2 = 1$, and $c_1 \oplus c_2 = 3$. Hence P does not satisfy (2.7.5). □

Table 2.2 A cancellative effect algebra

$\oplus$	0	1	2	3	4	5	6	7	8
0	0	1	2	3	4	5	6	7	8
1	1	8	3						
2	2	3	1	8	6			5	
3	3		8						
4	4		6		5	8		3	
5	5				8				
6	6							8	
7	7		5		3		8	6	
8	8								

Remark 2.7.14 Example 2.7.13 does not fit the intuition assigning to the symbol $\oplus$ the direct sum of (say) subspaces in a vector space. For example, in that example, $x \oplus x$ may be defined without x being zero (e.g., take $x = 1$).

An occurrence of the context of direct sums is given by the theory of *orthomodular lattices*. By definition, an orthomodular lattice (cf. Kalmbach [65]) is a lattice L, with a least element 0 and a largest element 1, endowed with an antitone, involutive unary operation $x \mapsto x^{\perp}$ such that $x \wedge x^{\perp} = 0$ and $x \vee x^{\perp} = 1$ for all $x \in L$ (we say that $x \mapsto x^{\perp}$ is an *orthocomplementation* on L), and $y = x \vee (x^{\perp} \wedge y)$ for all $x, y \in L$ with $x \leq y$. Then L can also be endowed a structure of a partial commutative monoid, by letting $z = x \oplus y$ hold if $z = x \vee y$ and $y \leq x^{\perp}$.

Elaborating on Greechie's construction of a finite orthomodular lattice with no non-trivial state (cf. Greechie [54]), Navara [89] and Weber [117] independently constructed finite orthomodular lattices with no non-trivial group-valued premeasure (a *premeasure* on L, with values in an Abelian group G, is just a homomorphism of partial commutative monoids from $(L, \oplus, 0)$ to $(G, +, 0)$).

Letting L be any orthomodular lattice with no non-trivial group-valued premeasure, we obtain that $(L, \oplus, 0)$ cannot be embedded into any cancellative commutative monoid (otherwise it could also be embedded into an Abelian group, and the corresponding embedding would be a group-valued premeasure).

This peculiar behavior of Navara's and Weber's orthomodular counterexamples is to be put in contrast with the following example, which, despite its apparent simplicity, is not devoid of further mysteries (see, in particular, Chap. 7, Problem 10).

Example 2.7.15 For any real or complex Hilbert space H, we endow the set $\operatorname{Sub} H$ of all closed subspaces of H with a structure of a partial commutative monoid, by letting $X \oplus Y$ denote the direct sum of X and Y whenever it is defined. We claim that $(\operatorname{Sub} H, \oplus, \{0\})$ embeds into an Abelian group. Recall that a *projection* of H is an idempotent, self-adjoint operator. The set $\operatorname{Proj} H$ of all projections of H can be endowed with a structure of a partial commutative monoid, by letting $r = p \oplus q$ if $r = p + q$ and $pq = 0$ (thus also $qp = 0$). The assignment that sends any projection p to its range defines an isomorphism of partial commutative monoids from $\operatorname{Proj} H$ onto $\operatorname{Sub} H$. The inverse isomorphism sends any closed subspace X to the orthogonal projection on X.

The inclusion map defines a premeasure from $\operatorname{Proj} H$ with values in the additive group $\operatorname{sAdj} H$ of all self-adjoint operators of H. Hence, $(\operatorname{Proj} H, \oplus, 0)$ *has a one-to-one embedding into the Abelian group* $(\operatorname{sAdj} H, +, 0)$. The submonoid $\operatorname{Proj}^{+} H$ generated by the range of that embedding consists exactly of the finite sums of projections of H. In particular, every element of $\operatorname{Proj}^{+} H$ is a positive self-adjoint operator of H. The study of those operators (finite sums of projections) was initiated in Fillmore [45].

Certainly, $\operatorname{Proj}^{+} H$ is a monoid quotient of the enveloping monoid $\mathrm{U}_{\mathrm{mon}}(\operatorname{Proj} H)$ of $\operatorname{Proj} H$. It turns out that $\operatorname{Proj}^{+} H$ is a *proper* quotient of $\mathrm{U}_{\mathrm{mon}}(\operatorname{Proj} H)$, whenever $\dim H \geq 2$. To see this, we may assume that $\dim H = 2$ and then we use the following expression of $(3/2)\mathrm{id}_H$ as a sum of three projections, given in matrix form in Rabanovich and Samoĭlenko [95, 2.6]:

$$(3/2)\mathrm{id}_H = p_0 + p_1 + p_2\,, \qquad \text{where}$$

$$p_0 = \begin{pmatrix} 0 & 0 \\ 0 & 1 \end{pmatrix}, \quad p_1 = \begin{pmatrix} 3/4 & \sqrt{3}/4 \\ \sqrt{3}/4 & 1/4 \end{pmatrix}, \quad p_2 = \begin{pmatrix} 3/4 & -\sqrt{3}/4 \\ -\sqrt{3}/4 & 1/4 \end{pmatrix}. \tag{2.7.6}$$

An important observation about (2.7.6) is that the projections p_i are pairwise non-commuting. In particular, they are pairwise non-orthogonal.

Now we abbreviate id_H by $\mathbf{1}$ and we write (2.7.6) in the form

$$2p_0 + 2p_1 + 2p_2 = 3 \cdot \mathbf{1}\,. \tag{2.7.7}$$

We claim that Eq. (2.7.7) does *not* hold in the enveloping monoid $\mathrm{U_{mon}}(\mathrm{Proj}\,H)$ (although we just observed that it holds in $\mathrm{Proj}^+ H$). In order to prove this, it suffices to prove that $u \not\equiv v$, where $u = 2\dot{p}_0 + 2\dot{p}_1 + 2\dot{p}_2$, $v = 3 \cdot \dot{\mathbf{1}}$, and $\equiv$ is the equivalence relation on the free commutative monoid $\mathrm{F_{mon}}(\mathrm{Proj}\,H)$ introduced just before Proposition 2.1.7, satisfying $\mathrm{U_{mon}}(\mathrm{Proj}\,H) = \mathrm{F_{mon}}(\mathrm{Proj}\,H)/{\equiv}$. Now observe that for each $i \in \{0, 1, 2\}$, p_i is a projection of rank one, thus the equality $p_i = x \oplus y$ in $\mathrm{Proj}\,H$ implies that either $x = 0$ or $y = 0$. Hence, as the p_i are pairwise non-orthogonal, it is easy to verify that $u \equiv w$ iff $u \stackrel{\circ}{=} w$, for all $w \in \mathrm{F_{mon}}(\mathrm{Proj}\,H)$. Since u and v are not equivalent modulo $\stackrel{\circ}{=}$, it follows that $u \not\equiv v$, that is, (2.7.7) does not hold in $\mathrm{U_{mon}}(\mathrm{Proj}\,H)$. Therefore, *the canonical homomorphism from* $\mathrm{U_{mon}}(\mathrm{Proj}\,H)$ *onto* $\mathrm{Proj}^+ H$ *is not one-to-one.*

For $\dim H = 2$, the lattice $\mathrm{Sub}\,H$ is infinite of length two. Denote by Δ the set of all atoms of $\mathrm{Sub}\,H$, that is, the one-dimensional subspaces of H. Hence, $\mathrm{U_{mon}}(\mathrm{Sub}\,H)$ is the commutative monoid defined by the generators e and m_X, for $X \in \Delta$, and the relations $m_X + m_{X^\perp} = e$ for $X \in \Delta$. It can be verified that this monoid is cancellative. We do not know whether the latter result extends to the case where $\dim H \geq 3$ (cf. Chap. 7, Problem 10).

2.8 Quotient of a Commutative Monoid Under a Group Action

Let us begin this section with some motivation. For a left action $(g, x) \mapsto g(x)$ of a group G on a ring R, the *crossed product* $R \rtimes G$ consists of all formal linear combinations $\sum_{i<n} x_i \cdot g_i$, where all $x_i \in R$ and all $g_i \in G$, where the multiplication is determined by $(x \cdot g)(y \cdot h) = xg(y) \cdot gh$, for all $x, y \in R$ and all $g, h \in G$.

In particular, if the action of G is trivial, then $R \rtimes G$ becomes the group ring $R[G]$.

The construction $M /\!/ G$, which we shall introduce now, intervenes at various places, which we will not attempt to enumerate. (For an example where M is the positive cone of $K_0(A)$, for a C*-algebra A, see Rainone [96, Definition 4.6].) We intend it here mainly to describe the effect of the crossed product on the nonstable K-theory. A more detailed form of that statement will be given in Proposition 2.8.4. Later on, we will examine that statement again in the context of Boolean inverse semigroups (cf. Theorem 4.1.10).

Definition 2.8.1 Let α be an action of a group G, by automorphisms, on a commutative monoid M. We set

$$x \sim_\alpha y \text{ if } (\exists g \in G)(y = \alpha_g(x)), \quad \text{for all } x, y \in M. \tag{2.8.1}$$

We denote by $\simeq_\alpha$ the monoid congruence generated by $\sim_\alpha$, and we denote the quotient monoid by $M/\!\!/\alpha = M/{\simeq_\alpha}$. We say that elements $x, y \in M$ are *α-equidecomposable* if $x \simeq_\alpha y$.

For any $x \in M$, we shall denote by $[x]_\alpha$ the equivalence class of x modulo $\simeq_\alpha$ and we let $\mu_\alpha : M \twoheadrightarrow M/\!\!/\alpha$, $x \mapsto [x]_\alpha$.

Notation 2.8.2 In the context of Definition 2.8.1, we shall write $\sim_G$, $\simeq_G$, $[x]_G$, μ_G, $M/\!\!/G$, and "G-equidecomposable", instead of $\sim_\alpha$, $\simeq_\alpha$, $[x]_\alpha$, μ_α, $M/\!\!/\alpha$, and "α-equidecomposable" whenever this does not arise confusion.

Moreover, if $G \cong \mathbb{Z}/2\mathbb{Z}$ and $\sigma = \alpha_1 \in \operatorname{Aut} M$ (this will be the case for all counterexamples of this section), then we shall identify the group action α with the involution σ, and thus we shall write $\sim_\sigma$, $\simeq_\sigma$, $[x]_\sigma$, μ_σ, $M/\!\!/\sigma$ instead of $\sim_\alpha$, $\simeq_\alpha$, $[x]_\alpha$, μ_α, $M/\!\!/\alpha$.

The structure $M/\!\!/G$ should not be confused with the orbit space M/G: for one thing, the former is a monoid, while the latter is just a set.

For a commutative monoid N, a monoid homomorphism $\mu : M \to N$ is *G-invariant* if $\mu(gx) = \mu(x)$ for every $(g, x) \in G \times M$. In such a case, μ is also an *invariant for G-equidecomposability*, that is, $x \simeq_G y$ implies $\mu(x) = \mu(y)$, for all $x, y \in M$. Moreover, we will say that μ is a *complete invariant for G-equidecomposability* if $\mu(x) = \mu(y)$ implies $x \simeq_G y$, for all $x, y \in M$.

By definition, the monoid homomorphism $\mu_G : M \twoheadrightarrow M/\!\!/G$ is a complete invariant for G-equidecomposability.

The following lemma is essentially a reformulation of Definition 2.8.1, with a straightforward proof.

Lemma 2.8.3 *The map μ_G is universal for the G-invariant monoid homomorphisms from M to a commutative monoid. That is, for every commutative monoid N and every G-invariant monoid homomorphism $\varphi : M \to N$, there exists a unique monoid homomorphism $\psi : M/\!\!/G \to N$ such that $\varphi = \psi \circ \mu_G$.*

Of course, ψ is defined by the rule $\psi([x]_G) = \varphi(x)$.

The connection between the crossed product construction for rings (i.e., $R \rtimes G$) and the construction $M/\!\!/G$ (for monoids) is given by the following result.

Proposition 2.8.4 *Let a group G act by automorphisms on a ring R. We endow the commutative monoid $\mathrm{V}(R)$ with the induced group action, that is, $g \cdot [x]_R = [g(x)]_R$, for every idempotent matrix x over R. Then there is a unique monoid homomorphism*

$\tau\colon \mathrm{V}(R)/\!\!/G \to \mathrm{V}(R \rtimes G)$ such that $\tau([[x]_R]_G) = [x]_{R\rtimes G}$ for every idempotent matrix x over R.

Proof The canonical map $R \hookrightarrow R \rtimes G$, $x \mapsto x \cdot 1$ induces a monoid homomorphism $\varphi\colon \mathrm{V}(R) \to \mathrm{V}(R \rtimes G)$. By Lemma 2.8.3, it suffices to prove that φ is G-invariant. Let a be an idempotent matrix over R and let $g \in G$. Setting $x = g(a) \cdot g$ and $y = a \cdot g^{-1}$, we get $g(a) \cdot 1 = xy$ and $a \cdot 1 = yx$; whence $a \cdot 1 \sim g(a) \cdot 1$ (Murray–von Neumann equivalence) within $R \rtimes G$. □

The homomorphism τ of Proposition 2.8.4 may not be an isomorphism. For example, in the conditions of Proposition 6.7.1, the canonical homomorphism $\tau\colon \mathrm{V}(\Bbbk)/\!\!/G \to \mathrm{V}(\Bbbk[G])$ is one-to-one, but not surjective, for any division ring $\Bbbk$.

Nonetheless, we will see in Theorem 4.1.10 that the analogue of Proposition 2.8.4, for so-called Boolean inverse semigroups (cf. Sect. 3.1), always yields an isomorphism.

Lemma 2.8.5 *Let a group G act by automorphisms on a commutative monoid M and let H be a subgroup of G. Then there is a unique monoid homomorphism $\mu_G^H\colon M/\!\!/H \to M/\!\!/G$ such that $\mu_G^H \circ \mu_H = \mu_G$. This map is surjective.*

Proof This follows trivially from μ_G being H-invariant, together with Lemma 2.8.3. □

In particular, if H is a subgroup of G, then $M/\!\!/G$ is a monoid quotient of $M/\!\!/H$. In case H is a normal subgroup of G, we shall now give more information. Thinking of Lemma 2.8.3 as the "First Isomorphism Theorem" for the structures $M/\!\!/G$, the "Second Isomorphism Theorem" is the following.

Proposition 2.8.6 *Let a group G act by automorphisms on a commutative monoid M and let H be a normal subgroup of G. The following statements hold:*

(1) *There is a unique group action of G/H on $M/\!\!/H$ sending every pair $(gH, [x]_H)$ to $[gx]_H$.*
(2) *The assignments $[x]_G \mapsto [[x]_H]_{G/H}$ and $[[x]_H]_{G/H} \mapsto [x]_G$ are well defined, and define mutually inverse isomorphisms between $M/\!\!/G$ and $(M/\!\!/H)/\!\!/(G/H)$. Therefore, $M/\!\!/G \cong (M/\!\!/H)/\!\!/(G/H)$.*

Proof (1) Let $x, y \in M$ with $x \sim_H y$ and let $g \in G$. By definition, $y = hx$ for some $h \in H$. Hence $gy = ghx = (ghg^{-1})gx$. Since H is a normal subgroup of G, $gy \sim_H gx$. It follows that $\sim_H$ is contained in the binary relation

$$\boldsymbol{\theta} = \{(x, y) \in M \times M \mid gx \simeq_H gy\}\,.$$

Since $\boldsymbol{\theta}$ is a monoid congruence of M containing $\sim_H$, it also contains $\simeq_H$. This means that for every $g \in G$, there is a unique map $\tau_g\colon M/\!\!/H \to M/\!\!/H$ such that $\tau_g([x]_H) = [gx]_H$ for all $x \in M$. Obviously, τ_g is a monoid homomorphism. Moreover, it is obvious that $\tau_g([x]_H)$ depends only on $Hg(= gH)$. Hence, the assignment $(gH, [x]_H) \mapsto [gx]_H$ is well defined. It is straightforward that it defines a group action of G/H on $M/\!\!/H$ by monoid automorphisms.

(2) For every $(g,x) \in G \times M$, $[gx]_H = (gH)[x]_H \sim_{G/H} [x]_H$. Hence $\sim_G$ is contained in the binary relation

$$\xi = \{(x,y) \in M \times M \mid [x]_H \simeq_{G/H} [y]_H\} .$$

Since ξ is a monoid congruence of M, it also contains $\simeq_G$. Hence, the map $\varphi: M/\!\!/G \to (M/\!\!/H)/\!\!/(G/H)$, $[x]_G \mapsto [[x]_H]_{G/H}$ is well defined. It is obviously a monoid homomorphism.

Conversely, for all $x, y \in M$ such that $[x]_H \sim_{G/H} [y]_H$, there is $g \in G$ such that $[y]_H = (gH)[x]_H$, that is, $gx \simeq_H y$, thus $x \simeq_G y$. Hence, the binary relation

$$\begin{aligned}\boldsymbol{\eta} &= \{(\boldsymbol{x},\boldsymbol{y}) \in (M/\!\!/H) \times (M/\!\!/H) \mid (\exists (x,y) \in \boldsymbol{x} \times \boldsymbol{y})(x \simeq_G y)\}\\ &= \{(\boldsymbol{x},\boldsymbol{y}) \in (M/\!\!/H) \times (M/\!\!/H) \mid (\forall (x,y) \in \boldsymbol{x} \times \boldsymbol{y})(x \simeq_G y)\}\end{aligned}$$

contains $\sim_{G/H}$. Since $\boldsymbol{\eta}$ is a monoid congruence of $M/\!\!/H$, it thus contains $\equiv_{G/H}$. Hence, the map $\psi: (M/\!\!/H)/\!\!/(G/H) \to M/\!\!/G$, $[[x]_H]_{G/H} \mapsto [x]_G$ is well defined. It is obviously a monoid homomorphism. It is trivial that φ and ψ are mutually inverse. □

The case of a group acting on a refinement monoid is especially straightforward.

Proposition 2.8.7 *Let a group G act on a refinement monoid M. Then $\simeq_G$ is a V-equivalence on M, and $M/\!\!/G$ is a refinement monoid. Furthermore, if M is conical, then so are the monoid $M/\!\!/G$ and the relation $\simeq_G$.*

Proof We first claim that $\sim_G$ is a V-equivalence: indeed, if $x_0 + x_1 \sim_G y$ in M, then $y = g(x_0 + x_1)$ for some $g \in G$, thus, setting $y_i = gx_i$, we get $y = y_0 + y_1$ with each $x_i \sim_G y_i$, thus proving our claim. By Lemma 2.4.3, it follows that the additive closure $\sim_G^+$ of $\sim_G$ in $M \times M$ is an additive V-equivalence on M. In particular, it is a monoid congruence of M. Since it contains $\sim_G$, it also contains $\simeq_G$. Since the converse containment is trivial, the relations $\simeq_G$ and $\sim_G^+$ are identical, and so $\simeq_G$ is an additive V-equivalence on M. By Lemma 2.4.4, $M/\!\!/G = M/{\simeq_G}$ is a refinement monoid.

The last statement, about conicality, is trivial. □

In case M is a refinement monoid, then the V-equivalence $\simeq_G$ is the additive closure of $\sim_G$ within $M \times M$, hence it can be explicitly defined as follows: for any $a, b \in M$, $a \simeq_G b$ holds iff there are decompositions $a = \sum_{i<n} a_i$ and $b = \sum_{i<n} g_i a_i$, where all $a_i \in M$ and $g_i \in G$. Let us highlight the following trivial consequence of Lemma 2.4.4(1).

Lemma 2.8.8 *Let a group G act on a refinement monoid M, let $\boldsymbol{a}, \boldsymbol{b} \in M/\!\!/G$, and let $c \in M$. If $[c]_G = \boldsymbol{a} + \boldsymbol{b}$, then there is $(a,b) \in \boldsymbol{a} \times \boldsymbol{b}$ such that $c = a + b$.*

As the following observations show, a significant class of examples arises from the construction $\mathbb{Z}^+\langle B\rangle = \mathrm{U}_{\mathrm{mon}}(B, \oplus, 0)$ given in Example 2.2.7.

Example 2.8.9 Let B be a Boolean ring. Observe that the automorphisms of the Boolean ring B are identical to the automorphisms of the partial refinement monoid $(B, \oplus, 0)$ (cf. Example 2.2.7). Since $\mathbb{Z}^+\langle B\rangle$ is the enveloping monoid of $(B, \oplus, 0)$, every automorphism α of B extends to a unique automorphism of the monoid $\mathbb{Z}^+\langle B\rangle$. This automorphism sends $\mathbf{1}_u$ to $\mathbf{1}_{\alpha(u)}$, for each $u \in B$. In particular, every action of a group G on the Boolean ring B extends to an action of G on the monoid $\mathbb{Z}^+\langle B\rangle$, thus making it possible to define the commutative monoid $\mathbb{Z}^+\langle B\rangle /\!\!/ G = \mathbb{Z}^+\langle B\rangle / {\simeq_G}$. The equivalence relation $\simeq_G$ is given by

$$u \simeq_G v \iff \Big(\exists \text{ decompositions } u = \sum_{i<n} \mathbf{1}_{a_i} \text{ and } v = \sum_{i<n} \mathbf{1}_{g_i a_i}\,, \quad \text{with all } a_i \in B \text{ and } g_i \in G\Big)\,.$$

An equivalent way to define $\mathbb{Z}^+\langle B\rangle /\!\!/ G$, taking advantage of $(B, \oplus, 0)$ being a partial refinement monoid, is the following. Since the orbital equivalence relation $\sim_G$ defines a V-equivalence on B, its additive closure $\sim_G^+$ *within* B is an additive V-equivalence on B (cf. Lemma 2.4.3). The relation $\sim_G^+$ is nothing else than the G-equidecomposability relation on B: namely,

$$a \sim_G^+ b \iff \Big(\exists \text{ decompositions } a = \bigoplus_{i<n} a_i \text{ and } b = \bigoplus_{i<n} g_i a_i\,, \quad \text{with all } a_i \in B \text{ and } g_i \in G\Big)\,.$$

By Lemma 2.4.4, this makes it possible to define the partial refinement monoid $B /\!\!/ G = B / {\sim_G^+}$. Furthermore, by Theorem 2.4.6, $\sim_G^+$ extends to a unique V-equivalence on $\mathrm{U_{mon}}(B, \oplus, 0) = \mathbb{Z}^+\langle B\rangle$, which is necessarily $\simeq_G$. By Theorem 2.4.6 again, $\mathbb{Z}^+\langle B\rangle /\!\!/ G \cong \mathrm{U_{mon}}(B /\!\!/ G)$.

We shall often write $[a]_G$ instead of $[\mathbf{1}_a]_G$, for $a \in B$. Since $\mathbb{Z}^+\langle B\rangle /\!\!/ G = \mathrm{U_{mon}}(B /\!\!/ G)$, every element of $\mathbb{Z}^+\langle B\rangle /\!\!/ G$ can be written in the form $\sum_{i<n} [a_i]_G$, where all $a_i \in B$.

2.9 Cancellation Properties of $M /\!\!/ G$

The question whether various cancellativity properties of M are inherited by $M /\!\!/ G$ will come up regularly in this work. It turns out that the study of that question leads to various nontrivial positive statements, but also to surprising counterexamples.

In this section we shall focus on positive statements, in the context of the cancellation properties of $M /\!\!/ G$, where a *finite* group G acts by automorphisms on a *refinement monoid* M. Although a few positive results can be proved, lots of surprising counterexamples arise.

Definition 2.9.1 Let a finite group G act by automorphisms on a commutative monoid M. The *G-trace* of an element $x \in M$ is defined as

$$\tau_G(x) = \sum_{g \in G} gx\,. \tag{2.9.1}$$

Proposition 2.9.2 *In the context of Definition* 2.9.1, *there is a unique monoid homomorphism $\overline{\tau}_G\colon M /\!\!/ G \to M$ such that $\overline{\tau}_G \mu_G = \tau_G$.*

Proof The map τ_G is a G-invariant monoid homomorphism from M to M. Apply Lemma 2.8.3. □

In particular, $a \simeq_G b$ implies that $\tau_G(a) = \tau_G(b)$, for all $a, b \in M$. The question whether τ_G is a *complete invariant* for $\simeq_G$, that is, $\tau_G(a) = \tau_G(b)$ implies $a \simeq_G b$, arises. This already leads us to an easy counterexample (we use Notation 2.8.2).

Example 2.9.3 An action σ of $\mathbb{Z}/2\mathbb{Z}$ by automorphisms on the additive group $\mathbb{Z}$ of all integers, such that τ_σ is not a complete invariant for $\simeq_\sigma$.

Proof Set $\sigma(x) = -x$, for any $x \in \mathbb{Z}$. Two integers x and y are σ-equidecomposable iff there are $u, v \in \mathbb{Z}$ such that $x = u + v$ and $y = u - v$, that is, iff $x - y$ is even. On the other hand, τ_σ is the constant zero map. In particular, $\tau_\sigma(0) = \tau_\sigma(1)$ while 0 and 1 are not σ-equidecomposable. □

Proposition 2.9.4 *Let a (not necessarily finite) group G act by automorphisms on an Abelian group M. Then the monoid $M /\!\!/ G$ is an Abelian group.*

Proof The canonical map $\mu_G\colon M \twoheadrightarrow M /\!\!/ G$ is a surjective monoid homomorphism. Since M is an Abelian group, so is $M /\!\!/ G$. □

By contrast, it follows from Example 2.9.3 that the power-cancellativity of M may not be inherited by $M /\!\!/ G$: for that example, $2[0]_G = 2[1]_G$ and $[0]_G \neq [1]_G$. Hence, power cancellativity is not propagated from M to $M /\!\!/ G$. Nevertheless, as the following result shows, stable finiteness propagates.

Proposition 2.9.5 *Let a finite group G act by automorphisms on a conical commutative monoid M. If M is stably finite, then so is $M /\!\!/ G$.*

Proof Let $\boldsymbol{a}, \boldsymbol{b} \in M /\!\!/ G$ such that $\boldsymbol{a} + \boldsymbol{b} = \boldsymbol{b}$. Write $\boldsymbol{a} = [a]_G$ and $\boldsymbol{b} = [b]_G$ for some $a, b \in M$. By applying the homomorphism $\overline{\tau}_G$ of Proposition 2.9.2 to the equation $\boldsymbol{a} + \boldsymbol{b} = \boldsymbol{b}$, we get the equation $\tau_G(a) + \tau_G(b) = \tau_G(b)$. Since M is stably finite, $\tau_G(a) = 0$. Since M is conical, $a = 0$, so $\boldsymbol{a} = 0$. □

The completeness of the trace invariant can be easily reached under an additional divisibility assumption.

Proposition 2.9.6 *Let a finite group G, with m elements, act by automorphisms on a uniquely m-divisible commutative monoid M (i.e., every element of M can be written in the form my for a unique $y \in M$). Then the trace function τ_G is a complete invariant for G-equidecomposability on M. In particular, if M is cancellative, then so is $M /\!\!/ G$.*

Proof Let $a, b, e \in M$ such that $e = \tau_G(a) = \tau_G(b)$. Denote by $(1/m)x$ the unique $y \in M$ such that $x = my$, for any $x \in M$. From $e = \sum_{g\in G} ga$ it follows that $(1/m)e = \sum_{g\in G} g((1/m)a)$. Since $a = \sum_{g\in G}(1/m)a$, it follows that $e \simeq_G a$. Likewise, $e \simeq_G b$. Therefore, $a \simeq_G b$, thus completing the proof that τ_G is a complete invariant. The statement about cancellativity follows trivially. □

In case of the action of a finite group on the positive cone of an Abelian lattice-ordered group, it turns out that the trace is also a complete invariant.

Theorem 2.9.7 *The following statements hold, for any action of a finite group G by automorphisms on an Abelian lattice-ordered group M:*

(1) *The G-trace map $\tau_G\colon M^+ \to M^+$ is a complete invariant for G-equidecomposability on M^+. In fact, $[a]_G = [b]_G$ iff $\tau_G(a) = \tau_G(b)$, and $[a]_G \leq^+ [b]_G$ iff $\tau_G(a) \leq \tau_G(b)$, for all $a, b \in M^+$.*
(2) *The monoid $M /\!/ G$ is the positive cone of an Abelian lattice-ordered group.*

Proof Since M is a dimension group (cf. Goodearl [49, Proposition 1.22]), M^+ is an unperforated, conical refinement monoid (cf. Goodearl [49, Proposition 2.1]).

(1) Let $a, b \in M^+$ such that $\tau_G(a) = \tau_G(b)$. We must prove that $[a]_G = [b]_G$. Set $m = \operatorname{card} G$ and fix an enumeration $G = \{g_1 = 1, g_2, \dots, g_m\}$. We argue by descending induction on the largest $k \leq m$, denoted by $\nu(a, b)$, such that $g_i a \wedge b = 0$ whenever $1 \leq i < k$. Suppose first that $k = m$, that is, $ga \wedge b = 0$ for all $g \in G$. Since M is lattice-ordered, $\tau_G(a) \wedge \tau_G(b) = 0$, hence, since $\tau_G(a) = \tau_G(b)$, we get $\tau_G(a) = \tau_G(b) = 0$, whence $a = b = 0$.

Suppose now that $k < m$ and set $a' = a - (a \wedge g_k^{-1} b)$ and $b' = b - (g_k a \wedge b)$. Then $0 \leq a' \leq a$ and $0 \leq b' \leq b$, thus $g_i a' \wedge b' = 0$ whenever $0 \leq i < k$. Since $g_k a' \wedge b' = 0$, we get $\nu(a', b') > \nu(a, b)$. Furthermore, $g_k a \wedge b = g_k(a \wedge g_k^{-1} b)$, thus $\tau_G(a') = \tau_G(b')$. Hence, by the induction hypothesis, $[a']_G = [b']_G$. Since $a = a' + a \wedge g_k^{-1} b$ and $b = b' + g_k(a \wedge g_k^{-1} b)$, it follows that $[a]_G = [b]_G$.

The proof that $\tau_G(a) \leq \tau_G(b)$ iff $[a]_G \leq^+ [b]_G$ is, *mutatis mutandis*, the same.

(2) It follows from (1) above that $M^+ /\!/ G$ is cancellative (*Proof*: if $[a]_G + [c]_G = [b]_G + [c]_G$, then $\tau_G(a) + \tau_G(c) = \tau_G(a + c) = \tau_G(b + c) = \tau_G(b) + \tau_G(c)$, thus $\tau_G(a) = \tau_G(b)$, and thus, by (1), $[a]_G = [b]_G$). Since M^+ is a conical refinement monoid, so is $M^+ /\!/ G$ (cf. Proposition 2.8.7).

Since $M^+ /\!/ G$ is cancellative, the canonical map $M^+ /\!/ G \to M /\!/ G$ that to $[x]_G$ (within M^+) associates $[x]_G$ (within M) is one-to-one. Hence we will identify any element of $M^+ /\!/ G$ with its image in $M /\!/ G$ via this embedding, and we will endow $M /\!/ G$ with the positive cone $M^+ /\!/ G$. We extend the map τ_G to the whole monoid M via the formula (2.9.1). By definition, this map is isotone.

To any elements $a, b \in M^+$, we associate finite sequences $(a_i \mid 0 \leq i \leq m)$ and $(b_i \mid 0 \leq i \leq m)$ of elements of M^+ by

$$a_0 = a\,, \qquad b_0 = b \tag{2.9.2}$$

$$a_{k+1} = a_k - (a_k \wedge g_{k+1}^{-1} b_k)\,, \quad b_{k+1} = b_k - (g_{k+1} a_k \wedge b_k) \quad \text{whenever } 0 \leq k < m\,. \tag{2.9.3}$$

Set $\overline{a} = \sum_{0 \le k < m}(a_k \wedge g_{k+1}^{-1} b_k)$ and $\overline{b} = \sum_{0 \le k < m}(g_{k+1}a_k \wedge b_k)$. From $a_k \wedge g_{k+1}^{-1} b_k = g_{k+1}(a_k \wedge g_{k+1}^{-1} b_k)$ whenever $0 \le k < m$, it follows that

$$\overline{a} \simeq_G \overline{b}\,. \tag{2.9.4}$$

Further, it follows easily from (2.9.2) and (2.9.3) that

$$a = \overline{a} + a_m \text{ and } b = \overline{b} + b_m\,. \tag{2.9.5}$$

By (2.9.5), it follows that $[\overline{a}]_G \le [a]_G, [b]_G$ in $M /\!\!/ G$. By using (2.9.3), we get

$$g_k a_k \wedge b_k = 0 \text{ whenever } 1 \le k \le m\,.$$

Since both finite sequences $(a_k \mid 0 \le k \le m)$ and $(b_k \mid 0 \le k \le m)$ are decreasing, we get $ga_m \wedge b_m = 0$ for every $g \in G$. Since M is an Abelian lattice-ordered group, this means that

$$\tau_G(a_m) \wedge \tau_G(b_m) = 0 \text{ within } M\,. \tag{2.9.6}$$

In order to prove that $[\overline{a}]_G = [a]_G \wedge [b]_G$, it suffices to prove that for any $c \in M$ such that $[c]_G \le [a]_G, [b]_G$, the inequality $[c]_G \le [\overline{a}]_G$ holds. From $[c]_G \le [a]_G, [b]_G$ it follows that $\tau_G(c) \le \tau_G(a), \tau_G(b)$, thus, by (2.9.5), $\tau_G(c - \overline{a}) \le \tau_G(a_m)$ and further, by using (2.9.4) and (2.9.5),

$$\tau_G(c - \overline{a}) = \tau_G(c) - \tau_G(\overline{a}) = \tau_G(c) - \tau_G(\overline{b}) = \tau_G(c - \overline{b}) \le \tau_G(b_m)\,.$$

By (2.9.6), it follows that $\tau_G(c - \overline{a}) \le 0$, so $\tau_G(c) \le \tau_G(\overline{a})$. By (1) above, it follows that $[c]_G \le [\overline{a}]_G$. □

Recall that a group is *locally finite* if every finitely generated subgroup is finite.

Corollary 2.9.8 *Let a locally finite group G act by automorphisms on an Abelian lattice-ordered group M. Then $M^+ /\!\!/ G$ is the positive cone of a structure of a dimension group on $M /\!\!/ G$.*

Proof Let $a, b, c \in M^+$ such that $a + c \simeq_G b + c$. There are decompositions $a + c = \sum_{i<n} x_i$ and $b + c = \sum_{i<n} g_i x_i$ in M^+, with all $g_i \in G$. Since G is locally finite, $\{g_i \mid i < n\}$ is contained in a finite subgroup H of G. Since $a + c \simeq_H b + c$ and by Theorem 2.9.7, it follows that $a \simeq_H b$, thus $a \simeq_G b$. This proves that $M^+ /\!\!/ G$ is cancellative. Since every Abelian lattice-ordered group is unperforated, a similar argument shows that $M^+ /\!\!/ G$ is unperforated.

Since $M^+ /\!\!/ G$ is cancellative, the canonical map from $M^+ /\!\!/ G$ to $M /\!\!/ G$ is a monoid embedding. Since the partially ordered Abelian group M, with its positive cone M^+, is directed, the partially ordered Abelian group $M /\!\!/ G$, with its positive cone $M^+ /\!\!/ G$, is also directed. □

It will follow from Theorem 5.1.10 that conversely, every positive cone of a dimension group, of cardinality at most $\aleph_1$, arises as $M^+/\!\!/G$ for a locally finite group G acting by automorphisms on an Abelian lattice-ordered group M.

2.10 Partially Ordered Abelian Groups and V-Equivalences with Bad Quotients

As we will see in Examples 2.10.7 and 2.10.8, Theorem 2.9.7 cannot be extended from lattice-ordered groups to dimension groups. We will also discover in this section a few additional positive results, notably involving the concept of weak comparability (cf. Sect. 1.6).

All the counterexamples of this section are based on the following construction.

Notation 2.10.1 For an arbitrary partially ordered Abelian group G and a nonnegative integer m, we shall denote by $E_{G,n}$ the set of all sequences $(x_n \mid n \in \mathbb{Z}^+)$ of elements of G such that

$$\text{For every integer } n \geq m\,, \quad x_{2n+2} = \sum_{k<2n} x_k \text{ and } x_{2n+3} = x_{2n} + x_{2n+1}\,.$$

We shall also set $E_G = \bigcup_{n\in\mathbb{Z}^+} E_{G,n}$.

It is trivial that the $E_{G,m}$ are all additive subgroups of $G^{\mathbb{Z}^+}$. Since they form an increasing chain (with respect to containment), E_G is also an additive subgroup of $G^{\mathbb{Z}^+}$. Hence the $E_{G,m}$ and E_G are all partially ordered subgroups of the partially ordered Abelian group $G^{\mathbb{Z}^+}$ (endowed with its componentwise ordering).

Lemma 2.10.2 *All partially ordered Abelian groups $E_{G,m}$ are finite powers of G. In particular, if G is a dimension group, then so are E_G and all $E_{G,n}$.*

Proof Every element of $E_{G,m}$ is determined by its first $2m+2$ coordinates, which yields $E_{G,m} \cong G^{2m+2}$. In particular, if G is a dimension group, then so are all $E_{G,m}$, thus so is E_G. □

Denote by $\rho_G\colon G \twoheadrightarrow G/2G$ the canonical projection map. The following lemma records a few elementary properties of the $E_{G,m}$ and E_G.

Lemma 2.10.3 *The following statements hold, for every nonnegative integer m, every $x \in E_{G,m}$, and every integer $n \geq m$:*

(1) *The equalities $x_{2n+4} = x_{2n+5}$ and $x_{2n+6} = x_{2n+7} = 2x_{2n+4}$ hold for every integer $n \geq m$.*
(2) $\sum_{k<n+2} \rho_G(x_{2k}) = \sum_{k<n+2} \rho_G(x_{2k+1})$.

Proof (1) From $x \in E_{G,m}$ it follows that $x_{2n+4} = \sum_{k<2n} x_k + x_{2n} + x_{2n+1}$, thus

$$x_{2n+5} = x_{2n+2} + x_{2n+3} = \sum_{k<2n} x_k + x_{2n} + x_{2n+1} = x_{2n+4}\,. \tag{2.10.1}$$

It follows that $x_{2n+7} = x_{2n+4} + x_{2n+5} = 2x_{2n+4}$. By using (2.10.1) with $n+1$ instead of n, we get $x_{2n+7} = x_{2n+6}$.

(2) By (1) above, $x_{2n+6} \in 2G$. Since $x_{2n+6} = \sum_{k<n+2} x_{2k} + \sum_{k<n+2} x_{2k+1}$, it follows that $\sum_{k<n+2} \rho_G(x_{2k}) + \sum_{k<n+2} \rho_G(x_{2k+1}) = 0$, which is equivalent to the desired conclusion. □

It follows from Lemma 2.10.3(1) that $x_n \in 2G$ (i.e., $\rho_G(x_n) = 0$), for all sufficiently large n, whenever $x \in E_G$. Hence, both expressions $\sum_{k\in\mathbb{Z}^+} \rho_G(x_{2k})$ and $\sum_{k\in\mathbb{Z}^+} \rho_G(x_{2k+1})$ are defined. By Lemma 2.10.3(2), these two elements of $G/2G$ are equal. We denote their common value by $\rho_G(x)$.

Now denote by σ the self-map of $\mathbb{Z}^+$ that interchanges $2n$ and $2n+1$, whenever $n \in \mathbb{Z}^+$.

Lemma 2.10.4 *The assignment* $\sigma_G : x \mapsto (x_{\sigma n} \mid n \in \mathbb{Z}^+)$ *defines an involutive automorphism of the dimension group* E_G.

Proof It follows immediately from Lemma 2.10.3(1) that σ_G maps $E_{G,m}$ into $E_{G,m+1}$. The other statements of Lemma 2.10.4 are trivial. □

The proof of the following lemma is straightforward and we leave it to the reader.

Lemma 2.10.5 *The map* ρ_G *is a* σ_G*-invariant homomorphism from* E_G *to* $G/2G$.

Theorem 2.10.6 *Suppose that G is a 2-power cancellative interpolation group. Then the pair* $(\mathrm{id} + \sigma_G, \rho_G)$ *is a complete invariant for* σ_G*-equidecomposability within* E_G^+. *In particular, the commutative monoid* $E_G^+ /\!\!/ \sigma_G$ *is cancellative.*

Of course, $\mathrm{id} + \sigma_G$ is the trace function associated to σ_G (cf. Definition 2.9.1).

Proof We must prove that any $a, b \in E_G^+$ such that $a + \sigma_G(a) = b + \sigma_G(b)$ and $\rho_G(a) = \rho_G(b)$ are σ_G-equidecomposable. By Lemma 2.10.3, there is a nonnegative integer m such that

$$a_{2n} = a_{2n+1} \text{ belongs to } 2G\,, \tag{2.10.2}$$

$$b_{2n} = b_{2n+1} \text{ belongs to } 2G\,, \tag{2.10.3}$$

$$a_{2n+2} = a_{2n+3} = \sum_{k<2n} a_k = 2a_{2n}\,, \tag{2.10.4}$$

$$b_{2n+2} = b_{2n+3} = \sum_{k<2n} b_k = 2b_{2n} \tag{2.10.5}$$

for every integer $n \geq m$. Since $a + \sigma_G(a) = b + \sigma_G(b)$, it follows from (2.10.2) and (2.10.3) that $2a_{2n} = 2b_{2n}$. Since G is 2-power cancellative, $a_{2n} = b_{2n}$, for every integer $n \geq m$. Hence, again by (2.10.2) and (2.10.3),

$$a_n = b_n \text{ belongs to } 2G\,, \quad \text{whenever } 2m \leq n \text{ in } \mathbb{Z}^+\,. \tag{2.10.6}$$

For every $k < m$, we pick any $x_{2k} \in G$ such that $\dfrac{0}{b_{2k} - a_{2k+1}} \leq x_{2k} \leq \dfrac{a_{2k}}{b_{2k}}$. Such an element certainly exists, because G has interpolation and $a + \sigma_G(a) = b + \sigma_G(b)$, thus $\dfrac{0}{b - \sigma_G(a)} \leq \dfrac{a}{b}$. Furthermore, we set

$$x_{2k+1} = x_{2k} + b_{2k+1} - a_{2k}\,, \quad \text{whenever } 0 \leq k < m\,. \tag{2.10.7}$$

Claim 1 The inequalities $\dfrac{0}{b_n - a_{\sigma(n)}} \leq x_n \leq \dfrac{a_n}{b_n}$ hold whenever $0 \leq n < 2m$.

Proof of Claim. For $n = 2k$, this follows from the definition of x_{2k}. By adding the term $b_{2k+1} - a_{2k} = -(b_{2k} - a_{2k+1})$ to the inequalities around x_{2k}, we obtain

$$\frac{0}{b_{2k+1} - a_{2k}} \leq x_{2k+1} \leq \frac{b_{2k+1}}{b_{2k} + b_{2k+1} - a_{2k}}\,,$$

which, since $b_{2k} + b_{2k+1} - a_{2k} = a_{2k+1}$, yields the desired inequalities. □ Claim 1.

Claim 2 $\sum_{k<2m} x_k$ belongs to $2G$.

Proof of Claim. For each $k < m$, $x_{2k} + x_{2k+1} = b_{2k+1} - a_{2k} + 2x_{2k} \equiv b_{2k+1} - a_{2k} \pmod{2G}$. Now by (2.10.2) and (2.10.3), $\sum_{k<m} \rho_G(a_{2k}) = \rho_G(a)$ and $\sum_{k<m} \rho_G(b_{2k+1}) = \rho_G(b)$. Hence, $\sum_{k<2m} \rho_G(x_k) = \sum_{k<m} \rho_G(x_{2k} + x_{2k+1}) = \rho_G(b) - \rho_G(a) = 0$. □ Claim 2.

It follows from Claims 1 and 2 that the element $x_{2m} = \frac{1}{2} \sum_{k<2m} x_k$ belongs to G^+. (We denote by $\frac{1}{2}x$ the unique $y \in G$ such that $x = 2y$, for any $x \in 2G$.) Set $x_{2m+1} = x_{2m}$.

Claim 3 The inequalities $0 \leq x_n \leq \dfrac{a_n}{b_n}$ hold whenever $0 \leq n \leq 2m + 1$.

Proof of Claim. For $0 \leq n < 2m$ this follows from Claim 1. For $n = 2m$, this follows from the inequalities

$$\begin{aligned} x_{2m} = \frac{1}{2} \sum_{k<2m} x_k \leq \frac{1}{2} \sum_{k<2m} a_k & \qquad \text{(use Claim 1)} \\ = a_{2m} & \qquad \text{[use (2.10.4)]}\,, \end{aligned}$$

and similarly, using (2.10.5), we get $x_{2m} \leq b_{2m}$. Since $x_{2m} = x_{2m+1}$ and by (2.10.2) and (2.10.3), we get the desired inequalities at $n = 2m + 1$. □ Claim 3.

We can now complete the construction of the x_n, by defining, inductively,

$$x_{2n+2} = \sum_{k<2n} x_k\,, \tag{2.10.8}$$

$$x_{2n+3} = x_{2n} + x_{2n+1}\,, \tag{2.10.9}$$

whenever $m \leq n$ in $\mathbb{Z}^+$. By construction, the sequence $x = (x_n \mid n \in \mathbb{Z}^+)$ belongs to E_G.

Claim 4 $0 \leq x \leq \dfrac{a}{b}$.

Proof of Claim. By Claim 3 together with (2.10.8) and (2.10.9) it follows, inductively, that $0 \leq x$. Let n be a nonnegative integer, we must prove that $0 \leq x_n \leq \dfrac{a_n}{b_n}$. If $n \leq 2m + 1$ this follows immediately from Claim 3. Suppose that $n = 2k + 2$, where $m \leq k$ in $\mathbb{Z}^+$. We compute

$$x_{2k+2} = \sum_{l<2k} x_l \leq \sum_{l<2k} a_l = a_{2k+2}\,,$$

and similarly, $x_{2k+2} \leq b_{2k+2}$. Since $x_{2k+2} = x_{2k+3}$, $a_{2k+2} = a_{2k+3}$, and $b_{2k+2} = b_{2k+3}$, the desired conclusion follows. □ Claim 4.

Claim 5 $x - \sigma_G(x) = b - \sigma_G(a)$.

Proof of Claim. We need to prove that $x_n - x_{\sigma(n)} = b_n - a_{\sigma(n)}$, for every nonnegative integer n. If $n < 2m$ this is taken care of by (2.10.7) together with the equation $b_{2k+1} - a_{2k} = a_{2k+1} - b_{2k}$. If $n \in \{2m, 2m+1\}$ this follows from the equations $a_{2m} = a_{2m+1} = b_{2m} = b_{2m+1}$ (cf. (2.10.2), (2.10.3), and (2.10.6)) together with the equations $x_{2m+1} = x_{2m} = \frac{1}{2}\sum_{k<2m} x_k$. Let $k \geq m$ such that $x_{2k} = x_{2k+1} = \frac{1}{2}\sum_{l<2k} x_l$. Then

$$x_{2k+3} = x_{2k} + x_{2k+1} = \sum_{l<2k} x_l = x_{2k+2}\,,$$

thus also $x_{2k+2} = x_{2k+3} = \frac{1}{2}\sum_{l<2k+2} x_l$. Therefore, the equations $x_{2k} = x_{2k+1} = \frac{1}{2}\sum_{l<2k} x_l$ hold for every integer $k \geq m$. Since $a_{2k} = b_{2k} = a_{2k+1} = b_{2k+1}$ for any such integer k, the desired conclusion follows. □ Claim 5.

The element $y = a - x$ belongs to E_G^+, and $a = x + y$. By Claim 5, $b = x + \sigma_G(y)$. Therefore, $a \simeq_G b$ within E_G^+.

Since the complete invariant $(1 + \sigma_G, \rho_G)$ takes its values in the cancellative monoid $E_G^+ \times (G/2G)$, it follows that $E_G^+ /\!\!/ \sigma_G$ is cancellative. □

By using Theorem 2.10.6, we can now construct the promised series of counterexamples.

Example 2.10.7 A simple dimension group E, with an action σ of $\mathbb{Z}/2\mathbb{Z}$, such that the monoid $E_G^+ /\!\!/ \sigma$ is cancellative, yet the trace function $\mathrm{id} + \sigma$ on E^+ is not a complete invariant for σ-equidecomposability.

It will turn out (cf. Corollary 2.10.11) that the cancellativity of $E_G^+ /\!\!/ \sigma$ is no surprise. Nevertheless, for the present example, we have a complete invariant, given by Theorem 2.10.6.

Proof Define G as the set of all fractions of the form p/q, where p and q are integers with q odd. Then G is a dense additive subgroup of the rationals, with $1 \in G$ and $1/2 \notin G$. Since G is 3-divisible (i.e., $3G = G$), so is E_G, thus every order-unit of E_G is the sum of two order-units. Since E_G is a dimension group and by Proposition 1.5.5, it follows that the structure $E = (E_G, \ll)$ [cf. (1.5.7)] is a dimension group. Moreover, σ_G restricts to an involutive automorphism σ of E. Denote by ρ the restriction of ρ_G to E. The element

$$e = (1, 1, 1, 1, 2, 2, 4, 4, 8, 8, 16, 16, \dots)$$

is an order-unit of E_G (and E), fixed under σ.

Claim The pair $(\mathrm{id} + \sigma, \rho)$ is a complete invariant for σ-equidecomposability within E^+. In particular, $E_G^+ /\!\!/ \sigma$ is cancellative.

Proof of Claim. Let $a, b \in E^+$ such that $a + \sigma(a) = b + \sigma(b)$ and $\rho(a) = \rho(b)$. We must prove that $a \simeq_\sigma b$ within E^+. If either a or b is zero then $a = b = 0$ and the statement is trivial. Suppose that a and b are both nonzero. Since a and b are both order-units, there is an integer n such that $2 \cdot 3^{-n} \cdot e \leq {}^{a}_{b}$. Since $\sigma_G(e) = e$, it follows that $a' = a - 3^{-n}e$ and $b' = b - 3^{-n}e$ are elements of E_G^+, with $a' + \sigma_G(a') = b' + \sigma_G(b')$ and $\rho_G(a') = \rho_G(b')$. By Theorem 2.10.6, it follows that $a' \simeq_{\sigma_G} b'$ within E_G^+, that is, there are $x, y \in E_G^+$ such that $a' = x + y$ and $b' = x + \sigma_G(y)$. Therefore, $a = (x + 3^{-n}e) + (y + 3^{-n}e)$ and $b = (x + 3^{-n}e) + \sigma(y + 3^{-n}e)$ are σ-equidecomposable within E^+. □ Claim.

Now we set

$$a = (1, 0, 1, 0, 1, 1, 2, 2, 4, 4, 8, 8, 16, 16, \dots), \tag{2.10.10}$$

$$b = (1, 0, 0, 1, 1, 1, 2, 2, 4, 4, 8, 8, 16, 16, \dots). \tag{2.10.11}$$

The elements a and b both belong to E_G^+, with $a + \sigma_G(a) = b + \sigma_G(b)$, $\rho_G(a) = 1$, and $\rho_G(b) = 0$. Hence, the elements $a' = a + e$ and $b' = b + e$ are order-units, with $a' + \sigma(a') = b' + \sigma(b')$ and $\rho(a') \neq \rho(b')$. In particular, $a' \not\simeq_\sigma b'$. □

We are now ready to introduce the most elaborate example of this section.

Example 2.10.8 A dimension group M, with an involutive automorphism σ, such that the monoid $M^+ /\!\!/ \sigma$ is not cancellative.

Proof Since $\mathbb{Z}^2$ and $E_\mathbb{Z}$ are both dimension groups, an easy application of Goodearl [49, Corollary 2.12] yields that the lexicographical product $F = \mathbb{Z}^2 \times_{\mathrm{lex}} E_\mathbb{Z}$ is a dimension group. We define an involutive automorphism σ of F, by setting

$$\sigma((m, n), x) = ((n, m), \sigma_\mathbb{Z}(x)), \quad \text{for all } ((m, n), x) \in F.$$

Consider the elements a and b of $E_{\mathbb{Z}}$ defined in (2.10.10) and (2.10.11), and set

$$d = a - b = (0, 0, 1, -1, 0, 0, 0, 0, \dots).$$

Denote by H the subgroup of F generated by $((1, -1), d)$. From $\sigma_{\mathbb{Z}}(d) = -d$ it follows that $\sigma((1, -1), d) = ((-1, 1), -d)$. Hence $\sigma[H] = H$.

Claim 1

(1) $H \cap F^+ = \{0\}$.
(2) H is a convex subgroup of F.
(3) $((0, 0), x) \in H$ iff $x = 0$, for every $x \in E_{\mathbb{Z}}$.

Proof of Claim. (1) Let $n \in \mathbb{Z}$. If $((n, -n), nd) \in F^+$, then $(n, -n) \in (\mathbb{Z}^2)^+ = \mathbb{Z}^+ \times \mathbb{Z}^+$, thus $n = 0$.

(2) follows trivially from (1).

(3) is trivial. □ Claim 1.

Since H is a convex subgroup of F, $M = F/H$ is a partially ordered Abelian group, with positive cone $M^+ = (F^+ + H)/H$. Since F is directed, so is M.

Set $[(m, n), x] = ((m, n), x) + H \in F/H$, for every $((m, n), x) \in F$. From $\sigma[H] = H$ it follows that σ induces an involutive automorphism $\overline{\sigma}$ of M, defined by

$$\overline{\sigma}(t + H) = \sigma(t) + H \quad \text{for every } t \in F.$$

The value of $m + n$ is unchanged by adding any element of H to $((m, n), x)$. This enables us to define $\ell([(m, n), x]) = m + n$, for all $((m, n), x) \in F$.

Claim 2

(1) Every element $\boldsymbol{x} \in M$ can be written in the form $[(\ell(\boldsymbol{x}), 0), x]$ for a unique $x \in E_{\mathbb{Z}}$.
(2) The map ℓ is a positive homomorphism from M onto $\mathbb{Z}$.
(3) $\ell(\boldsymbol{x}) > 0$ implies that $\boldsymbol{x} \in M^{++}$, for every $\boldsymbol{x} \in M$.

Proof of Claim. (1) Write $\boldsymbol{x} = [(m, n), y]$. Then $\boldsymbol{x} = [(m, n), y] + [(n, -n), nd] = [(\ell(\boldsymbol{x}), 0), x]$ where we put $x = y + nd$. The uniqueness statement on x follows from Claim 1(3).

(2) follows from $\ell \restriction_{F^+} \geq 0$ together with $\ell \restriction_H = 0$.

(3) follows trivially from (1). □ Claim 2.

Claim 3 $[(m, n), x] \in M^+$ iff one of the following statements holds:

(i) $m + n > 0$.
(ii) $m + n = 0$, $-x_3 \leq m \leq x_2$, and $x_k \geq 0$ for every nonnegative integer $k \notin \{2, 3\}$ (Here and elsewhere in this proof, we write $x = \big(x_n \mid n \in \mathbb{Z}^+\big)$, for any $x \in E_{\mathbb{Z}}$).

Consequently, M is unperforated.

Proof of Claim. Since $m+n>0$ implies that $[(m,n),x]\in M^+$, which implies in turn that $m+n\geq 0$ (cf. Claim 2), the only case we need to consider is the one where $m+n=0$. Then $[(m,n),x]\in M^+$ iff $((m,-m),x)\in F^+ + H$, iff there is $k\in\mathbb{Z}$ such that $((m-k,k-m),x-kd)\in F^+$. Since the latter condition implies that $k=m$, it follows that $[(m,n),x]\in M^+$ iff $x-md\in E_{\mathbb{Z}}^+$, which is easily seen to be equivalent to the given condition.

Since Conditions (i) and (ii) are unchanged by positive scaling, M is unperforated. □ Claim 3.

Claim 4 M is a dimension group.

Proof of Claim. By our previous claims, it suffices to verify that M has the interpolation property. Let $\boldsymbol{x},\boldsymbol{y}_0,\boldsymbol{y}_1\in M$ such that $\begin{matrix}0\\ \boldsymbol{x}\end{matrix}\leq\begin{matrix}\boldsymbol{y}_0\\ \boldsymbol{y}_1\end{matrix}$, we must find $\boldsymbol{z}\in M$ such that $\begin{matrix}0\\ \boldsymbol{x}\end{matrix}\leq\boldsymbol{z}\leq\begin{matrix}\boldsymbol{y}_0\\ \boldsymbol{y}_1\end{matrix}$. We may assume that the pairs $(0,\boldsymbol{x})$ and $(\boldsymbol{y}_0,\boldsymbol{y}_1)$ are both incomparable in M. By Claim 2, $\ell(\boldsymbol{x})=0$ and $\ell(\boldsymbol{y}_1)=\ell(\boldsymbol{y}_2)\geq 0$. Set $m=\ell(\boldsymbol{y}_1)$. By Claim 2, there are (unique) $x,y_0,y_1\in E_{\mathbb{Z}}$ such that $\boldsymbol{x}=[(0,0),x]$ and $\boldsymbol{y}_i=[(m,0),y_i]$ whenever $i\in\{0,1\}$. We separate cases.

Case 1. $m>0$. Since $E_{\mathbb{Z}}$ is directed, there is $z\in E_{\mathbb{Z}}$ such that $\begin{matrix}0\\ x\end{matrix}\leq z$. Then $\boldsymbol{z}=[(0,0),z]$ is as required.

Case 2. $m=0$. It follows from Claim 3 that $\begin{matrix}0\\ x\end{matrix}\leq\begin{matrix}y_0\\ y_1\end{matrix}$. Since $E_{\mathbb{Z}}$ is an interpolation group, there is $z\in E_{\mathbb{Z}}$ such that $\begin{matrix}0\\ x\end{matrix}\leq z\leq\begin{matrix}y_0\\ y_1\end{matrix}$. Then $\boldsymbol{z}=[(0,0),z]$ is as required. □ Claim 4.

Now we can conclude the proof of Example 2.10.8. We define elements of M^{++} by

$$\boldsymbol{a}=[(0,0),a]\,,\quad \boldsymbol{b}=[(0,0),b]\,,\quad \boldsymbol{c}=[(1,0),0]\,.$$

Observe that $\boldsymbol{a}+\boldsymbol{c}=[(1,0),a]$ and $\boldsymbol{b}+\sigma(\boldsymbol{c})=[(0,1),b]$. By the definition of H, we get $\boldsymbol{a}+\boldsymbol{c}=\boldsymbol{b}+\sigma(\boldsymbol{c})$, thus $\boldsymbol{a}+\boldsymbol{c}\simeq_\sigma\boldsymbol{b}+\boldsymbol{c}$. Suppose that $\boldsymbol{a}\simeq_\sigma\boldsymbol{b}$. There are $\boldsymbol{x},\boldsymbol{y}\in M^+$ such that $\boldsymbol{a}=\boldsymbol{x}+\boldsymbol{y}$ and $\boldsymbol{b}=\boldsymbol{x}+\sigma(\boldsymbol{y})$. From $\ell(\boldsymbol{a})=\ell(\boldsymbol{b})=0$ it follows that $\ell(\boldsymbol{x})=\ell(\boldsymbol{y})=0$, thus (cf. Claim 2) there are $x,y\in E_{\mathbb{Z}}^+$ such that $\boldsymbol{x}=[(0,0),x]$ and $\boldsymbol{y}=[(0,0),y]$. It follows that $a=x+y$ and $b=x+\sigma_{\mathbb{Z}}(y)$, so $a\simeq_{\sigma_{\mathbb{Z}}}b$, thus (cf. Lemma 2.10.5) $\rho_{\mathbb{Z}}(a)=\rho_{\mathbb{Z}}(b)$, a contradiction since $\rho_{\mathbb{Z}}(a)=1$ and $\rho_{\mathbb{Z}}(b)=0$. □

The dimension group constructed in Example 2.10.8 is not simple. In view of Example 2.10.7, this raises the question whether Example 2.10.8 could be achieved with M a simple dimension group. As shown by the following Theorems 2.10.9 and 2.10.10, together with Corollary 2.10.11, this is not the case. The crucial concepts are the ones of strict unperforation and weak comparability (cf. Sect. 1.6).

Theorem 2.10.9 *Let a finite group G act by automorphisms on a conical commutative monoid M. If M satisfies weak comparability, then so does $M /\!/ G$.*

Proof Since M is conical, the statements $x = 0$, $[x]_G = 0$, and $\tau_G(x) = 0$ are equivalent, for all $x \in M$. Hence we must prove that given $e, y \in M \setminus \{0\}$, the set $\operatorname{comp}([y]_G : [e]_G)$ is nonempty. Since M has weak comparability, $\tau_G(e)$ belongs to the weak comparability set of M, thus $\operatorname{comp}(y : \tau_G(e))$ is nonempty. Let k be an element of that set and let $\boldsymbol{x} \in M /\!/ G$ such that $k\boldsymbol{x} \leq^+ [e]_G$. Write $\boldsymbol{x} = [x]_G$, where $x \in M$. By applying the homomorphism $\overline{\tau}_G$ of Proposition 2.9.2, it follows that $k\tau_G(x) \leq^+ \tau_G(e)$, thus, since $k \in \operatorname{comp}(y : \tau_G(e))$, we get $\tau_G(x) \leq^+ y$, thus, a fortiori, $x \leq^+ y$, and thus $\boldsymbol{x} \leq^+ [y]_G$. Therefore, k belongs to $\operatorname{comp}([y]_G : [e]_G)$. □

Theorem 2.10.10 *Let a finite group G act by automorphisms on a commutative monoid M. If M is simple, strictly unperforated, and cancellative, then so is $M /\!/ G$.*

Proof If M is not conical, then it is an Abelian group (cf. Proposition 1.6.6) and everything is trivial. Suppose from now on that M is conical. We already know that $M /\!/ G$ is a conical refinement monoid. Furthermore, the simplicity of M obviously carries over to $M /\!/ G$. Since M is strictly unperforated, it has weak comparability (cf. Proposition 1.6.8). By Theorem 2.10.9, $M /\!/ G$ also has weak comparability. Since M is cancellative, it is stably finite, thus, by Proposition 2.9.5, $M /\!/ G$ is also stably finite. By Ara and Pardo [11, Corollary 1.8], $M /\!/ G$ is cancellative. By Proposition 1.6.8 again, $M /\!/ G$ is strictly unperforated. □

Corollary 2.10.11 *Let a finite group G act by automorphisms on a simple dimension group M. Then the monoid $M^+ /\!/ G$ is cancellative.*

Recall from Example 2.10.7 that $M^+ /\!/ G$ may not be power cancellative (for that example, $M^+ /\!/ G$ is not 2-power cancellative).

The final counterexample of this section, although not directly related to problems of equidecomposability, is also obtained via the construction E_G. This example solves Problem 31 of Goodearl [49], in the negative.

Example 2.10.12 A dimension group E, with an involutive automorphism σ, such that the subgroup $E^\sigma = \{x \in E \mid \sigma(x) = x\}$ does not satisfy the interpolation property.

Proof Consider again the dimension group $E = E_{\mathbb{Z}}$, now endowed with the involutive automorphism σ defined by

$$\sigma(x) = (x_0, x_1, x_3, x_2, x_5, x_4, x_7, x_6, \dots)\,, \quad \text{whenever } x \in E_{\mathbb{Z}}\,.$$

It follows that E^σ consists of all the sequences $x = \left(x_n \mid n \in \mathbb{Z}^+\right)$ of integers such that $x_{2n} = x_{2n+1}$ whenever $n \in \mathbb{N}$, and $x_{2n+2} = 2x_{2n} = x_0 + x_1 + 2\sum_{1 \leq k < n} x_{2k}$ for all sufficiently large n. In particular,

$$x_0 + x_1 \text{ is even, for every } x \in E^\sigma . \tag{2.10.12}$$

We consider the following interpolation problem in E^σ.

$$\begin{matrix}(0,0,0,0,0,0,\ldots) \\ (1,-1,0,0,0,0,\ldots)\end{matrix} \leq x \leq \begin{matrix}(2,0,1,1,2,2,4,4,\ldots) \\ (1,1,1,1,2,2,4,4,\ldots)\end{matrix}$$

Suppose that this problem has a solution $x \in E^\sigma$. Necessarily,

$$\begin{matrix}(0,0) \\ (1,-1)\end{matrix} \leq (x_0, x_1) \leq \begin{matrix}(2,0) \\ (1,1)\end{matrix},$$

thus $(x_0, x_1) = (1, 0)$, in contradiction with (2.10.12). □

Chapter 3
Boolean Inverse Semigroups and Additive Semigroup Homomorphisms

Tarski investigates in [109] partial commutative monoids constructed from *partial bijections* on a given set. In Kudryavtseva et al. [71], this study is conveniently formalized in the language of *inverse semigroups*. Further connections can be found in works on K-theory of rings, such as Ara and Exel [7].

By definition, a partial bijection on a set Ω is a bijection from a subset of Ω onto another subset of Ω. Partial bijections can be composed, by letting $(g \circ f)(x)$ be defined if $f(x)$ is defined and belongs to the domain of g. Instead of forming a group, the partial bijections on Ω form an *inverse semigroup*. Moreover, two partial bijections f and g, with disjoint domains and ranges, can be added, by defining their orthogonal join $f \oplus g$ as the smallest common extension of f and g. This brings us naturally to the widely studied concept of a *Boolean inverse semigroup*, in particular Lawson [74, 75], Lawson and Lenz [76] (the original definition of a Boolean inverse semigroup got subsequently extended by no longer assuming the existence of finite meets; see Sect. 3.1.2 for more detail). While the literature contains a number of interesting weakenings of the concept of Boolean inverse semigroup, most notably the one of *distributive inverse semigroup* (cf. Lawson [74], Lawson and Scott [77]), Boolean inverse semigroups will take up most of our discussion, mostly due to our ring-theoretical emphasis and the results of Sect. 6.1.

In Sect. 3.1 we recall a few basic results on inverse semigroups and Boolean inverse semigroups, in particular emphasizing with Proposition 3.1.9 that they are distributive, and beginning the discussion of additivity in Sect. 3.1.3.

In Sect. 3.2, we prove that the category of all Boolean inverse semigroups, with additive semigroup homomorphisms, is identical to a variety of algebras (in the sense of universal algebra) that we call *biases*, with their homomorphisms.

In Sect. 3.3, we discuss the faithfulness of Exel's regular representation, defined for any inverse semigroup, with emphasis on the class of all Boolean inverse semigroups. We also present a variant of this representation which is valid for all distributive inverse semigroups, as a specialization of a duality theorem due to Lawson and Lenz [76].

F. Wehrung, *Refinement Monoids, Equidecomposability Types, and Boolean Inverse Semigroups*, Lecture Notes in Mathematics 2188,
DOI 10.1007/978-3-319-61599-8_3

In Sect. 3.4 we study the bias congruences of a given Boolean inverse semigroup, in terms of the semigroup operations and the orthogonal join. We also describe the congruence associated with an additive ideal.

Section 3.5 is of preparatory nature, and it introduces a minor extension of the concept of *generalized rook matrices* introduced in Kudryavtseva et al. [71]. The results of that section are applied in Sect. 3.6 to an extension, to the class of all Boolean inverse semigroups, of the ring-theoretical concept of *crossed product*.

Section 3.7 introduces some material on two subclasses of Boolean inverse semigroups, called *fundamental Boolean inverse semigroups* and *Boolean inverse meet-semigroups*.

Section 3.8 is devoted to a brief study of *inner automorphisms* of a Boolean inverse semigroup, which can be defined even in the non-unital case.

Our main textbook references on inverse semigroups will be Howie [60] and Lawson [73].

3.1 Boolean Inverse Semigroups

3.1.1 Arbitrary Inverse Semigroups

We first recall a few classical definitions. Let S be a semigroup (i.e., a set endowed with an associative binary operation). For $x, y \in S$, we say that y is a *quasi-inverse* (resp., an *inverse*) of x if $x = xyx$ (resp., $x = xyx$ and $y = yxy$).

Recall (cf. Howie [60]) that S is

- a *regular semigroup* if every element of S has a quasi-inverse (this is consistent with Definition 1.5.1),
- an *inverse semigroup* if every $x \in S$ has a unique inverse, then denoted by x^{-1}. The assignment $x \mapsto x^{-1}$ is the *inversion map* of S.

Every semigroup homomorphism between inverse semigroups is also a homomorphism of inverse semigroups. We denote by $\operatorname{Idp} S$ the set of all idempotent elements of S. A regular semigroup S is an inverse semigroup iff all the idempotent elements of S commute (cf. Howie [60, Theorem V.1.2]). In that case, $(xy)^{-1} = y^{-1}x^{-1}$ for all $x, y \in S$, and e idempotent implies that xex^{-1} is also idempotent (cf. Howie [60, Proposition V.1.4]).

For the remainder of this section we shall fix an inverse semigroup S. We set $XY = \{xy \mid (x, y) \in X \times Y\}$, $aX = \{a\}X$, $Xa = X\{a\}$, and $X^{-1} = \{x^{-1} \mid x \in X\}$, for all $a \in S$ and all $X, Y \subseteq S$.

We set $\mathbf{d}(x) = x^{-1}x$ (the *domain* of x) and $\mathbf{r}(x) = xx^{-1}$ (the *range* of x), for any $x \in S$. Both $\mathbf{d}(x)$ and $\mathbf{r}(x)$ are idempotent.

Recall that *Green's relations* $\mathscr{L}$, $\mathscr{R}$, $\mathscr{D}$, $\mathscr{H}$, and $\mathscr{J}$ can be defined on S by

$$x \mathscr{L} y \quad \text{if} \quad \mathbf{d}(x) = \mathbf{d}(y)\,;$$
$$x \mathscr{R} y \quad \text{if} \quad \mathbf{r}(x) = \mathbf{r}(y)\,,$$

$\mathscr{D} = \mathscr{L} \circ \mathscr{R} = \mathscr{R} \circ \mathscr{L}$ (cf. Howie [60, Proposition II.1.3]), $\mathscr{H} = \mathscr{L} \cap \mathscr{R}$, and

$$x \mathscr{J} y \quad \text{if} \quad SxS = SyS\,.$$

Every congruence of S with respect to the semigroup structure is also a congruence with respect to the inverse semigroup structure.

The following very useful lemma, contained in Schein [101], yields an alternate characterization of inverse semigroups. We include a proof for convenience.

Lemma 3.1.1 *Let $(S, \cdot)$ be a semigroup and let $i \colon S \to S$ be a map satisfying the following conditions:*

(I1) $x = x \cdot i(x) \cdot x$ *for all* $x \in S$.
(I2) $i(x \cdot y) = i(y) \cdot i(x)$ *for all* $x, y \in S$.
(I3) $i(i(x)) = x$ *for all* $x \in S$.
(I4) $i(x) \cdot x \cdot x \cdot i(x) = x \cdot i(x) \cdot i(x) \cdot x$ *for all* $x \in S$.

Then S is an inverse semigroup, with inversion map i.

Proof By applying (I1) to $i(x)$, we obtain, by virtue of (I3),

$$i(x) \cdot x \cdot i(x) = i(x)\,, \quad \text{for all } x \in S\,. \tag{3.1.1}$$

By (I1) and (3.1.1), $i(x)$ is an inverse of x.

We claim that $i(e) = e$, for every idempotent element e of S. Indeed, by applying (I4), we obtain, by virtue of (I3),

$$i(e) \cdot e \cdot e \cdot i(e) = e \cdot i(e) \cdot i(e) \cdot e\,,$$

hence, as e and $i(e)$ are both idempotent (use (I2)),

$$i(e) \cdot e \cdot i(e) = e \cdot i(e) \cdot e\,.$$

By (I1) and (3.1.1), this means that $i(e) = e$, thus proving our claim.

Since every element of S has an inverse [use (I1) and (3.1.1)], it suffices, in order to reach the desired conclusion, to prove that any idempotent elements a and b of S commute. By (I2) together with the claim above,

$$i(a \cdot b) = b \cdot a\,. \tag{3.1.2}$$

By applying (I1) to $x = a \cdot b$, we thus obtain that $a \cdot b = a \cdot b \cdot b \cdot a \cdot a \cdot b$, that is, $a \cdot b = a \cdot b \cdot a \cdot b$, which means that $a \cdot b$ is idempotent. By the claim above, $i(a \cdot b) = a \cdot b$. By (3.1.2), it thus follows that $a \cdot b = b \cdot a$. □

Let $x \leq y$ hold if $x = y\,\mathbf{d}(x)$, for all elements x and y in an inverse semigroup S. This relation is a partial ordering on S, called the *natural ordering of* S. It is compatible with the multiplication and the inversion operation on S (cf. Howie [60, Proposition V.2.4], or Lemma 1.4.6 and Proposition 1.4.7 in Lawson [73]). Various statements equivalent to $x \leq y$ can be found in Howie [60, Proposition V.2.2] or Lawson [73, Proposition 1.4.6]: $x = \mathbf{r}(x)y$; $x = ye$ for some idempotent e; $x = ey$ for some idempotent e; $\mathbf{r}(x) = yx^{-1}$; $\mathbf{d}(x) = y^{-1}x$; $x = xy^{-1}x$. The set $\operatorname{Idp} S$ of all idempotents of S is a lower subset of $(S, \leq)$.

For $x, y \in S$, let $x \sim y$ hold (we say that x and y are *compatible*) if $x^{-1}y$ and xy^{-1} are both idempotent. Equivalently (cf. Lawson [73, Lemma 1.4.11]), the meet $x \wedge y$ exists in S, $\mathbf{d}(x \wedge y) = \mathbf{d}(x)\,\mathbf{d}(y)$, and $\mathbf{r}(x \wedge y) = \mathbf{r}(x)\,\mathbf{r}(y)$. In that case (see, for example, Lawson [73, Lemma 1.4.12]),

$$x \wedge y = \mathbf{r}(x)y = y\,\mathbf{d}(x) = xy^{-1}x\,. \tag{3.1.3}$$

If $\{x, y\}$ is bounded above, then $x \sim y$; the converse fails for easy examples. A subset A of S is compatible if any two elements of A are compatible.

Definition 3.1.2 Let S be an inverse semigroup with zero and let $x, y \in S$.

(1) We say that x and y are *left orthogonal*, in notation $x \perp_{\mathrm{lt}} y$, if $xy^{-1} = 0$; equivalently, $\mathbf{d}(x)\,\mathbf{d}(y) = 0$.
(2) We say that x and y are *right orthogonal*, in notation $x \perp_{\mathrm{rt}} y$, if $x^{-1}y = 0$; equivalently, $\mathbf{r}(x)\,\mathbf{r}(y) = 0$.
(3) We say that x and y are *orthogonal*, in notation $x \perp y$, if $x \perp_{\mathrm{lt}} y$ and $x \perp_{\mathrm{rt}} y$.

A subset A of S is orthogonal if any two distinct elements of A are orthogonal.

In particular, $x \perp y$ (orthogonality) implies that $x \sim y$ (compatibility).

For a congruence relation $\boldsymbol{\theta}$ on S and elements $x, y \in S$, let $x \leq_{\boldsymbol{\theta}} y$ hold if $x \equiv_{\boldsymbol{\theta}} y\,\mathbf{d}(x)$. Equivalently, $\theta(x) \leq \theta(y)$, where $\theta\colon S \twoheadrightarrow S/\boldsymbol{\theta}$ denotes the canonical projection. Observe, in particular, that since $S/\boldsymbol{\theta}$ is an inverse semigroup, $x \leq_{\boldsymbol{\theta}} y$ and $y \leq_{\boldsymbol{\theta}} x$ iff $x \equiv_{\boldsymbol{\theta}} y$, for all $x, y \in S$.

For any $a \in S$, let $\lambda_a\colon S \to S, x \mapsto ax$ and $\rho_a\colon S \to S, x \mapsto xa$. The following lemma is well known but we could not trace it back to any particular source. It enables us to reduce order properties of an inverse semigroup to its semilattice of idempotents. We include a proof for convenience.

Lemma 3.1.3 (Folklore) *The following statements hold, for any* $a \in S$.

(1) *The maps* λ_a *and* $\lambda_{a^{-1}}$ *restrict to mutually inverse, domain-preserving order-isomorphisms, from* $S \downarrow \mathbf{d}(a)$ *onto* $S \downarrow a$ *and from* $S \downarrow a$ *onto* $S \downarrow \mathbf{d}(a)$*, respectively. The graphs of those maps are all contained in* $\mathscr{L}$.
(2) *The maps* ρ_a *and* $\rho_{a^{-1}}$ *restrict to mutually inverse, range-preserving order-isomorphisms, from* $S \downarrow \mathbf{r}(a)$ *onto* $S \downarrow a$ *and from* $S \downarrow a$ *onto* $S \downarrow \mathbf{r}(a)$*, respectively. The graphs of those maps are all contained in* $\mathscr{R}$.

Furthermore, all the isomorphisms above preserve orthogonality, and also all existing meets and joins, evaluated in S.

Proof It is clear that $\lambda_a, \lambda_{a^{-1}}, \rho_a, \rho_{a^{-1}}$ are all isotone.

(1) Any $x \in S \downarrow \mathbf{d}(a)$ satisfies $x = \mathbf{d}(a)\,\mathbf{d}(x)$ (so x is idempotent), thus $\lambda_a(x) = a\,\mathbf{d}(x) \leq a$. Furthermore, $\lambda_{a^{-1}}\lambda_a(x) = \mathbf{d}(a)x = x$, and further, by using the idempotence of x, $\mathbf{d}(\lambda_a(x)) = x^{-1}a^{-1}ax = x$, so $(x, \lambda_a(x)) \in \mathscr{L}$. This proves that $\lambda_a[S \downarrow \mathbf{d}(a)] \subseteq S \downarrow a$, $\lambda_{a^{-1}}\lambda_a \upharpoonright_{S\downarrow\mathbf{d}(a)} = \mathrm{id}_{S\downarrow\mathbf{d}(a)}$, and the graph of λ_a is contained in $\mathscr{L}$.

Any $y \in S \downarrow a$ satisfies $y = a\,\mathbf{d}(y)$, thus $\lambda_{a^{-1}}(y) = \mathbf{d}(a)\,\mathbf{d}(y) \leq \mathbf{d}(a)$. Furthermore, $\lambda_a\lambda_{a^{-1}}(y) = aa^{-1}a\,\mathbf{d}(y) = a\,\mathbf{d}(y) = y$, and $\lambda_{a^{-1}}(y) = a^{-1}a\,\mathbf{d}(y) = \mathbf{d}(y)$, so $(y, \lambda_{a^{-1}}(y)) \in \mathscr{L}$. This proves that $\lambda_{a^{-1}}[S \downarrow a] \subseteq S \downarrow \mathbf{d}(a)$, $\lambda_a\lambda_{a^{-1}} \upharpoonright_{S\downarrow a} = \mathrm{id}_{S\downarrow a}$, and the graph of $\lambda_{a^{-1}}$ is contained in $\mathscr{L}$. This completes the proof of (1).

The proof of (2) is symmetric.

For all $x, y, z \in S$, if $x \perp y$, then $(xz)^{-1}yz = z^{-1}(x^{-1}y)z = 0$ and $xz(yz)^{-1} = xzz^{-1}y^{-1} \leq xy^{-1} = 0$, thus $xz \perp yz$. Symmetrically, $zx \perp zy$. In particular, all maps $\lambda_a, \lambda_{a^{-1}}, \rho_a, \rho_{a^{-1}}$ preserve orthogonality. By virtue of Lawson [73, Proposition 1.4.19], all those maps preserve all existing meets evaluated in S.

Let $u \in S$ and let $X \subseteq S \downarrow \mathbf{d}(a)$ such that $u = \bigvee X$ within S. In particular, $u \in S \downarrow \mathbf{d}(a)$. Let $y \in S$ such that $aX \leq y$. Then $x = a^{-1}ax \leq a^{-1}y$, for each $x \in X$. It follows that $X \leq a^{-1}y$, so $u \leq a^{-1}y$, and so $au \leq aa^{-1}y \leq y$. This proves that $au = \bigvee(aX)$ within S.

Let $u \in S$ and let $X \subseteq S \downarrow a$ such that $u = \bigvee X$. In particular, $u \in S \downarrow a$. Let $y \in S$ such that $a^{-1}X \leq y$. Then $x = aa^{-1}x \leq ay$, for each $x \in X$. It follows that $u \leq ay$, so $a^{-1}u \leq a^{-1}ay \leq y$. This proves that $a^{-1}u = \bigvee(a^{-1}X)$ within S.

Therefore, both λ_a and $\lambda_{a^{-1}}$ preserve all existing joins from S. The proofs for ρ_a and $\rho_{a^{-1}}$ are symmetric. □

In particular, since $\mathbf{d}(x) = a^{-1}x$ and $\mathbf{r}(x) = xa^{-1}$ whenever $x \in S \downarrow a$, we obtain the following result, contained in Schein [102, Lemma 1.12]; see also Lawson [73, Proposition 1.4.17].

Lemma 3.1.4 *The maps* **d** *and* **r** *both preserve all existing meets and joins in S.*

Since the map **d** preserves all existing meets and joins, it follows that Idp *S is closed under all existing meets and joins in S.*

3.1.2 Boolean Inverse Semigroups

Definition 3.1.5 (Orthogonal Join in an Inverse Semigroup with Zero) For elements x, y, z in an inverse semigroup S with zero, let $z = x \oplus y$ hold if $z = x \vee y$ in S and $x \perp y$.

Definition 3.1.6 An inverse semigroup S is

- *distributive* if $\mathrm{Idp}\,S$ is a distributive lattice and $x \vee y$ exists for all compatible $x, y \in \mathrm{Idp}\,S$;

– *Boolean* if Idp S is a generalized Boolean lattice and $x \vee y$ exists for all compatible $x, y \in \operatorname{Idp} S$.

Although distributive inverse semigroups will be met on an occasional basis throughout the present work, Boolean inverse semigroups will be given the lion's share.

It is well known that an inverse semigroup S is Boolean iff Idp S is a generalized Boolean lattice and $x \vee y$ exists for all *orthogonal* $x, y \in \operatorname{Idp} S$ (thus $x \vee y = x \oplus y$). The original definition of a Boolean inverse semigroup (cf. Lawson [74]) assumed that the natural ordering is a meet-semilattice. This definition got subsequently relaxed, by dropping the meet-semilattice assumption (cf. Lawson [75]). In the latter paper, inverse semigroups for which the natural ordering is a meet-semilattice are called *inverse* $\wedge$*-semigroups* (cf. Definition 3.7.7). Proposition 3.1.9 shows, in particular, that our Boolean inverse semigroups are identical to Lawson's Boolean inverse semigroups from [75].

As the following example shows, those concepts are stronger than the eponymous one introduced in Exel [41]. For further discussion about this, see Sect. 3.2.

Example 3.1.7 Table 3.1 describes a finite commutative inverse monoid S with zero, such that Idp S is the Boolean semilattice with two atoms, but the atoms of S have no join.

The atoms of S are 1 and 2. They are both idempotent, and they join to 4 in Idp S. However, they do not have a join in S.

Another inverse monoid, which is Boolean in Exel's sense but not in ours, is the final example in Exel [41]. This will be discussed further in Sect. 3.2.

Example 3.1.8 The semigroup $\mathfrak{I}_X$, of all partial one-to-one functions between subsets of a set X, is a Boolean inverse monoid, the so-called *symmetric inverse monoid on* X. It has a zero element, namely the function with empty domain (and range). Its unit element is the identity function on X. For $u, v \in \mathfrak{I}_X$, the inequality $u \leq v$ holds iff v extends u. Furthermore, $u \sim v$ iff u and v agree on the intersection of the domains of u and v, and $u \perp v$ iff $\operatorname{dom}(u) \cap \operatorname{dom}(v) = \operatorname{rng}(u) \cap \operatorname{rng}(v) = \varnothing$. If $X = [n] = \{1, \dots, n\}$, for a nonnegative integer n, then we shall write $\mathfrak{I}_n$ instead of $\mathfrak{I}_{[n]}$.

The inverse monoid $\mathfrak{I}_X$ has the additional property that every collection $\mathcal{F}$ of elements of $\mathfrak{I}_X$ has a meet (with respect to the natural ordering), whose domain is the set of all elements of X on which all members of $\mathcal{F}$ agree. In particular, $\mathfrak{I}_X$ is an *inverse meet-semigroup* as introduced further (cf. Definition 3.7.7).

Table 3.1 A non-Boolean inverse monoid with zero, with Boolean semilattice of idempotents

S	0	1	2	3	4
0	0	0	0	0	0
1	0	1	0	1	1
2	0	0	2	2	2
3	0	1	2	4	3
4	0	1	2	3	4

The monoid $\mathfrak{I}_X$ can be viewed as a "skeleton" of $X \times X$ matrix rings. In particular, for $i, j \in X$, the unique function $e_{i,j}$ with domain $\{j\}$ and range $\{i\}$ belongs to $\mathfrak{I}_X$. Denoting by $\delta_{x,y}$ the Kronecker symbol and interpreting $0 \cdot f$ as the empty function (which is the zero element of $\mathfrak{I}_X$), the $e_{i,j}$ satisfy the following relations:

$$e_{i,j}e_{k,l} = \delta_{j,k}e_{i,l}\,, \tag{3.1.4}$$

$$e_{i,j}^{-1} = e_{j,i}\,, \tag{3.1.5}$$

for all $i, j, k, l \in X$. We call the $e_{i,j}$ the *matrix units* of $\mathfrak{I}_X$.

The following result is at the basis of many calculations in distributive and Boolean inverse semigroups.

Proposition 3.1.9 *The following statements hold for any distributive inverse semigroup S, with a zero element required in* (2)–(4)*:*

(1) *For every nonempty finite compatible subset* $\{b_1, \ldots, b_n\}$ *of S, the join* $\bigvee_{i=1}^n b_i$ *exists and the following statements hold:*
   (i) $a \cdot \bigvee_{i=1}^n b_i = \bigvee_{i=1}^n (a \cdot b_i)$ *and* $\left(\bigvee_{i=1}^n b_i\right) \cdot a = \bigvee_{i=1}^n (b_i \cdot a)$*, for every* $a \in S$.
   (ii) *For every* $a \in S$, $a \wedge \bigvee_{i=1}^n b_i$ *exists iff each* $a \wedge b_i$ *exists, and then* $a \wedge \bigvee_{i=1}^n b_i = \bigvee_{i=1}^n (a \wedge b_i)$.
(2) *The partial operation* $(x, y) \mapsto x \oplus y$ *endows S with a structure of a conical partial refinement monoid.*
(3) *The results of* (1) *above extend to the orthogonal join* $\oplus$ *in place of the join* $\vee$.
(4) *If* $a \oplus c = b \oplus c$ *in S, then* $a = b$.

Proof (1i) follows from the finite version of Lawson [73, Proposition 1.4.20].

(1ii) follows from the finite version of Resende [98].

(2) Let $u = (x \oplus y) \oplus z$ in S. From $x \oplus y \perp z$ it follows that $x \perp z$ and $y \perp z$. Since S is distributive, it follows that $y \oplus z$ is defined. By (1ii), it follows that $x \perp y \oplus z$. Since S is distributive, $x \oplus (y \oplus z)$ is defined, with value $x \vee (y \vee z) = (x \vee y) \vee z = u$. Hence, $\oplus$ is associative, so $(S, \oplus, 0)$ is a partial commutative monoid.

Let $x \oplus x' = y \oplus y'$ in S. Applying (1ii), we get the following refinement matrix:

	y	y'
x	$x \wedge y$	$x \wedge y'$
x'	$x' \wedge y$	$x' \wedge y'$

within $(S, \oplus)$. (3.1.6)

Item (2) follows. If $a \oplus c = b \oplus c$, then, taking $x = y = c$, $x' = a$, and $y' = b$ yields $x \wedge y' = x' \wedge y = 0$, thus $x' = y'$. Item (4) follows.

Finally, it is straightforward to obtain (3) from (1). □

As an immediate application of Proposition 3.1.9(1ii), we record the following.

Corollary 3.1.10 *Let S be a distributive inverse semigroup, let m and n be positive integers, and let $a_1, \dots, a_m, b_1, \dots, b_n \in S$. Then $\left(\bigvee_{i=1}^m a_i\right) \wedge \left(\bigvee_{j=1}^n b_j\right)$ exists iff each $a_i \wedge b_j$ exists, and then*

$$\left(\bigvee_{i=1}^{m} a_i\right) \wedge \left(\bigvee_{j=1}^{n} b_j\right) = \bigvee_{1\leq i\leq m,\ 1\leq j\leq n} (a_i \wedge b_j)\,. \tag{3.1.7}$$

Furthermore, if S has a zero element, then the analogue of (3.1.7)*, with $\vee$ replaced by $\oplus$, also holds.*

It is well known (see, for example, Lawson and Lenz [76, Lemma 3.27]) that for any elements x and y in a Boolean inverse semigroup S, if $x \leq y$, then there exists a unique $z \in S$ such that $y = x \oplus z$. Consequently, the natural ordering $\leq$ of S is also the algebraic ordering $\leq^{\oplus}$ of the partial commutative monoid $(S, \oplus, 0)$.

Notation 3.1.11 For any elements x and y in a Boolean inverse semigroup S such that $x \leq y$, we denote by $y \smallsetminus x$ the unique $z \in S$ such that $y = x \oplus z$. The range of this symbol is extended to all pairs (x, y) such that $x \wedge y$ exists, by defining $x \smallsetminus y = x \smallsetminus (x \wedge y)$.

A direct application of Proposition 3.1.9 yields the following result, whose easy proof we omit.

Lemma 3.1.12 *The following statements hold, for every Boolean inverse semigroup S and all $x, y, z \in S$ such that $x \leq y$:*

(1) $z(y \smallsetminus x) = (zy) \smallsetminus (zx)$ *and* $(y \smallsetminus x)z = (yz) \smallsetminus (xz)$.
(2) *If $y \wedge z$ exists, then $x \wedge z$ exists and* $(y \smallsetminus x) \wedge z = (y \wedge z) \smallsetminus (x \wedge z)$.

We will repeatedly use the following easy fact.

Lemma 3.1.13 *Let S be an inverse subsemigroup of a Boolean inverse semigroup T. If S is closed under finite orthogonal joins and $a \smallsetminus b \in S$ whenever $a, b \in \operatorname{Idp} S$ with $b \leq a$, then S is a Boolean inverse semigroup, and $x \smallsetminus y \in S$ whenever $x, y \in S$ with $y \leq x$.*

Proof It follows from our assumptions, together with the identity $\mathsf{x} \smallsetminus \mathsf{y} = \mathsf{x} \smallsetminus \mathsf{xy}$ (in generalized Boolean algebras), that $\operatorname{Idp} S$ is a subsemigroup of $\operatorname{Idp} T$, closed under the operation $(x, y) \mapsto x \smallsetminus y$ and under finite orthogonal joins. Since the latter is a Boolean ring, so is the former.

By assumption, S contains the empty sum 0. Hence, the orthogonality relation on S is the restriction to S of the orthogonality relation on T. By our assumption, $x \oplus y$ exists in S whenever x and y are orthogonal elements of S. Therefore, S is Boolean.

Let $x, y \in S$ with $y \leq x$. From $y \leq x$ it follows that $y = x\,\mathbf{d}(y)$, whence $x \smallsetminus y = x(\mathbf{d}(x) \smallsetminus \mathbf{d}(y))$. By assumption, $\mathbf{d}(x) \smallsetminus \mathbf{d}(y) \in S$; whence $x \smallsetminus y \in S$. □

The following example shows that the additional assumption, in Lemma 3.1.13, that $\operatorname{Idp} S$ be closed under $(x, y) \mapsto x \smallsetminus y$, cannot be dropped.

Example 3.1.14 The powerset algebra T of $\{0, 1\}$, endowed with set intersection, is a Boolean inverse semigroup. The subset $S = \{\varnothing, \{0\}, \{0, 1\}\}$ is an inverse subsemigroup of T, closed under finite orthogonal joins. However, S is not a Boolean inverse semigroup.

3.1.3 Additivity in Boolean Inverse Semigroups

The definition of additivity given below is the restriction, to Boolean inverse semigroups, of a definition by Lawson and Lenz [76].

Definition 3.1.15 Let S and T be Boolean inverse semigroups.

- A semigroup homomorphism $f: S \to T$ is *additive* if the equality $f(x \oplus y) = f(x) \oplus f(y)$ holds whenever x and y are orthogonal elements in S.
- A one-to-one map $f: S \hookrightarrow T$ is a *lower semigroup embedding* if it is an additive semigroup homomorphism and $f[S]$ is a lower subset of T with respect to the natural ordering.

In particular, every lower semigroup embedding is also a V-embedding (cf. Definition 2.1.2). Further, it is well known, and an easy exercise, that any additive semigroup homomorphism between Boolean inverse semigroups preserves finite compatible joins. We postpone a more complete characterization of additive semigroup homomorphisms until Theorem 3.2.5.

Definition 3.1.16 An inverse subsemigroup S of an inverse semigroup T is an *ideal of T* (resp., a *quasi-ideal of T*) if $TST \subseteq S$ (resp., $STS \subseteq S$).

It is easy to verify that every ideal is a quasi-ideal. Our concept of quasi-ideal is the restriction, to inverse semigroups, of the classical one (cf. Lawson [72]), defined for arbitrary semigroups. The five-element inverse semigroup T represented in Table 3.2, with the subsemigroup $S = \{0, 1\}$, shows that the assumption that S is an inverse subsemigroup of T is not redundant in Definition 3.1.16.

Definition 3.1.17 A subset S in a Boolean inverse semigroup T is

- an *additive inverse subsemigroup of T* if S is an inverse subsemigroup of T, $\mathrm{Idp}\, S$ is closed under the operation $(x, y) \mapsto x \smallsetminus y$, and S is closed under finite orthogonal joins in T;

Table 3.2 The subsemigroup $S = \{0, 1\}$ of T has $STS = S$ and $S^{-1} \neq S$

T	0	1	2	3	4
0	0	0	0	0	0
1	0	0	3	0	1
2	0	4	0	2	0
3	0	1	0	3	0
4	0	0	2	0	4

- a *lower inverse subsemigroup* of T if S is an additive inverse subsemigroup of T and S is a lower subset of T with respect to the natural ordering;
- an *additive ideal* of T if S is an ideal of T and S is closed under finite orthogonal joins in T.

Our additive ideals are called $\vee$*-ideals* in Kudryavtseva et al. [71]. The following result shows that the concepts introduced above occur only between Boolean inverse semigroups.

Proposition 3.1.18 *The following implications hold, for any subset S in a Boolean inverse semigroup T:*

$$S \text{ additive ideal of } T \Rightarrow S \text{ additive quasi-ideal of } T \Rightarrow$$

$$S \text{ lower inverse subsemigroup of } T \Rightarrow S \text{ additive inverse subsemigroup of } T \Rightarrow$$
$$S \text{ Boolean inverse semigroup}\,.$$

Proof It is trivial that any ideal of T is also a quasi-ideal of T. Now suppose that S is a quasi-ideal of T and let $x \in T$ and $y \in S$ such that $x \leq y$. Then $x = \mathbf{r}(y)x\,\mathbf{d}(y) \in STS \subseteq S$. Hence, S is a lower subset of T. Now if S is a lower inverse subsemigroup of T, then $\operatorname{Idp} S$ is a lower subset of $\operatorname{Idp} T$, thus it is closed under the operation $(x, y) \mapsto x \smallsetminus y$. The final implication follows immediately from Lemma 3.1.13. □

Proposition 3.1.19 *Every additive inverse subsemigroup S of a Boolean inverse semigroup T is closed under finite compatible joins.*

Proof Let $x, y \in S$ with $x \sim y$. Using (3.1.3), we get $x \wedge y = y\,\mathbf{d}(x) \in S$. Further, by Lemma 3.1.13, $x \smallsetminus y = x \smallsetminus (x \wedge y)$ belongs to S. It follows that $x \vee y = (x \smallsetminus y) \oplus y \in S$. □

Proposition 3.1.20 *Let X be a subset in a Boolean inverse semigroup S. Then $(SXS)^{\oplus}$ is the smallest additive ideal of S containing X.*

Proof For each $x \in X$, $x = \mathbf{r}(x)x\,\mathbf{d}(x) \in SXS \subseteq (SXS)^{\oplus}$; thus $X \subseteq (SXS)^{\oplus}$. Moreover, SXS is an ideal of S, thus, using Proposition 3.1.9, it follows that $(SXS)^{\oplus}$ is also an ideal of S. This ideal is obviously additive in S. □

Definition 3.1.21 A Boolean inverse semigroup T is an *additive enlargement* of a quasi-ideal S if $T = (TST)^{\oplus}$.

Our concept of additive enlargement is obtained, from the one of enlargement introduced in Lawson [72], by replacing the condition $T = TST$ by the weaker condition $T = (TST)^{\oplus}$. For an interpretation of additive enlargements in terms of the type monoid introduced in Definition 4.1.3, see Theorem 4.2.2. An important class of additive enlargements is given by the following result.

Proposition 3.1.22 *Let e be an idempotent element in a Boolean inverse semigroup S. Then eSe is an additive quasi-ideal of S, and $(SeS)^{\oplus}$ is an additive enlargement of eSe.*

Proof Set $B = \operatorname{Idp} S$. It is trivial that eSe is an inverse subsemigroup of S. Clearly, $\operatorname{Idp}(eSe) = B \downarrow e$ is Boolean. Furthermore, it follows from Proposition 3.1.9 that eSe is closed under finite orthogonal joins. Hence, eSe is an additive inverse subsemigroup of S. It is trivial that $(eSe)S(eSe) \subseteq eSe$. Thus, eSe is an additive quasi-ideal of S.

By Proposition 3.1.20, $(SeS)^{\oplus}$ is the additive ideal of S generated by e. Setting $S' = eSe$ and $T' = (SeS)^{\oplus}$, it is straightforward to verify that $(T'S'T')^{\oplus} = T'$. □

The two following examples show that none of the converse implications in Proposition 3.1.18 holds.

Example 3.1.23 Let $T = \mathfrak{I}_2$ (cf. Example 3.1.8). Then $S = \{\varnothing, \mathrm{id}_{\{1\}}\}$ is an additive quasi-ideal of T, but not an ideal of T.

The following result gives a convenient characterization of lower inverse subsemigroups.

Proposition 3.1.24 *An inverse subsemigroup S of a Boolean inverse semigroup T is a lower inverse subsemigroup of T iff S is closed under finite orthogonal joins and* $\operatorname{Idp} S$ *is a lower subset of* $\operatorname{Idp} T$.

Proof It is sufficient to prove that if $\operatorname{Idp} S$ is a lower subset of $\operatorname{Idp} T$, then S is a lower subset of T. Let $t \leq s$ where $t \in T$ and $s \in S$. Since S is an inverse subsemigroup of T, $\mathbf{d}(s) \in S$. Since $\mathbf{d}(t) \leq \mathbf{d}(s)$ and by assumption, it follows that $\mathbf{d}(t) \in S$. Therefore, $t = s\,\mathbf{d}(t) \in S$. □

In particular, $\operatorname{Idp} S$ is a lower inverse subsemigroup of S, for every Boolean inverse semigroup S.

Example 3.1.25 Let $T = \mathfrak{I}_2$. Then $S = \operatorname{Idp} T = \{\varnothing, \mathrm{id}_{\{1\}}, \mathrm{id}_{\{2\}}, \mathrm{id}_{\{1,2\}}\}$ is a lower inverse subsemigroup of T, but not a quasi-ideal. In fact, $STS = T$.

3.2 The Concept of Bias: An Equational Definition of Boolean Inverse Semigroups

Due to their formulation in terms of the partial operation $\oplus$, the original defining axioms of the class of all Boolean inverse semigroups are not identities in the usual sense of universal algebra. For example, the formula $\mathsf{x}(\mathsf{y} \oplus \mathsf{z}) = (\mathsf{xy}) \oplus (\mathsf{xz})$ makes sense only in case $\mathsf{y} \oplus \mathsf{z}$ and $(\mathsf{xy}) \oplus (\mathsf{xz})$ are both defined. This causes confusion when it comes to handling standard concepts of universal algebra, such as homomorphisms, colimits, or free algebras.

In general, a *similarity type* is a set of "operation symbols" (or just, using a standard abuse of language, "operations"), each one given a nonnegative natural number called the *arity*. Operations with arity zero are usually called *constants*. Formal compositions of operations, starting with variables, are called *terms*. An *identity* is a formal expression of the form $\mathsf{p} = \mathsf{q}$, where p and q are terms. A *variety* is the class of all algebras that satisfy a given set of identities. For more detail, see McKenzie et al. [82].

We shall provide in this section an alternative characterization of Boolean inverse semigroups by a finite set of identities. This characterization will be formulated in the language of inverse semigroups (i.e., a binary operation for the product, a unary operation for the inversion, and a constant for the zero), enriched by two additional binary operations $\otimes$ and ∇ (cf. Definition 3.2.1). This will enable us to define natural concepts such as homomorphisms, congruences, or free objects, within Boolean inverse semigroups, and more generally, study that class from the vantage point of universal algebra.

Our definition of the operations $\otimes$ and ∇ is inspired by Leech [78, Example 1.7(c)].

Definition 3.2.1 Let S be a Boolean inverse semigroup. We set

$$x \otimes y = (\mathbf{r}(x) \smallsetminus \mathbf{r}(y))x(\mathbf{d}(x) \smallsetminus \mathbf{d}(y)) \text{ and } x \mathbin{\nabla} y = (x \otimes y) \oplus y\,, \quad \text{for all } x, y \in S\,. \tag{3.2.1}$$

We shall call $x \otimes y$ the *skew difference* and $x \mathbin{\nabla} y$ the *left skew join*—from now on *skew join*[1] —of x and y.

Since B is a Boolean inverse semigroup and $\mathbf{d}(x)$, $\mathbf{d}(y)$, $\mathbf{r}(x)$, $\mathbf{r}(y)$ are all idempotent, both differences $\mathbf{d}(x) \smallsetminus \mathbf{d}(y)$ and $\mathbf{r}(x) \smallsetminus \mathbf{r}(y)$ are always defined, thus $x \otimes y$ is always defined. Furthermore, $\mathbf{r}(y)(x \otimes y) = (x \otimes y)\,\mathbf{d}(y) = 0$, thus $x \otimes y \perp y$, and thus $x \mathbin{\nabla} y$ is also always defined.

Notation 3.2.2 We denote by $\mathcal{L}_{\mathrm{IS}}$ the similarity type of inverse semigroups. It is thus defined as $\mathcal{L}_{\mathrm{IS}} = (0, {}^{-1}, \cdot)$, where 0 is a symbol of constant, ${}^{-1}$ is a symbol of unary operation, and $\cdot$ is a symbol of binary operation.[2]

We also denote by $\mathcal{L}_{\mathrm{BIS}}$ the similarity type obtained by enriching $\mathcal{L}_{\mathrm{IS}}$ with two binary operation symbols $\otimes$ and ∇.

As the sequel of the present section will involve relatively complicated identities, we shall use a number of abbreviations, such as $\mathbf{d}(\mathsf{x}) = \mathsf{x}^{-1}\mathsf{x}$, $\mathbf{r}(\mathsf{x}) = \mathsf{x}\mathsf{x}^{-1}$, $\mathsf{x}^2 = \mathsf{x} \cdot \mathsf{x}$, $\mathsf{x} \leq \mathsf{y}$ instead of $\mathsf{x} = \mathsf{y}\,\mathbf{d}(\mathsf{x})$, $\mathsf{x} \perp \mathsf{y}$ instead of $\mathsf{x}^{-1}\mathsf{y} = \mathsf{x}\mathsf{y}^{-1} = 0$, and so on. For instance, the identity $\mathsf{x}^{-1}\mathsf{x}\mathsf{y}^{-1}\mathsf{y} = (\mathsf{x}\mathsf{y})^{-1}\mathsf{x}\mathsf{y}$ (which is not valid in all inverse semigroups!) will be abbreviated by $\mathbf{d}(\mathsf{x}\mathsf{y}) = \mathbf{d}(\mathsf{x})\,\mathbf{d}(\mathsf{y})$.

The characterization of the class of all Boolean inverse semigroups by a set of identities will be performed via the following concept.

Definition 3.2.3 A *bias* is a $\mathcal{L}_{\mathrm{BIS}}$-structure $(S, 0, {}^{-1}, \cdot, \otimes, \nabla)$, that is, a set S together with a distinguished element $0 \in S$, a unary operation $x \mapsto x^{-1}$ on S, and binary operations $(x, y) \mapsto x \cdot y$, $(x, y) \mapsto x \otimes y$, $(x, y) \mapsto x \mathbin{\nabla} y$ on S, subject to the following

[1]The right skew join of x and y could of course be defined as $x \oplus (y \otimes x)$, that is, $y \mathbin{\nabla} x$.

[2]Although strictly speaking, the operation symbols should not be denoted the same way as their interpretations (in a given structure), that confusion is widespread and harmless.

(finite) collection of identities:

(InvSem) Any set of identities defining inverse semigroups with zero. For example, state that $\cdot$ is associative, 0 is a zero element with respect to $\cdot$, $\mathsf{x} = \mathsf{x}\mathsf{x}^{-1}\mathsf{x}$, $(\mathsf{x}^{-1})^{-1} = \mathsf{x}$, and $\mathbf{d}(\mathsf{x})\,\mathbf{d}(\mathsf{y}) = \mathbf{d}(\mathsf{y})\,\mathbf{d}(\mathsf{x})$.

($\mathbf{GBa}_{\otimes,\triangledown}$) All defining identities (1.4.3) of generalized Boolean algebras, with $\wedge$ changed to the product operation $\cdot$, $\smallsetminus$ changed to $\otimes$, $\vee$ changed to $\triangledown$, and x, y, z respectively replaced by $\mathbf{d}(\mathsf{x})$, $\mathbf{d}(\mathsf{y})$, $\mathbf{d}(\mathsf{z})$. For example, the identity $\mathbf{d}(\mathsf{x}) = (\mathbf{d}(\mathsf{x}) \otimes \mathbf{d}(\mathsf{y}))\ \triangledown\ (\mathbf{d}(\mathsf{x})\,\mathbf{d}(\mathsf{y}))$, which is the translation of the identity $\mathsf{x} = (\mathsf{x} \smallsetminus \mathsf{y}) \vee (\mathsf{x} \wedge \mathsf{y})$, belongs to the list.

($\mathbf{Idp}_{\otimes,\triangledown}$) $(\mathbf{d}(\mathsf{x}) \otimes \mathbf{d}(\mathsf{y}))^2 = \mathbf{d}(\mathsf{x}) \otimes \mathbf{d}(\mathsf{y})$ and $(\mathbf{d}(\mathsf{x})\ \triangledown\ \mathbf{d}(\mathsf{y}))^2 = \mathbf{d}(\mathsf{x})\ \triangledown\ \mathbf{d}(\mathsf{y})$. This says that the set of all idempotents is closed under both operations $\otimes$ and $\triangledown$.

($\mathbf{Distr}_{\otimes,\triangledown}$) $\mathsf{z}\Big(\big(\mathbf{d}(\mathsf{x}) \otimes \mathbf{d}(\mathsf{y})\big)\ \triangledown\ \mathbf{d}(\mathsf{y})\Big) = \mathsf{z}\big(\mathbf{d}(\mathsf{x}) \otimes \mathbf{d}(\mathsf{y})\big)\ \triangledown\ \mathsf{z}\,\mathbf{d}(\mathsf{y})$. This states a certain restricted distributivity of the product $\cdot$ on the skew join $\triangledown$.

($\mathbf{Maj}_{\otimes,\triangledown}$) $\mathsf{x} \otimes \mathsf{y} \leq \mathsf{x}\ \triangledown\ \mathsf{y}$ and $\mathsf{y} \leq \mathsf{x}\ \triangledown\ \mathsf{y}$.

($\mathbf{Dom}_{\triangledown}$) $\mathbf{d}(\mathsf{x}\ \triangledown\ \mathsf{y}) = \mathbf{d}(\mathsf{x} \otimes \mathsf{y})\ \triangledown\ \mathbf{d}(\mathsf{y})$.

($\mathbf{Def}_{\otimes}$) $\mathsf{x} \otimes \mathsf{y} = \big(\mathbf{r}(\mathsf{x}) \otimes \mathbf{r}(\mathsf{y})\big)\mathsf{x}\big(\mathbf{d}(\mathsf{x}) \otimes \mathbf{d}(\mathsf{y})\big)$.

The equivalence between the concept of bias on the one hand, and the one of Boolean inverse semigroup on the other hand, is achieved by the following result.

Theorem 3.2.4

(1) *Every Boolean inverse semigroup* $(S, 0, ^{-1}, \cdot)$ *expands, via the operations* $\otimes$ *and* $\triangledown$ *defined in* (3.2.1), *to a bias.*

(2) *For every bias* $(S, 0, ^{-1}, \otimes, \triangledown)$, *the reduct* $(S, 0, ^{-1}, \cdot)$ *to the similarity type* $\mathcal{L}_{\mathrm{IS}}$ *is a Boolean inverse semigroup.*

(3) *Any two biases on* S *with the same inverse semigroup reduct are equal. In particular, the two operations of expansion and reduction, defined in* (1) *and* (2) *above, are mutually inverse.*

Proof (1) The identities **(InvSem)** and **($\mathbf{Maj}_{\otimes,\triangledown}$)** are both satisfied by definition. For all idempotents $a, b \in S$, $a \cdot b = a \wedge b$, $a \otimes b = a \smallsetminus b$, and $a\ \triangledown\ b = a \vee b$, thus, since $\operatorname{Idp} S$ is a generalized Boolean algebra, the identities **($\mathbf{GBa}_{\otimes,\triangledown}$)** and **($\mathbf{Idp}_{\otimes,\triangledown}$)** are satisfied. It follows that **($\mathbf{Def}_{\otimes}$)** holds as well.

In order to verify **($\mathbf{Distr}_{\otimes,\triangledown}$)**, we just need to observe that $z(a \oplus b) = (za) \oplus (zb)$, for all $z \in S$ and all orthogonal $a, b \in \operatorname{Idp} S$. (Indeed, whenever $x, y \in S$, the elements $a = \mathbf{d}(x) \otimes \mathbf{d}(y) = \mathbf{d}(x) \smallsetminus \mathbf{d}(y)$ and $b = \mathbf{d}(y)$ are orthogonal idempotents, thus $a\ \triangledown\ b = a \oplus b$.)

Now we verify **($\mathbf{Dom}_{\triangledown}$)**. Observe first that whenever $x, y \in S$, the elements $x' = x \otimes y$ and y are orthogonal, thus $x'\ \triangledown\ y = x' \oplus y$. Further, $\mathbf{d}(x')$ and $\mathbf{d}(y)$ are orthogonal, and $\mathbf{d}(x' \oplus y) = \mathbf{d}(x') \oplus \mathbf{d}(y)$.

(2) It follows from **(InvSem)** that $(S, 0, ^{-1}, \cdot)$ is an inverse semigroup. Further, it follows from **($\mathbf{GBa}_{\otimes,\triangledown}$)** and **($\mathbf{Idp}_{\otimes,\triangledown}$)** that $\operatorname{Idp} S$, endowed with the restriction $\wedge$ of $\cdot$, the restriction $\smallsetminus$ of $\otimes$, and the restriction $\vee$ of $\triangledown$, is a generalized Boolean algebra.

Now let $x, y \in S$ be orthogonal elements. Since $\mathbf{d}(x)$ and $\mathbf{d}(y)$ are orthogonal idempotents, $\mathbf{d}(x) \otimes \mathbf{d}(y) = \mathbf{d}(x) \smallsetminus \mathbf{d}(y) = \mathbf{d}(x)$, and similarly, $\mathbf{r}(x) \otimes \mathbf{r}(y) = \mathbf{r}(x)$.

Further, it follows from $(\mathbf{Def}_{\obslash})$ that $x \obslash y = \mathbf{r}(x)x\,\mathbf{d}(x) = x$. By $(\mathbf{Maj}_{\obslash,\triangledown})$, this implies that $\begin{smallmatrix}x\\y\end{smallmatrix} \leq x \triangledown y$. By using $(\mathbf{Dom}_{\triangledown})$, we get $\mathbf{d}(x \triangledown y) = \mathbf{d}(x) \triangledown \mathbf{d}(y)$.

Now let $z \in S$ such that $\begin{smallmatrix}x\\y\end{smallmatrix} \leq z$. We compute:

$$
\begin{aligned}
z\,\mathbf{d}(x \triangledown y) &= z\big(\mathbf{d}(x) \triangledown \mathbf{d}(y)\big) && \text{(by the above)}\\
&= z\big((\mathbf{d}(x) \obslash \mathbf{d}(y)) \triangledown \mathbf{d}(y)\big) && \text{(because } \mathbf{d}(x) \obslash \mathbf{d}(y) = \mathbf{d}(x))\\
&= z(\mathbf{d}(x) \obslash \mathbf{d}(y)) \triangledown z\,\mathbf{d}(y) && \text{(by } (\mathbf{Distr}_{\obslash,\triangledown}))\\
&= z\,\mathbf{d}(x) \triangledown z\,\mathbf{d}(y) && \text{(because } \mathbf{d}(x) \obslash \mathbf{d}(y) = \mathbf{d}(x))\\
&= x \triangledown y && \text{(because } x \leq z \text{ and } y \leq z)\,,
\end{aligned}
$$

so $x \triangledown y \leq z$. This completes the proof that $x \triangledown y$ is the orthogonal join of $\{x, y\}$ in S. Therefore, S is a Boolean inverse semigroup.

(3) We need to prove that in the presence of the bias identities, the operations $\obslash$ and $\triangledown$ are necessarily given by (3.2.1). Observe from the start that by (2) above, S is a Boolean inverse semigroup.

Due to $(\mathbf{GBa}_{\obslash,\triangledown})$ and $(\mathbf{Idp}_{\obslash,\triangledown})$, this certainly holds on $\operatorname{Idp} S$: that is, $a \obslash b = a \smallsetminus b$ and $a \triangledown b = a \vee b$ (within $\operatorname{Idp} S$), for any $a, b \in \operatorname{Idp} S$. Due to $(\mathbf{Def}_{\obslash})$, it follows that the operation $\obslash$ is given by (3.2.1); thus it is uniquely determined.

Now let $x, y \in S$. We must prove that $x \triangledown y = (x \obslash y) \oplus y$. Since S is a Boolean inverse semigroup and by $(\mathbf{Maj}_{\obslash,\triangledown})$, $(x \obslash y) \oplus y \leq x \triangledown y$. Further, it follows from $(\mathbf{Dom}_{\triangledown})$ that $\mathbf{d}(x \triangledown y) = \mathbf{d}(x \obslash y) \triangledown \mathbf{d}(y)$. Since $x \obslash y \perp y$ and since the restriction of $\triangledown$ to the idempotents is the join within $\operatorname{Idp} S$, it follows that $\mathbf{d}(x \triangledown y) = \mathbf{d}(x \obslash y) \oplus \mathbf{d}(y)$. Therefore, we get

$$
\begin{aligned}
(x \obslash y) \oplus y &= (x \triangledown y)\,\mathbf{d}\big((x \obslash y) \oplus y\big) && \text{(because } (x \obslash y) \oplus y \leq x \triangledown y)\\
&= (x \triangledown y)\,\mathbf{d}(x \triangledown y)\\
&= x \triangledown y\,,
\end{aligned}
$$

so the operation $\triangledown$ is given by (3.2.1).

The second statement of (3) follows immediately. □

In particular, given a Boolean inverse semigroup S, Theorem 3.2.4 enables us to define the *Boolean inverse subsemigroup of S generated by a subset X*, as the sub-bias of S generated by X.

The following result, crucial despite the easiness of its proof, identifies the homomorphisms on Boolean inverse semigroups, with respect to the structure of bias.

Theorem 3.2.5 *Let S and T be Boolean inverse semigroups and let $f: S \to T$ be a semigroup homomorphism. The following are equivalent:*

(i) *f is a bias homomorphism.*
(ii) *The domain-range restriction of f from* Idp S *to* Idp T *is a homomorphism of Boolean rings.*
(iii) *$c = a \oplus b$ implies that $f(c) = f(a) \oplus f(b)$, for all $a, b, c \in$* Idp S.
(iv) *f is additive.*

Proof (i)⇒(ii). Since f is a semigroup homomorphism, it sends Idp S into Idp T. Since the bias operations $\oslash$ and $\triangledown$ restrict, on the idempotents, to the difference $(x, y) \mapsto x \smallsetminus y$ and the join $(x, y) \mapsto x \vee y$, (ii) follows.

(ii)⇒(iii) is trivial.

(iii)⇒(iv) Let $z = x \oplus y$ in S, we must prove that $f(z) = f(x) \oplus f(y)$ in T. Since f is a homomorphism of inverse semigroups with zero, $f(x) \perp f(y)$; whence $f(x) \oplus f(y) \leq f(z)$. It follows from Lemma 3.1.4 that $\mathbf{d}(z) = \mathbf{d}(x) \oplus \mathbf{d}(y)$, thus, by assumption and since f is a homomorphism of inverse semigroups,

$$\mathbf{d}(f(z)) = f(\mathbf{d}(z)) = f(\mathbf{d}(x)) \oplus f(\mathbf{d}(y)) = \mathbf{d}(f(x)) \oplus \mathbf{d}(f(y)) = \mathbf{d}(f(x) \oplus f(y)) .$$

Since $f(x) \oplus f(y) \leq f(z)$, it follows that $f(x) \oplus f(y) = f(z)$.

(iv)⇒(i) Suppose that f is an additive semigroup homomorphism from S to T. For all $a, b \in$ Idp S, it follows from the additivity of f together with the equation $a = (a \smallsetminus b) \oplus ab$ that $f(a) = f(a \smallsetminus b) \oplus f(ab) = f(a \smallsetminus b) \oplus f(a)f(b)$. It follows that $f(a \smallsetminus b) = f(a) \smallsetminus f(b)$. Since f is an inverse semigroup homomorphism, it follows that $f(x \oslash y) = f(x) \oslash f(y)$, for all $x, y \in S$. Since f is additive, it follows that $f(x \triangledown y) = f(x) \triangledown f(y)$, for all $x, y \in S$. □

Corollary 3.2.6 *Let S and T be Boolean inverse semigroups and let $f: S \to T$ be an additive semigroup homomorphism. Then $f[S]$ is a sub-bias of T. In particular, it is a Boolean inverse semigroup.*

The following result relates Theorem 3.2.5 with the concept of additive inverse subsemigroup introduced in Definition 3.1.17.

Corollary 3.2.7 *An inverse subsemigroup S of a Boolean inverse semigroup T is a sub-bias of T iff it is an additive inverse subsemigroup of T.*

Proof It is trivial that every sub-bias is an additive inverse subsemigroup. Suppose, conversely, that S is an additive inverse subsemigroup of T. By Lemma 3.1.13, S is a Boolean inverse semigroup. The desired conclusion follows then from Theorem 3.2.5. □

In particular, a Boolean inverse subsemigroup S of a Boolean inverse semigroup T is an additive inverse subsemigroup iff S is a sub-bias of T. Even more particularly, an ideal I of S is a sub-bias of S iff it is an additive ideal of S.

By Theorems 3.2.4 and 3.2.5, the category of all Boolean inverse semigroups and additive semigroup homomorphisms is identical to the category of all biases and bias homomorphisms. In particular, this category is a variety of algebras (in the sense of universal algebra).

3.3 The Prime Spectrum Representation of a Distributive Inverse Semigroup

Cayley's Theorem states that every group embeds into some symmetric group, and the Wagner[3]-Preston Theorem (cf. Lawson [73, Theorem 1.5.4]) states that every inverse semigroup embeds into some symmetric inverse semigroup. As observed in Exel [41], the implied embedding does not preserve finite joins as a rule, even starting with a Boolean inverse semigroup.

The following theorem is an analogue of those results for distributive inverse semigroups and embeddings preserving finite joins and meets. Although it is not explicitly stated there, most of it can, in principle, deduced from results of Lawson and Lenz [76] via elementary arguments: ε being one-to-one is essentially contained in the combination of Lemma 3.6, Propositions 3.12, and 3.19 in [76], and ε preserving existing meets can be deduced from Lemma 2.16 and Corollary 2.18 in [76]. Since the required translations involve the digestion of a fair number of nontrivial definitions, we provide direct proofs for convenience.

Theorem 3.3.1 *Let S be a distributive inverse semigroup with zero. Then there are a set Ω and a zero-preserving semigroup embedding $\varepsilon\colon S \hookrightarrow \mathfrak{I}_\Omega$ such that the following conditions hold for every positive integer n and all $x_1, \dots, x_n \in S$:*

(i) *$\bigvee_{i=1}^n x_i$ exists in S iff $\bigvee_{i=1}^n \varepsilon(x_i)$ exists in $\mathfrak{I}_\Omega$, and then*

$$\bigvee_{i=1}^{n} \varepsilon(x_i) = \varepsilon\Big(\bigvee_{i=1}^{n} x_i\Big). \tag{3.3.1}$$

(ii) *If $\bigwedge_{i=1}^n x_i$ exists in S, then*

$$\bigwedge_{i=1}^{n} \varepsilon(x_i) = \varepsilon\Big(\bigwedge_{i=1}^{n} x_i\Big). \tag{3.3.2}$$

Note Remember that every subset of $\mathfrak{I}_\Omega$ has a meet (cf. Example 3.1.8).

Proof Let us recall the definition of the *prime spectrum* $\mathsf{G}_{\mathsf{P}}(S)$ of S, as considered in Lawson and Lenz [76]. By definition, a nonempty subset $\mathfrak{p}$ of S is a *filter* of S if it is a downward directed, upper subset of S, with respect to the natural ordering of S. In addition, we say that $\mathfrak{p}$ is *prime* if $x \vee y \in \mathfrak{p}$ implies that either $x \in \mathfrak{p}$ or $y \in \mathfrak{p}$, for all $x, y \in S$ such that $x \vee y$ is defined. By definition, $\Omega = \mathsf{G}_{\mathsf{P}}(S)$ is[4] the set of all prime filters of S. Set $D = \operatorname{Idp} S$ and $\Omega_e = \{\mathfrak{p} \in \Omega \mid e\mathfrak{p} \subseteq \mathfrak{p}\}$, for every $e \in D$.

---

[3]Often transliterated as "Vagner".

[4]This set can be endowed with a well studied structure of topological groupoid, which will however not be of concern in the present work.

For all $x \in S$ and all $\mathfrak{p} \in \Omega_{\mathbf{d}(x)}$, we define $\varepsilon(x)(\mathfrak{p}) = \uparrow x\mathfrak{p}$ (where $\uparrow X$ is shorthand for $S \uparrow X$). If $\mathfrak{p} \notin \Omega_{\mathbf{d}(x)}$, let $\varepsilon(x)(\mathfrak{p})$ be undefined.

Claim 1 Let $x \in S$ and let $\mathfrak{p} \in \Omega_{\mathbf{d}(x)}$. Then $\varepsilon(x)(\mathfrak{p})$ is a prime filter of S. Moreover, $\varepsilon(x)(\mathfrak{p}) \in \Omega_{\mathbf{r}(x)}$, and $\varepsilon(x^{-1})\big(\varepsilon(x)(\mathfrak{p})\big) = \mathfrak{p}$.

Proof of Claim. If is obvious that $\varepsilon(x)(\mathfrak{p})$ is a proper filter of S. Let $y_0, y_1 \in S$ be compatible such that $y_0 \vee y_1 \in \varepsilon(x)(\mathfrak{p})$, so $xp \leq y_0 \vee y_1$ for some $p \in \mathfrak{p}$. Since $\mathbf{d}(x)\mathfrak{p} \subseteq \mathfrak{p}$, we may assume that $p = \mathbf{d}(x)p$. Since S is distributive, it follows from Proposition 3.1.9 that

$$xp = xp \wedge (y_0 \vee y_1) = (xp \wedge y_0) \vee (xp \wedge y_1)\,,$$

thus, by applying again Proposition 3.1.9, $p = \mathbf{d}(x)p = p_0 \vee p_1$ where each $p_i = x^{-1}(xp \wedge y_i)$. Since $p \in \mathfrak{p}$ and $\mathfrak{p}$ is prime, there is $i \in \{0, 1\}$ such that $p_i \in \mathfrak{p}$. Since $xp_i = \mathbf{r}(x)(xp \wedge y_i) \leq y_i$, it follows that $y_i \in \uparrow x\mathfrak{p}$, thus completing the proof that $\varepsilon(x)(\mathfrak{p})$ is prime.

The proofs of the relations $\varepsilon(x)(\mathfrak{p}) \in \Omega_{\mathbf{r}(x)}$ and $\varepsilon(x^{-1})\big(\varepsilon(x)(\mathfrak{p})\big) = \mathfrak{p}$ are routine and we omit them. □ Claim 1.

It follows from Claim 1 that ε takes its values in $\mathfrak{I}_\Omega$.

Denote by $\overline{\Omega}$ the prime spectrum of D. An argument, similar to the one of the proof of Claim 1, yields the following claim.

Claim 2 Let $\mathfrak{p} \in \overline{\Omega}$. Then $\uparrow\mathfrak{p} \in \Omega$, and $\mathfrak{p} = D \cap \uparrow\mathfrak{p}$.

Claim 3 Let $a, b \in D$. Then $\Omega_{a \wedge b} = \Omega_a \cap \Omega_b$ and $\Omega_{a \vee b} = \Omega_a \cup \Omega_b$. Furthermore, $\Omega_a = \Omega_b$ implies that $a = b$.

Proof of Claim. The relation $\Omega_{a \wedge b} = \Omega_a \cap \Omega_b$ follows immediately from the distributivity of the multiplication on the meet in S, while the relation $\Omega_{a \vee b} = \Omega_a \cup \Omega_b$ follows immediately from Proposition 3.1.9.

Now suppose that $\Omega_a = \Omega_b$. By Claim 2, it follows that $a \cdot (\uparrow\mathfrak{p}) \subseteq \uparrow\mathfrak{p}$ iff $b \cdot (\uparrow\mathfrak{p}) \subseteq \uparrow\mathfrak{p}$, for every $\mathfrak{p} \in \overline{\Omega}$; that is, $a \in \mathfrak{p}$ iff $b \in \mathfrak{p}$, for every $\mathfrak{p} \in \overline{\Omega}$. By Proposition 1.4.1, this implies that $a = b$. □ Claim 3.

The proof of the following claim is routine (and it does not require distributivity), and we omit it.

Claim 4 The map ε is a semigroup homomorphism from S to $\mathfrak{I}_X$.

Claim 5 The map ε is one-to-one.

Proof of Claim. Let $x, y \in S$ such that $\varepsilon(x) = \varepsilon(y)$. By equating the domains of the two sides, we get $\Omega_{\mathbf{d}(x)} = \Omega_{\mathbf{d}(y)}$, thus, by Claim 3, $\mathbf{d}(x) = \mathbf{d}(y)$. Set $e = \mathbf{d}(x)$. The filter $\uparrow\mathfrak{p}$ belongs to Ω_e, for every $\mathfrak{p} \in \overline{\Omega}(e)$ (cf. Claim 2). Hence, it follows from our assumption $\varepsilon(x) = \varepsilon(y)$ that $\uparrow x\mathfrak{p} = \uparrow y\mathfrak{p}$. This implies easily that for every $\mathfrak{p} \in \overline{\Omega}(e)$, there exists $p \in \mathfrak{p} \downarrow e$ such that $xp = yp$. Setting $\Delta = \{p \in D \downarrow e \mid xp = yp\}$, this means that

$$\overline{\Omega}(e) \subseteq \bigcup \big(\overline{\Omega}(p) \mid p \in \Delta\big)\,.$$

Since $\overline{\Omega}(e)$ is compact and all $\overline{\Omega}(p)$ are open within $\overline{\Omega}$ (cf. Theorem 1.4.2), there is a finite subset X of Δ such that

$$\overline{\Omega}(e) \subseteq \bigcup \left(\overline{\Omega}(p) \mid p \in X\right) .$$

By Proposition 1.4.1, this means that $e \leq \bigvee X$. Since $xp = yp$ for every $p \in X$, it follows from the distributivity of S that $xe = ye$, that is, $x = y$. □ Claim 5.

Now we know that ε is a semigroup embedding. Trivially, this embedding maps 0 to the empty function.

Let us prove (i). Since S and $\mathfrak{I}_\Omega$ are both distributive inverse semigroups and since compatibility can be expressed equationally, $\bigvee_{i=1}^{n} \varepsilon(x_i)$ is defined iff $\bigvee_{i=1}^{n} x_i$ is defined. Suppose that this holds and set $y = \bigvee_{i=1}^{n} x_i$. Obviously,

$$\bigvee_{i=1}^{n} \varepsilon(x_i) \leq \varepsilon(y) . \tag{3.3.3}$$

Furthermore, by using Lemma 3.1.4 together with Claim 3, we obtain the relations

$$\operatorname{dom} \varepsilon(y) = \Omega_{\mathbf{d}(y)} = \bigcup_{i=1}^{n} \Omega_{\mathbf{d}(x_i)} = \bigcup_{i=1}^{n} \operatorname{dom} \varepsilon(x_i) = \operatorname{dom}\Big(\bigvee_{i=1}^{n} \varepsilon(x_i)\Big) .$$

By (3.3.3), it follows that $\bigvee_{i=1}^{n} \varepsilon(x_i) = \varepsilon(y)$, thus completing the proof of (i).

Finally, suppose that $z = \bigwedge_{i=1}^{n} x_i$ exists in S. Obviously,

$$\varepsilon(z) \leq \bigwedge_{i=1}^{n} \varepsilon(x_i) . \tag{3.3.4}$$

Thus, in order to complete the proof of (ii), it suffices to prove that the domain of the right hand side of (3.3.4) is contained in the domain of its left hand side. That is, for every element $\mathfrak{p}$ of the domain of $\bigwedge_{i=1}^{n} \varepsilon(x_i)$, we must prove that $\mathbf{d}(z)\mathfrak{p} \subseteq \mathfrak{p}$. Let $p \in \mathfrak{p}$. For all $i, j \in [n]$, $\varepsilon(x_i)(\mathfrak{p}) = \varepsilon(x_j)(\mathfrak{p})$, thus there is $q_{i,j} \in \mathfrak{p} \downarrow p$ such that $x_i q_{i,j} \leq x_j q_{i,j}$. Pick $q \in \mathfrak{p}$ such that $q \leq q_{i,j}$ for all $i, j \in [n]$; since $\mathbf{d}(x_1)\mathfrak{p} \subseteq \mathfrak{p}$, we may assume that $q = \mathbf{d}(x_1)q$. Then $x_i q = x_j q$ for all $i, j \in [n]$, whence $x_1 q = \bigwedge_{i=1}^{n} x_i q = zq$. From $z \leq x_1$ it follows that $x_1^{-1} z = \mathbf{d}(z)$. Therefore, $q = \mathbf{d}(x_1)q = x_1^{-1} x_1 q = x_1^{-1} z q = \mathbf{d}(z) q \leq \mathbf{d}(z) p$, so $\mathbf{d}(z) p \in \mathfrak{p}$, as desired. □

Specializing Theorem 3.3.1 to Boolean inverse semigroups, we obtain immediately the following result.

Corollary 3.3.2 *Every Boolean inverse semigroup S has an additive semigroup embedding into some symmetric inverse semigroup $\mathfrak{I}_\Omega$, preserving all existing nonempty finite meets. In particular, S is a sub-bias of $\mathfrak{I}_\Omega$.*

Remark 3.3.3 The set Ω of Corollary 3.3.2 is identical to the one of Theorem 3.3.1, that is, it is the prime spectrum of S. In the context of Corollary 3.3.2 (i.e., S is

Boolean), more can be said: the prime filters of S are exactly the *ultrafilters* of S, that is, the maximal elements of the set of all filters of S with respect to set inclusion (cf. Lawson and Lenz [76, Lemma 3.20]).

For an arbitrary inverse semigroup S, the canonical semigroup homomorphism $\lambda\colon S \to \mathfrak{I}_{\Omega'}$ introduced in Exel [41], where Ω' is the set of all ultrafilters of S (denoted by $\mathsf{G}_{\mathrm{M}}(S)$ in Lawson and Lenz [76]), is tight in Exel's sense. As in [41], λ will be called the *regular representation of* S. Although Exel's concept of a Boolean inverse semigroup is less restrictive than ours, it follows from Exel [41, Proposition 6.2], together with Theorem 3.2.5, that his concept of a tight homomorphism extends our concept of an additive semigroup homomorphism. Moreover, for a Boolean inverse semigroup S, $\mathsf{G}_{\mathrm{M}}(S) = \mathsf{G}_{\mathrm{P}}(S)$ and the canonical embedding $\varepsilon\colon S \hookrightarrow \mathfrak{I}_{\mathsf{G}_{\mathrm{P}}(S)}$ of Theorem 3.3.1 is identical to Exel's regular representation λ.

On the other hand, $\mathsf{G}_{\mathrm{M}}(S) \not\cong \mathsf{G}_{\mathrm{P}}(S)$ for most distributive inverse semigroups S (consider the three-element chain), so there are examples where $\varepsilon \neq \lambda$.

Remark 3.3.4 Say that an inverse semigroup is *Exel-Boolean* if its semilattice of idempotents is Boolean (not necessarily unital). The final example of Exel [41] is an Exel-Boolean inverse semigroup with no additive semigroup embedding into any symmetric inverse semigroup. Of course, by Corollary 3.3.2, such an inverse semigroup cannot be Boolean. A much easier example, serving the same purpose, is the one of Example 3.1.7: in that example, the ultrafilters of S are $\mathfrak{p}_i = \{i, 3, 4\}$, for $i \in \{1, 2\}$; and $3\mathfrak{p}_i = 4\mathfrak{p}_i = \mathfrak{p}_i$, whenever $i \in \{1, 2\}$. In particular, $\lambda(3) = \lambda(4)$, with $3 \neq 4$.

We say that two elements x and y in an inverse semigroup S with zero *essentially coincide*, in notation $x \equiv y$, if $\mathbf{d}(x) = \mathbf{d}(y)$ and for every nonzero idempotent $e \leq \mathbf{d}(x)$ there exists a nonzero idempotent $a \leq e$ such that $xa = ya$. We say that S is *continuous* if $x \equiv y$ implies that $x = y$, for all $x, y \in S$. Exel proved in [41, Theorem 7.5] that every continuous Exel-Boolean inverse semigroup embeds tightly (in his sense) into some symmetric inverse monoid. He also asks, just before the statement of [41, Theorem 7.5], whether $x \equiv y$ implies $\lambda(x) = \lambda(y)$. The following example, whose construction is inspired by the final counterexample of Exel [41], shows that this is not the case as a rule. This example turns out to be Boolean.

Example 3.3.5 A Boolean inverse monoid S, with unit element 1_S and an element x such that $1_S \equiv x$ and $\lambda(1_S) \neq \lambda(x)$. In particular, S is not continuous.

Proof We denote by $\mathcal{B}$ the Boolean algebra of all subsets of $\mathbb{Z}^+$ that are either finite or cofinite, and we fix a nontrivial group G. For each $x \in \mathcal{B}$, we set $N_x = G$ if x is finite, and $N_x = \{1\}$ if x is cofinite. The semigroup $\mathcal{B} \times G$ is an inverse monoid, and the binary relation $\sim$ on S defined by the rule

$$(x, g) \sim (y, h) \quad \text{if} \quad x = y \text{ and } g \equiv h \pmod{N_x}\,, \qquad \text{for any } (x, g), (y, h) \in \mathcal{B} \times G\,,$$

is a monoid congruence on $S \times G$. The quotient monoid $S = (\mathcal{B} \times G)/{\sim}$ is an inverse monoid with zero. Denoting by $[x, g]$ the equivalence class of (x, g) modulo $\sim$, the zero element of S is $[\varnothing, 1] = [\varnothing, g]$ (for all $g \in G$) and the unit element of S

is $1_S = [\mathbb{Z}^+, 1]$. Easy calculations show that $[x, g]^{-1} = [x, g^{-1}]$ and $\mathbf{d}([x, g]) = \mathbf{r}([x, g]) = [x, 1]$ whenever $(x, g) \in \mathcal{B} \times G$. Two elements $[x_0, g_0]$ and $[x_1, g_1]$ of S are orthogonal if $x_0 \cap x_1 = \varnothing$ (thus one of x_0 and x_1 needs to be finite), and then their orthogonal sum is $[x_0 \cup x_1, g_{1-i}]$ if x_i is finite. The semilattice of all idempotents of S is $B = \{[x, 1] \mid x \in \mathcal{B}\}$, which is isomorphic to $\mathcal{B}$. Therefore, S is a Boolean inverse monoid.

Pick $g \in G \setminus \{1\}$ and set $x = [\mathbb{Z}^+, g]$. Every nonzero idempotent of S contains an idempotent of the form $e_n = [\{n\}, 1]$, where $n \in \mathbb{Z}^+$; and $1_S e_n = x e_n = e_n$. This proves that $1_S \equiv x$.

However, since S is Boolean, it follows from Corollary 3.3.2 that Exel's regular representation λ of S (cf. Remark 3.3.3) is one-to-one; whence $\lambda(1_S) \neq \lambda(x)$. □

3.4 Additive Congruences of Boolean Inverse Semigroups

In this section we shall investigate in our context the crucial universal-algebraic concept of a *congruence*, in particular by describing bias congruences in terms of the semigroup operations and the orthogonal join operation $\oplus$.

Proposition 3.4.1 *Let S be a Boolean inverse semigroup. An equivalence relation $\boldsymbol{\theta}$ on S is a bias congruence iff $\boldsymbol{\theta}$ is a semigroup congruence and the following condition holds:*

For all $x \in S$ and all $a, b \in \operatorname{Idp} S$ orthogonal,

$$(xa \equiv_{\boldsymbol{\theta}} a \text{ and } xb \equiv_{\boldsymbol{\theta}} b) \Rightarrow x(a \oplus b) \equiv_{\boldsymbol{\theta}} a \oplus b\,. \tag{3.4.1}$$

Proof We prove the non-trivial direction. Let $\boldsymbol{\theta}$ be a semigroup congruence of S (thus also an inverse semigroup congruence) satisfying (3.4.1).

The assumption (3.4.1) means that for all orthogonal idempotents a and b, from $\begin{matrix} a \\ b \end{matrix} \leq_{\boldsymbol{\theta}} x$ it follows that $a \oplus b \leq_{\boldsymbol{\theta}} x$, for each $x \in S$. (Recall that $x \leq_{\boldsymbol{\theta}} y$ is shorthand for $x \equiv_{\boldsymbol{\theta}} y\mathbf{d}(x)$.) Denoting by $\theta \colon S \twoheadrightarrow S/\boldsymbol{\theta}$ the canonical projection, this means that $\theta(a \oplus b) = \theta(a) \oplus \theta(b)$ within $S/\boldsymbol{\theta}$.

Claim 1 $\theta(a \vee b)$ is the join of $\{\theta(a), \theta(b)\}$ within $S/\boldsymbol{\theta}$, for any idempotents a and b of S. Hence, $\boldsymbol{\theta}$ is compatible with the operations $\wedge$, $\vee$, and $\smallsetminus$ on idempotents.

Proof of Claim. Any upper bound, within $S/\boldsymbol{\theta}$, of $\{\theta(a), \theta(b)\}$ is also an upper bound of the set $\{\theta(a \smallsetminus b), \theta(b)\}$, thus, by (3.4.1), it is an upper bound of $\theta((a \smallsetminus b) \oplus b) = \theta(a \vee b)$. Hence, $\theta(a \vee b)$ is the join of $\{\theta(a), \theta(b)\}$ within $S/\boldsymbol{\theta}$, and hence $\boldsymbol{\theta}$ is compatible with the $\vee$ operation on $\operatorname{Idp} S$. Since $\boldsymbol{\theta}$ is also a congruence with respect to the product operation, its restriction to the generalized Boolean algebra $\operatorname{Idp} S$ is a congruence with respect to join and meet, thus it is also a congruence with respect to the difference operation $\smallsetminus$. □ Claim 1.

Claim 2 The equivalence relation $\boldsymbol{\theta}$ is compatible with the operation $\oslash$ on S.

Proof of Claim. Since $\boldsymbol{\theta}$ is compatible with the product operation, it is also compatible with the operations $\mathbf{d}$ and $\mathbf{r}$, thus, by Claim 1, it is also compatible with the operations $(x, y) \mapsto \mathbf{r}(x) \smallsetminus \mathbf{r}(y)$ and $(x, y) \mapsto \mathbf{d}(x) \smallsetminus \mathbf{d}(y)$. Since it is compatible with the product operation, the desired conclusion follows. □ Claim 2.

Claim 3 Let $x_0, y_0, x_1, y_1 \in S$ such that $x_0 \equiv_{\boldsymbol{\theta}} y_0$, $x_1 \equiv_{\boldsymbol{\theta}} y_1$, $x_0 \perp x_1$, and $y_0 \perp y_1$. Then $x_0 \oplus x_1 \equiv_{\boldsymbol{\theta}} y_0 \oplus y_1$.

Proof of Claim. Set $x = x_0 \oplus x_1$ and $y = y_0 \oplus y_1$. Then

$$\mathbf{d}(x_i) = x_i^{-1}x_i = x^{-1}x_i \equiv_{\boldsymbol{\theta}} x^{-1}y_i \leq_{\boldsymbol{\theta}} x^{-1}y\,, \quad \text{for each } i \in \{0,1\}\,.$$

Thus, by our assumption (3.4.1), $\mathbf{d}(x_0) \oplus \mathbf{d}(x_1) \leq_{\boldsymbol{\theta}} x^{-1}y$, that is, $\mathbf{d}(x) \leq_{\boldsymbol{\theta}} x^{-1}y$, and thus $x = x\,\mathbf{d}(x) \leq_{\boldsymbol{\theta}} xx^{-1}y$, and so $x \leq_{\boldsymbol{\theta}} y$. Symmetrically, $y \leq_{\boldsymbol{\theta}} x$, and therefore, since $\boldsymbol{\theta}$ is an inverse semigroup congruence, $x \equiv_{\boldsymbol{\theta}} y$. □ Claim 3.

Let $x_0, x_1, y_0, y_1 \in S$ such that $x_0 \equiv_{\boldsymbol{\theta}} y_0$ and $x_1 \equiv_{\boldsymbol{\theta}} y_1$. It follows from Claim 2 that $x_0 \oslash x_1 \equiv_{\boldsymbol{\theta}} y_0 \oslash y_1$. Since $x_0 \oslash x_1 \perp x_1$, $y_0 \oslash y_1 \perp y_1$, and $y_0 \equiv_{\boldsymbol{\theta}} y_1$, it follows from Claim 3 that $(x_0 \oslash x_1) \oplus x_1 \equiv_{\boldsymbol{\theta}} (y_0 \oslash y_1) \oplus y_1$, that is, $x_0 \triangledown x_1 \equiv_{\boldsymbol{\theta}} y_0 \triangledown y_1$. Therefore, $\boldsymbol{\theta}$ is compatible with the operation $\triangledown$. □

Define an *additive congruence* of a Boolean inverse semigroup S as a semigroup congruence satisfying (3.4.1). Proposition 3.4.1 says that the concepts of additive congruence and bias congruence are equivalent.

It would be nicer if, within the statement of Proposition 3.4.1, the assumption (3.4.1) could be replaced by the weaker assumption that the restriction of $\boldsymbol{\theta}$ to $\operatorname{Idp} S$ is a ring congruence. The following example shows that this cannot be done, even for idempotent-separating congruences. (A congruence $\boldsymbol{\theta}$ of S is *idempotent-separating* if $a \equiv_{\boldsymbol{\theta}} b$ implies that $a = b$, for all $a, b \in \operatorname{Idp} S$. By Howie [60, Proposition II.4.8], this is equivalent to saying that $\boldsymbol{\theta} \subseteq \mathscr{H}$.)

Example 3.4.2 Denote the two-element group $G = \mathbb{Z}/2\mathbb{Z}$ multiplicatively, so $G = \{1, u\}$ with $u^2 = 1$. The inverse semigroup $S = G^{\sqcup 0} \times \{0, 1\}$ is a Boolean inverse monoid. The equivalence relation $\boldsymbol{\theta}$ on S, defined as the union of the diagonal of S with the set $\{((u,0), (1,0)), ((1,0), (u,0))\}$, is an idempotent-separating semigroup congruence of S.

This congruence is not additive, for $(u, 0) \equiv_{\boldsymbol{\theta}} (1, 0)$ while $(u, 1) \not\equiv_{\boldsymbol{\theta}} (1, 1)$, the latter meaning that $(u, 0) \oplus (0, 1) \not\equiv_{\boldsymbol{\theta}} (1, 0) \oplus (0, 1)$.

Observe the contrast between Example 3.4.2 and Theorem 3.2.5. The point is that the quotient inverse semigroup $S/\boldsymbol{\theta}$, in Example 3.4.2, is not Boolean.

Notation 3.4.3 We set

$$x\,\langle y\rangle = xyx^{-1}\,, \quad \text{for all } x\,,\ y \text{ in any inverse semigroup}\,. \tag{3.4.2}$$

Recall that if y is idempotent, then so is $x\langle y\rangle$. The following observation will be used repeatedly without mentioning, throughout this work.

Lemma 3.4.4 *Let x, u, v be elements in an inverse semigroup, with either u or v idempotent. Then $x\langle uv\rangle = x\langle u\rangle \cdot x\langle v\rangle$.*

Proof If, for example, u is idempotent, then u and $x^{-1}x$ commute (they are both idempotent), thus $x\langle uv\rangle = xx^{-1}xuvx^{-1} = xux^{-1}xvx^{-1} = x\langle u\rangle \cdot x\langle v\rangle$. □

As our next result shows, the somewhat irregular-looking behavior witnessed by Example 3.4.2 does not occur for the largest idempotent-separating congruence of a Boolean inverse semigroup.

Proposition 3.4.5 *Let S be a Boolean inverse semigroup. Then the largest idempotent-separating congruence μ of S is an additive congruence of S. In particular, S/μ is a Boolean inverse semigroup and the canonical projection $S \twoheadrightarrow S/\mu$ is an additive semigroup homomorphism.*

Proof Recall (cf. Howie [60, Theorem V.3.2]) that μ can be described by

$$\mu = \{(x, y) \in S \times S \mid (\forall e \in \operatorname{Idp} S)(x\langle e\rangle = y\langle e\rangle)\} \tag{3.4.3}$$

(cf. Notation 3.4.3). Now let $a, b \in \operatorname{Idp} S$ be orthogonal and let $x \in S$ with $xa \equiv_\mu a$ and $xb \equiv_\mu b$. By (3.4.3), this means that $x\langle ae\rangle = ae$ and $x\langle be\rangle = be$ for every $e \in \operatorname{Idp} S$. Now for every $e \in \operatorname{Idp} S$,

$$x\langle (a \oplus b)e\rangle = x\langle ae \oplus be\rangle = x\langle ae\rangle \oplus x\langle be\rangle = ae \oplus be = (a \oplus b)e\,,$$

so $x(a \oplus b) \equiv_\mu a \oplus b$. By Proposition 3.4.1, it follows that μ is a bias congruence. The last part of Proposition 3.4.5 follows immediately. □

Proposition 3.4.6 *Let I be an additive ideal in a Boolean inverse semigroup S. Then the binary relation $\equiv_I$ on S, defined by the rule*

$$x \equiv_I y \Leftrightarrow (\exists z)\big(z \leq x \text{ and } z \leq y \text{ and } \{x \smallsetminus z, y \smallsetminus z\} \subseteq I\big)\,, \quad \text{for all } x, y \in S\,, \tag{3.4.4}$$

is the least additive congruence of S for which the equivalence class of 0 contains I. Moreover, $0/{\equiv_I} = I$.

Proof It is obvious that $0/{\equiv_I} = I$ and that every additive congruence of S, for which the equivalence class of 0 contains I, contains $\equiv_I$. Hence, it suffices to prove that $\equiv_I$ is an additive congruence of S. It is trivial that $\equiv_I$ is both reflexive and symmetric. Let $x, y, z \in S$ such that $x \equiv_I y$ and $y \equiv_I z$. There are $u, v \in S$ with $u \leq x$, $v \leq z$, and $\genfrac{}{}{0pt}{}{u}{v} \leq y$, such that $x \smallsetminus u$, $y \smallsetminus u$, $y \smallsetminus v$, and $z \smallsetminus v$ all belong to I. From $\genfrac{}{}{0pt}{}{u}{v} \leq y$ it follows that $u \sim v$ and $y \smallsetminus u \sim y \smallsetminus v$. The latter relation implies that $(y \smallsetminus u) \vee (y \smallsetminus v)$ exists in S. Since $\{y \smallsetminus u, y \smallsetminus v\} \in I$, it follows that $(y \smallsetminus u) \vee (y \smallsetminus v) \in I$ (cf. Proposition 3.1.19), that is, $y \smallsetminus (u \wedge v) \in I$. (All statements, such as $y \smallsetminus (u \wedge v) = (y \smallsetminus u) \vee (y \smallsetminus v)$, can easily be proved by reduction to the idempotent case, via Lemma 3.1.3.) By meeting that relation with u, we get $u \smallsetminus (u \wedge v) \in I$ and $v \smallsetminus (u \wedge v) \in I$. Therefore, $x \smallsetminus (u \wedge v) = (x \smallsetminus u) \oplus (u \smallsetminus (u \wedge v)) \in I$, and, similarly, $z \smallsetminus (u \wedge v) \in I$, so $x \equiv_I z$.

Let $x, y, z \in S$ with $x \equiv_I y$. There exists $u \in S$ such that $u \leq \begin{smallmatrix} x \\ y \end{smallmatrix}$ and $\{x \smallsetminus u, y \smallsetminus u\} \subseteq I$. By Lemma 3.1.12 and since I is an ideal of S, it follows that $\{xz \smallsetminus uz, yz \smallsetminus uz\} \subseteq I$ and $\{zx \smallsetminus zu, zy \smallsetminus zu\} \subseteq I$, thus $xz \equiv_I yz$ and $zx \equiv_I zy$. Therefore, $\equiv_I$ is a semigroup congruence of S.

In order to verify that $\equiv_I$ is an additive congruence, it suffices to verify (3.4.1). Let $a, b \in \operatorname{Idp} S$ be orthogonal idempotents and let $x \in S$ such that $xa \equiv_I a$ and $xb \equiv_I b$. There are $u \leq \begin{smallmatrix} xa \\ a \end{smallmatrix}$ and $v \leq \begin{smallmatrix} xb \\ b \end{smallmatrix}$ such that $xa \smallsetminus u$, $a \smallsetminus u$, $xb \smallsetminus v$, and $b \smallsetminus v$ all belong to I. From $u \leq a$ and $v \leq b$ it follows that u and v are both idempotent. Further, $u \oplus v \leq \begin{smallmatrix} x(a \oplus b) \\ a \oplus b \end{smallmatrix}$, and

$$(a \oplus b) \smallsetminus (u \oplus v) = (a \smallsetminus u) \oplus (b \smallsetminus v) \in I\,,$$
$$x(a \oplus b) \smallsetminus (u \oplus v) = (xa \oplus xb) \smallsetminus (u \oplus v) = (xa \smallsetminus u) \oplus (xb \smallsetminus v) \in I\,,$$

so $x(a \oplus b) \equiv_I a \oplus b$. Therefore, $\equiv_I$ is an additive congruence of S. By virtue of Theorem 3.2.4 and Proposition 3.4.1, the final statement of Proposition 3.4.6 follows immediately. □

For an additive ideal I in a Boolean inverse semigroup S, we will denote by x/I the equivalence class of x with respect to $\equiv_I$, for each $x \in S$. Observe that $0/I = I$.

In the context of Proposition 3.4.6, $\equiv_I$ is a bias congruence of S (cf. Proposition 3.4.1), thus the quotient structure $S/I = S/{\equiv_I}$ is a Boolean inverse semigroup.

Our next group of results introduces an alternate way to view additive ideals of S, by focusing attention on the idempotents of S.

Definition 3.4.7 Let S be a Boolean inverse semigroup. An ideal I of the Boolean ring $\operatorname{Idp} S$ is *$\mathscr{D}$-closed* if for all $a, b \in \operatorname{Idp} S$, $a \mathrel{\mathscr{D}_S} b$ and $a \in I$ implies that $b \in I$.

Our next result shows that additive ideals (of Boolean inverse semigroups) are essentially the same concept as $\mathscr{D}$-closed ideals (in Boolean rings of idempotents).

Proposition 3.4.8 *Let S be a Boolean inverse semigroup and set $B = \operatorname{Idp} S$. The following statements hold:*

(1) *For any additive ideal J of S, the intersection $J \cap B$ is a $\mathscr{D}$-closed ideal of B.*
(2) *For any $\mathscr{D}$-closed ideal I of the Boolean ring B, the equality $\mathbf{d}^{-1}[I] = \mathbf{r}^{-1}[I]$ holds. Furthermore, this set is the ideal of S generated by I, and it is also an additive ideal.*
(3) *The two transformations described in* (1) *and* (2) *above are mutually inverse.*

Proof (1) From $0 \in J$ it follows that $0 \in J \cap B$. Let $a \in B$ and $b \in J \cap B$ with $a \leq b$. Then $a = ab \in SJ \subseteq J$, so $a \in J \cap B$, and so $J \cap B$ is a lower subset of B. Since J is closed under finite orthogonal joins, so is $J \cap B$. Hence $J \cap B$ is an ideal of

B. Now let $a, b \in B$ with $a \mathscr{D} b$. Pick $x \in S$ with $\mathbf{d}(x) = a$ and $\mathbf{r}(x) = b$. In particular, $b = xax^{-1}$, hence $a \in I$ implies that $b \in I$.

(2) Since $\mathbf{d}(x) \mathrel{\mathscr{D}} \mathbf{r}(x)$ for all $x \in S$, the equality $\mathbf{d}^{-1}[I] = \mathbf{r}^{-1}[I]$ is obvious. It is then easy to verify, in particular by using Lemma 3.1.4, that this set is an additive ideal of S. It obviously contains I. Let J be an ideal of S containing I. For any $x \in \mathbf{d}^{-1}[I]$, the element $\mathbf{d}(x)$ belongs to I, thus to J, thus $x = x\mathbf{d}(x) \in J$; whence $\mathbf{d}^{-1}[I] \subseteq J$.

(3) Let J be an additive ideal of S and set $I = J \cap B$. We claim that $J = \mathbf{d}^{-1}[I]$. For each $x \in J$, $\mathbf{d}(x) = x^{-1}x \in J$, thus $\mathbf{d}(x) \in I$, that is, $x \in \mathbf{d}^{-1}[I]$. Conversely, let $x \in \mathbf{d}^{-1}[I]$. Then $\mathbf{d}(x) \in J$ as well, so $x = x\mathbf{d}(x) \in J$, thus completing the proof of our claim.

Finally, it is trivial that $I = \mathbf{d}^{-1}[I] \cap B$, for any $\mathscr{D}$-closed ideal I of B. □

Proposition 3.4.9 *Let S and T be Boolean inverse semigroups and let $f \colon S \to T$ be an additive map. Then the set* $\ker f = f^{-1}\{0\}$ *is an additive ideal of S. Furthermore, denoting by $p \colon S \twoheadrightarrow S/\ker f$ the canonical projection, there is a unique additive semigroup homomorphism $\overline{f} \colon S/\ker f \to T$ such that $f = \overline{f} \circ p$.*

Note The set $\ker f = f^{-1}\{0\}$, which is a subset of the domain of f, should not be confused with the kernel $\operatorname{Ker} f$ of f, which is an equivalence relation on the domain of f (cf. Sect. 1.3).

Proof It is straightforward to verify that the subset $I = \ker f$ is an additive ideal of S.

Since biases form a variety of algebras, the standard concepts of universal algebra apply to the category of biases and bias homomorphisms. This is the case, in particular, for the First Isomorphism Theorem. Since bias homomorphisms and additive semigroup homomorphisms are the same concept (cf. Theorem 3.2.5), in order to prove the final statement of Proposition 3.4.9, it suffices to prove that $p(x) = p(y)$ (i.e., $x \equiv_I y$) implies that $f(x) = f(y)$, for all $x, y \in S$. Let $z \in S$ witness $x \equiv_I y$, that is, $z \leq {x \atop y}$ and $\{x \smallsetminus z, y \smallsetminus z\} \subseteq I$. Since $x = (x \smallsetminus z) \oplus z$ and f is additive, we get $f(x) = f(x \smallsetminus z) \oplus f(z) = f(z)$. Similarly, $f(y) = f(z)$, so $f(x) = f(y)$. □

Say that a congruence $\boldsymbol{\theta}$ of S is *ideal-induced* if $\boldsymbol{\theta}$ is equal to $\equiv_I$ for some additive ideal I of S. As the following example shows, a Boolean inverse semigroup may have many additive congruences that are not ideal-induced. This example also shows that the map $\overline{f}$ of the statement of Proposition 3.4.9 may not be one-to-one.

Example 3.4.10 For any group G, the inverse semigroup $G^{\sqcup 0}$ is a Boolean inverse semigroup, where $x \perp y$ iff either $x = 0$ or $y = 0$. If an additive congruence $\boldsymbol{\theta}$ of $G^{\sqcup 0}$ identifies 0 with some nonzero element, then $\boldsymbol{\theta} = G^{\sqcup 0} \times G^{\sqcup 0}$ is the largest congruence. If $\boldsymbol{\theta}$ does not identify 0 with any nonzero element, then $\boldsymbol{\theta}$ is the congruence $\boldsymbol{\theta}_H$ associated with a normal subgroup H of G, in the sense that $x \equiv_{\boldsymbol{\theta}} y$ iff either $x = y = 0$ or $x, y \neq 0$ and $x^{-1}y \in H$. Observe that $\boldsymbol{\theta}_H$ is ideal-induced iff $H = \{1\}$.

It follows, in particular, that the lattice of all additive congruences of $G^{\sqcup 0}$ is isomorphic to the normal subgroup lattice NSub G of G, with a top element added. In particular, taking for G the Klein group $(\mathbb{Z}/2\mathbb{Z}) \times (\mathbb{Z}/2\mathbb{Z})$, the lattice of all additive congruences of $G^{\sqcup 0}$ is the five-element modular non distributive lattice M_3, with a top element added. Thus we get the following observation: *The lattice of all additive congruences of a Boolean inverse semigroup may not be distributive.*

On the other hand, it is well know that the lattice NSub G is modular, for any group G. Hence, the lattice of all additive congruences of $G^{\sqcup 0}$ is modular. We shall now see that this observation can be extended to any Boolean inverse semigroup.

To this end, let us introduce the following ternary term m, in the similarity type $\mathcal{L}_{\mathrm{BIS}}$ of all biases (cf. Notation 3.2.2; recall that $\mathbf{d}(\mathsf{x})$ and $\mathbf{r}(\mathsf{z})$ are shorthand for $\mathsf{x}^{-1}\mathsf{x}$ and $\mathsf{x}\mathsf{x}^{-1}$, respectively):

$$\mathsf{m}(\mathsf{x},\mathsf{y},\mathsf{z}) = \Big(\mathsf{x}\big(\mathbf{d}(\mathsf{x}) \otimes \mathbf{d}(\mathsf{y})\big) \mathbin{\triangledown} \mathsf{x}\mathsf{y}^{-1}\mathsf{z}\Big) \mathbin{\triangledown} \big(\mathbf{r}(\mathsf{z}) \otimes \mathbf{r}(\mathsf{y})\big)\mathsf{z}\,. \tag{3.4.5}$$

It is worthwhile noticing that the right hand side of (3.4.5) contains, as a subterm, the group-theoretical term $\mathsf{x}\mathsf{y}^{-1}\mathsf{z}$, which is the standard Mal'cev term for groups.

Recall that a variety $\mathbf{V}$ of algebras is *congruence-permutable* if $\alpha \circ \beta = \beta \circ \alpha$ for any congruences α and β of any algebra in $\mathbf{V}$. We also say that $\mathbf{V}$ is *congruence-modular* if the lattice of all congruences of any algebra $A \in \mathbf{V}$ is modular, that is, $\alpha \cap (\beta \vee (\alpha \cap \gamma)) = (\alpha \cap \beta) \vee (\alpha \cap \gamma)$ for any congruences α, β, γ of A. It is well known that every congruence-permutable variety is congruence-modular (for every lattice of pairwise commuting equivalence relations is modular, and even Arguesian; this originates in Jónsson [63], see also Grätzer [53, Theorem 410]).

Theorem 3.4.11 *The term* m *is a Mal'cev term for the variety of all biases; that is, the equations* $\mathsf{m}(\mathsf{x},\mathsf{x},\mathsf{y}) = \mathsf{m}(\mathsf{y},\mathsf{x},\mathsf{x}) = \mathsf{y}$ *hold identically in every bias. Therefore, the variety of all biases is congruence-permutable, thus also congruence-modular.*

Proof Let S be a bias. It is straightforward to verify that $x \mathbin{\triangledown} 0 = 0 \mathbin{\triangledown} x = x$, for every $x \in S$. Since the operations $\otimes$ and $\smallsetminus$ agree on the idempotents of S, while $\triangledown$ and $\oplus$ agree on orthogonal pairs, we can compute

$$\begin{aligned}
\mathsf{m}(x,x,y) &= \Big(x\big(\mathbf{d}(x) \smallsetminus \mathbf{d}(x)\big) \mathbin{\triangledown} xx^{-1}y\Big) \mathbin{\triangledown} \big(\mathbf{r}(y) \smallsetminus \mathbf{r}(x)\big)y \\
&= \mathbf{r}(x)y \oplus \big(\mathbf{r}(y) \smallsetminus \mathbf{r}(x)\big)y \\
&= \big(\mathbf{r}(x) \vee \mathbf{r}(y)\big)y \\
&= y,
\end{aligned}$$

and

$$\begin{aligned}
\mathsf{m}(y,x,x) &= \Big(y\big(\mathbf{d}(y) \smallsetminus \mathbf{d}(x)\big) \mathbin{\triangledown} yx^{-1}x\Big) \mathbin{\triangledown} \big(\mathbf{r}(x) \smallsetminus \mathbf{r}(x)\big)x \\
&= y\big(\mathbf{d}(y) \smallsetminus \mathbf{d}(x)\big) \mathbin{\triangledown} y\,\mathbf{d}(x)
\end{aligned}$$

$$
\begin{aligned}
&= y(\mathbf{d}(y) \vee \mathbf{d}(x)) \\
&= y\,.
\end{aligned}
$$

Hence m is a Mal'cev term for biases. It is well known since Mal'cev [80] (cf. McKenzie et al. [82, Theorem 4.141]) that this implies the congruence-permutability result, whence the congruence-modularity result. □

Theorem 3.4.11 marks a crucial difference between Boolean inverse semigroups on the one hand, and inverse semigroups on the other hand. Indeed, it is well known that there is no lattice identity satisfied by the congruence lattices of all semilattices (cf. Freese and Nation [48]), thus, a fortiori, by the congruence lattices of all inverse semigroups.

3.5 Generalized Rook Matrices over Boolean Inverse Semigroups

The following concept is taken from Wallis [116, § 4.5], see also Kudryavtseva et al. [71]. It extends the one of a *rook matrix* introduced in Solomon [103]. Solomon's rook matrices are identical to generalized rook matrices taken over the two-element inverse semigroup.

Recall that left and right orthogonality are both introduced in Definition 3.1.2.

Definition 3.5.1 Let S be an inverse semigroup with zero and let Ω be a (possibly infinite) set. A square matrix $a = \bigl(a_{i,j} \mid (i,j) \in \Omega \times \Omega\bigr)$, with all $a_{i,j} \in S$, is a $\Omega \times \Omega$ *generalized rook matrix over S* if any two distinct rows (resp., columns) of S are left orthogonal (resp., right orthogonal). In formula,

$$a_{i,j} \perp_{\mathrm{rt}} a_{i,k} \text{ and } a_{j,i} \perp_{\mathrm{lt}} a_{k,i}\,, \quad \text{for all } i,j,k \in \Omega \text{ with } j \neq k\,,$$

or, equivalently,

$$a_{i,j}^{-1} a_{i,k} = a_{j,i} a_{k,i}^{-1} = 0\,, \quad \text{for all } i,j,k \in \Omega \text{ with } j \neq k\,.$$

We denote by $\mathrm{R}_{\Omega}^{\oplus}(S)$ the set of all $\Omega \times \Omega$ generalized rook matrices over S. We also consider the following subsets of $\mathrm{R}_{\Omega}^{\oplus}(S)$:

- the set $\mathrm{B}_{\Omega}^{\oplus}(S)$ of all generalized rook matrices a that are both *row-finite* (i.e., for each $i \in \Omega$, $a_{i,j} = 0$ for all but finitely many $j \in \Omega$) and *column-finite* (i.e., for each $j \in \Omega$, $a_{i,j} = 0$ for all but finitely many $i \in \Omega$);
- the set $\mathrm{M}_{\Omega}^{\oplus}(S)$ of all generalized rook matrices a such that $a_{i,j} = 0$ for all but finitely many $(i,j) \in \Omega \times \Omega$.

If $\Omega = [n]$, for $n \in \mathbb{N}$, we will write $\mathrm{M}_{n}^{\oplus}(S) = \mathrm{B}_{n}^{\oplus}(S) = \mathrm{R}_{n}^{\oplus}(S) = \mathrm{R}_{[n]}^{\oplus}(S)$.

The basic properties of generalized rook matrices over S are summed up in Wallis [116, § 4.5], Kudryavtseva et al. [71, Proposition 3.5]. Since we are dealing with a slightly more general context (due to the possibility that Ω be infinite), we include proofs for convenience.

In what follows, for any family $(a_i \mid i \in I)$ of elements in a Boolean inverse semigroup S, we say that the orthogonal join $\bigoplus_{i \in I} a_i$ is *defined* if the a_i are pairwise orthogonal and $a_i = 0$ for all but finitely many $i \in I$.

Lemma 3.5.2 *Let S be an inverse semigroup with zero and let Ω be a set. The following statements hold, for any generalized rook matrices $a = \big(a_{i,j} \mid (i,j) \in \Omega \times \Omega\big)$ and $b = \big(b_{i,j} \mid (i,j) \in \Omega \times \Omega\big)$ over S:*

(1) *For any $i, j \in \Omega$, the elements $a_{i,k}b_{k,j}$, where $k \in \Omega$, are pairwise orthogonal.*
(2) *If S is Boolean inverse and all elements $c_{i,j} = \bigoplus_{k \in \Omega} a_{i,k}b_{k,j}$, for $i, j \in \Omega$, are defined* (in which case we say that the matrix ab is defined), *then $c = \big(c_{i,j} \mid (i,j) \in \Omega \times \Omega\big)$ is a generalized rook matrix over S.*
(3) *If S is Boolean inverse, $a, b \in \mathrm{R}_\Omega^\oplus(S)$, and either a is row-finite or b is column-finite, then ab is defined. Furthermore, if a is row-finite and b is column-finite, then ab is both row-finite and column-finite.*
(4) *If S is Boolean inverse, then $\mathrm{M}_\Omega^\oplus(S)$ is an ideal of $\mathrm{B}_\Omega^\oplus(S)$.*

In the context of Lemma 3.5.2(2), we say that c is the product of a and b, and we write $c = ab$.

Proof (1) For any distinct $k, l \in \Omega$, from $b_{k,j}b_{l,j}^{-1} = 0$ it follows that

$$a_{i,k}b_{k,j}(a_{i,l}b_{l,j})^{-1} = a_{i,k}b_{k,j}b_{l,j}^{-1}a_{i,l}^{-1} = 0\,,$$

so $a_{i,k}b_{k,j} \perp_{\mathrm{lt}} a_{i,l}b_{l,j}$. Similarly, from $a_{i,k}^{-1}a_{i,l} = 0$ it follows that

$$(a_{i,k}b_{k,j})^{-1}a_{i,l}b_{l,j} = b_{k,j}^{-1}a_{i,k}^{-1}a_{i,l}b_{l,j} = 0\,.$$

so $a_{i,k}b_{k,j} \perp_{\mathrm{rt}} a_{i,l}b_{l,j}$. Hence, $a_{i,k}b_{k,j} \perp a_{i,l}b_{l,j}$.

(2) Suppose that the matrix $c = ab$ is defined. Let $i, j, k \in \Omega$ with $j \neq k$. We claim that $c_{i,j} \perp_{\mathrm{rt}} c_{i,k}$ and $c_{j,i} \perp_{\mathrm{lt}} c_{k,i}$. In order to verify the first statement, it suffices to verify that $a_{i,p}b_{p,j} \perp_{\mathrm{rt}} a_{i,q}b_{q,k}$, that is, $b_{p,j}^{-1}a_{i,p}^{-1}a_{i,q}b_{q,k} = 0$, for all $p, q \in \Omega$. If $p \neq q$, then this follows from $a_{i,p}^{-1}a_{i,q} = 0$. If $p = q$, then $a_{i,p}^{-1}a_{i,q} = \mathbf{d}(a_{i,p})$ is idempotent, thus, since $b_{p,j}^{-1}b_{p,k} = 0$, we get

$$b_{p,j}^{-1}a_{i,p}^{-1}a_{i,q}b_{q,k} \leq b_{p,j}^{-1}b_{p,k} = 0\,,$$

thus $b_{p,j}^{-1}a_{i,p}^{-1}a_{i,q}b_{q,k} = 0$, as desired. The proof of the relation $c_{j,i} \perp_{\mathrm{lt}} c_{k,i}$ is similar.

(3) Suppose first that a is row-finite and let $i, j \in \Omega$. By assumption, the set $X = \{k \in \Omega \mid a_{i,k} \neq 0\}$ is finite. It follows that $\bigoplus_{k \in \Omega} a_{i,k}b_{k,j} = \bigoplus_{k \in X} a_{i,k}b_{k,j}$ is defined. Hence ab is defined. The argument is similar in case b is column-finite.

Now suppose that a is row-finite and b is column-finite. By the paragraph above, $c = ab$ is defined. Let $i \in \Omega$. Since a is row-finite, the set $X = \{k \in \Omega \mid a_{i,k} \neq 0\}$ is finite. Since b is column-finite, the set $Y = \{j \in \Omega \mid (\exists k \in X)(b_{k,j} \neq 0)\}$ is finite. We shall prove that $a_{i,k}b_{k,j} = 0$, for any $j \in \Omega \setminus Y$ and any $k \in \Omega$. If $k \notin X$, then $a_{i,k} = 0$ and we are done. If $k \in X$, then, since $j \notin Y$, we get $b_{k,j} = 0$. In any case, we are done. This proves that $c_{i,j} = 0$ whenever $j \in \Omega \setminus Y$, thus completing the proof that c is row-finite. The proof that c is column-finite is symmetric.

A similar type of argument yields (4). □

Proposition 3.5.3 *The following statements hold, for any Boolean inverse semigroup S and every set* Ω*:*

(1) *The multiplication,* $(a, b) \mapsto ab$*, defined in the statement of Lemma* 3.5.2*, endows* $\mathrm{B}_{\Omega}^{\oplus}(S)$ *with a structure of an inverse semigroup, for which the inverse of a matrix* $a = (a_{i,j} \mid (i,j) \in \Omega \times \Omega)$ *is given by* $a^{-1} = \left(a_{j,i}^{-1} \mid (i,j) \in \Omega \times \Omega\right)$*. The idempotent elements of* $\mathrm{B}_{\Omega}^{\oplus}(S)$ *are the diagonal matrices with idempotent entries.*
(2) *Let* $a, b \in \mathrm{B}_{\Omega}^{\oplus}(S)$*. Then* $a \leq b$ *iff* $a_{i,j} \leq b_{i,j}$ *for all* $i, j \in \Omega$*.*
(3) *Two matrices* $a, b \in \mathrm{B}_{\Omega}^{\oplus}(S)$ *are left orthogonal (resp., right orthogonal) iff any row of a is left orthogonal to any row of b (resp., any column of a is right orthogonal to any column of b). Furthermore, if a and b are orthogonal, then their orthogonal join* $a \oplus b$ *is defined, and*

$$a \oplus b = (a_{i,j} \oplus b_{i,j} \mid (i,j) \in \Omega \times \Omega) .$$

(4) $\mathrm{B}_{\Omega}^{\oplus}(S)$ *is a Boolean inverse semigroup, in which* $\mathrm{M}_{\Omega}^{\oplus}(S)$ *is an additive ideal.*

Proof (1) The proof of the associativity of the matrix multiplication on $\mathrm{B}_{\Omega}^{\oplus}(S)$, given in the statement of Lemma 3.5.2, is identical, *mutatis mutandis* (and using Proposition 3.1.9), to the one of the associativity of the matrix multiplication over any ring, so we omit it.

Now set $\iota(a) = \left(a_{j,i}^{-1} \mid (i,j) \in \Omega \times \Omega\right)$, for any generalized rook matrix a over S. A straightforward calculation yields that $a \cdot \iota(a)$ is the diagonal matrix with diagonal entries $\bigoplus_{j \in \Omega} \mathbf{r}(a_{i,j})$, for $i \in \Omega$. A further easy calculation yields $a \cdot \iota(a) \cdot a = a$. In particular, any matrix of the form $a \cdot \iota(a)$ is diagonal with idempotent diagonal entries. Hence, any two such matrices commute. Since the map ι is obviously involutive, it follows from Lemma 3.1.1 that $\mathrm{B}_{\Omega}^{\oplus}(S)$ is an inverse semigroup, with inversion map ι. Further, the zero matrix is the zero element of $\mathrm{B}_{\Omega}^{\oplus}(S)$.

(2) As observed in the proof of (1), $\mathbf{r}(a)$ is the diagonal matrix with entries $e_i = \bigoplus_{j \in \Omega} \mathbf{r}(a_{i,j})$, for $i \in \Omega$. Hence, $\mathbf{r}(a)b = \left(b'_{i,j} \mid (i,j) \in \Omega \times \Omega\right)$ where we set $b'_{i,j} = e_i b_{i,j}$ whenever $i, j \in \Omega$. In particular, if $a \leq b$, that is, $a = \mathbf{r}(a)b$, then $a_{i,j} \leq b_{i,j}$ for all $i, j \in \Omega$. Suppose, conversely, that $a_{i,j} \leq b_{i,j}$ for all $i, j \in \Omega$. Let $k \in \Omega \setminus \{j\}$. From $\mathbf{r}(b_{i,k})b_{i,j} = 0$ and $a_{i,k} \leq b_{i,k}$ it follows that $\mathbf{r}(a_{i,k})b_{i,j} = 0$. Since $\mathbf{r}(a_{i,j})b_{i,j} = a_{i,j}$, a direct application of Proposition 3.1.9 yields that $e_i b_{i,j} = a_{i,j}$. Hence, $a = \mathbf{r}(a)b$, that is, $a \leq b$.

(3) For all $i, j \in \Omega$, the (i,j)-th entry of ab^{-1} is $\bigoplus_{k\in\Omega} a_{i,k}b_{j,k}^{-1}$. Hence, $ab^{-1} = 0$ iff $a_{i,k}b_{j,k}^{-1} = 0$ for each $k \in \Omega$, that is, any row of a is left orthogonal to any row of b. The proof of the statement about right orthogonality is similar.

Now suppose that $a \perp b$. Let $i, j, k \in \Omega$ with $j \neq k$. Since a and b are both generalized rook matrices over S, $a_{i,j} \perp_{\mathrm{rt}} a_{i,k}$ and $b_{i,j} \perp_{\mathrm{rt}} b_{i,k}$. Moreover, by the paragraph above, $a_{i,j} \perp_{\mathrm{rt}} b_{i,k}$ and $b_{i,j} \perp_{\mathrm{rt}} a_{i,k}$. Therefore, $a_{i,j} \oplus b_{i,j} \perp_{\mathrm{rt}} a_{i,k} \oplus b_{i,k}$. The proof of the relation $a_{j,i} \oplus b_{j,i} \perp_{\mathrm{lt}} a_{k,i} \oplus b_{k,i}$ is similar. It follows that the matrix $c = \big(a_{i,j} \oplus b_{i,j} \mid (i,j) \in \Omega \times \Omega\big)$ is a generalized rook matrix over S. An easy application of (2) yields then that c is the orthogonal join of $\{a, b\}$.

(4) By (1) above, $\operatorname{Idp} \mathrm{B}_{\Omega}^{\oplus}(S)$ is isomorphic to $(\operatorname{Idp} S)^{\Omega}$ endowed with the componentwise ordering. By (3) above, it follows that $\operatorname{Idp} \mathrm{B}_{\Omega}^{\oplus}(S)$ is Boolean. Hence, $\mathrm{B}_{\Omega}^{\oplus}(S)$ is a Boolean inverse semigroup. The subset $\mathrm{M}_{\Omega}^{\oplus}(S)$ is an ideal (cf. Lemma 3.5.2), closed under finite orthogonal sum by (3) above, so it is an additive ideal. □

For a Boolean inverse semigroup S and a set Ω, denote by $x_{(i,j)}$ the matrix with (i,j)-th entry x and all other entries 0, for all $x \in S$ and all $(i,j) \in \Omega \times \Omega$. It follows from Proposition 3.5.3 that every element of $\mathrm{M}_{\Omega}^{\oplus}(S)$ is a finite orthogonal join of elements of the form $x_{(i,j)}$. The $x_{(i,j)}$ behave essentially like matrix units:

$$x_{(i,j)} \cdot y_{(k,l)} = \delta_{j,k} \cdot (xy)_{(i,l)}\,, \qquad \text{for all } x, y \in S \text{ and all } i, j, k, l \in \Omega\,, \tag{3.5.1}$$

$$(x_{(i,j)})^{-1} = (x^{-1})_{(j,i)}\,, \qquad \text{for all } x \in S \text{ and all } i, j \in \Omega\,, \tag{3.5.2}$$

where $\delta_{j,k}$ denotes the Kronecker symbol. In particular,

$$e_{(i,i)} = e_{i,j}(e_{(i,j)})^{-1} \text{ and } e_{(j,j)} = (e_{(i,j)})^{-1}e_{(i,j)}\,, \quad \text{for all } e \in \operatorname{Idp} S \text{ and all } i, j \in \Omega\,, \tag{3.5.3}$$

so $e_{(i,i)} \mathrel{\mathscr{D}} e_{(j,j)}$ within $\mathrm{M}_{\Omega}^{\oplus}(S)$.

Corollary 3.5.4 *Let S be a Boolean inverse semigroup, let Ω be a set, and let $o \in \Omega$. Then the map $\eta\colon S \hookrightarrow \mathrm{M}_{\Omega}^{\oplus}(S)$, $x \mapsto x_{(o,o)}$ is a lower semigroup embedding and $\mathrm{M}_{\Omega}^{\oplus}(S)$ is an additive enlargement of $\eta[S]$.*

Proof It is straightforward to verify from Proposition 3.5.3 that η is an additive semigroup embedding. Set $\overline{S} = \eta[S]$ and $\overline{T} = \mathrm{M}_{\Omega}^{\oplus}(S)$. Then $\overline{S}$ consists of all matrices with all entries, with the possible exception of the (o,o)-th, zero. By the definition of the multiplication in $\overline{T}$, we obtain easily that $\overline{S}\,\overline{T}\,\overline{S} = \overline{S}$. Since $\overline{S}^{-1} = \overline{S}$, it follows that $\overline{S}$ is an additive quasi-ideal of $\overline{T}$.

Finally, it follows from Proposition 3.5.3(3) that the orthogonal joins in $\overline{T}$ are evaluated componentwise, thus every element of $\overline{T}$ is a finite orthogonal join of elements of the form $x_{(i,j)}$, where $x \in S$ and $(i,j) \in \Omega \times \Omega$. From $x_{(i,j)} = x_{(i,o)}x_{(o,o)}x_{(o,j)}$ it follows that $x_{(i,j)} \in \overline{T}\,\overline{S}\,\overline{T}$. Therefore, $\overline{T} = (\overline{T}\,\overline{S}\,\overline{T})^{\oplus}$. □

It is interesting to compare the results of this section, especially Proposition 3.5.3, to the corresponding results in ring theory. A unital ring R is an *exchange ring* if for every $x \in R$, there is an idempotent $e \in R$ such that $eR \subseteq xR$ and $(1-e)R \subseteq (1-x)R$.

Every von Neumann regular ring is an exchange ring, but the converse fails. A C*-algebra is an exchange ring iff it has real rank zero (cf. Ara et al. [13, Theorem 7.2]). O'Meara proves in [91] that the ring $\mathrm{B}(R)$, of all countably infinite, row-finite, and column-finite matrices over a regular ring R, is an exchange ring. He also observes there that for an arbitrary exchange ring R, $\mathrm{B}(R)$ may not be an exchange ring. On the other hand, it is well known that the ring $\mathrm{B}(R)$ is not regular unless R is trivial (if s is the matrix of the shift operator, then $1-s$ has no quasi-inverse in $\mathrm{B}(R)$).

3.6 Crossed Product of a Boolean Inverse Semigroup by a Group Action

The goal of this section is to extend, to Boolean inverse semigroups, the classical construction of the crossed product of a ring by a group action (cf. Sect. 2.8).

Let a group G act by automorphisms on a Boolean inverse semigroup S. We denote the group action by $(g,x)\mapsto g(x)$. We set $x\cdot g = \bigl(\delta_{p,gq}p^{-1}(x) \mid (p,q)\in G\times G\bigr)$, for any $(x,g)\in S\times G$, where $\delta_{p,q}$ denotes the Kronecker symbol. The set $S\cdot G=\{x\cdot g \mid (x,g)\in S\times G\}$ is a subset of the Boolean inverse semigroup $\mathrm{B}_G^{\oplus}(S)$ of row-finite and column-finite $G\times G$ generalized rook matrices over S (cf. Proposition 3.5.3). The following lemma records a few elementary properties of the elements $x\cdot g$. Its proof is straightforward and we leave it to the reader.

Lemma 3.6.1 *The following statements hold, for any $x,y\in S$ and $g,h\in G$:*

(1) $x\cdot g = y\cdot h$ *iff* $x=y$ *and either* $g=h$ *or* $x=0$;
(2) $(x\cdot g)(y\cdot h)=\bigl(xg(y)\bigr)\cdot gh$;
(3) $(x\cdot g)^{-1}=g^{-1}(x^{-1})\cdot g^{-1}$;
(4) $\mathbf{d}(x\cdot g)=g^{-1}(\mathbf{d}(x))\cdot 1$;
(5) $\mathbf{r}(x\cdot g)=\mathbf{r}(x)\cdot 1$.

In particular, $S\cdot G$ is an inverse subsemigroup of $\mathrm{B}_G^{\oplus}(S)$, and the idempotent elements of $S\cdot G$ are the $e\cdot 1$, where $e\in \operatorname{Idp} S$.

Definition 3.6.2 The *crossed product of S by* (*the action of*) G, denoted by $S\rtimes G$, is the closure of $S\cdot G$ under finite orthogonal joins, within $\mathrm{B}_G^{\oplus}(S)$.

Hence the elements of $S\rtimes G$ are the orthogonal joins of the form

$$x=\bigoplus_{i=1}^{n}(x_i\cdot g_i)\,,\quad \text{where } n\in\mathbb{Z}^+ \text{ and each } (x_i,g_i)\in S\times G\,. \tag{3.6.1}$$

The orthogonality, within $\mathrm{B}_G^{\oplus}(S)$, of the finite sequence $(x_i\cdot g_i \mid i\in[n])$ is, by Lemma 3.6.1, equivalent to the orthogonality, within $\operatorname{Idp} S$, of both finite sequences $\bigl(g_i^{-1}(\mathbf{d}(x_i)) \mid i\in[n]\bigr)$ and $(\mathbf{r}(x_i)\mid i\in[n])$.

Proposition 3.6.3 *Let a group G act by automorphisms on a Boolean inverse semigroup S. Then $S \rtimes G$ is an additive inverse subsemigroup of $\mathrm{B}_G^{\oplus}(S)$. In particular, it is a Boolean inverse semigroup. Furthermore,* $\mathrm{Idp}(S \rtimes G) = (\mathrm{Idp}\, S) \cdot 1$, *and the canonical map $\varepsilon \colon S \hookrightarrow S \rtimes G$, $x \mapsto x \cdot 1$ is a lower semigroup embedding.*

Proof It follows from the definition of $S \rtimes G$, together with Lemma 3.6.1 and Proposition 3.1.9, that $S \rtimes G$ is an inverse subsemigroup of $\mathrm{B}_G^{\oplus}(S)$, closed under finite orthogonal sums. Any element $x \in S \rtimes G$ can be written in the form (3.6.1), and then, using Lemma 3.6.1, we get $\mathbf{r}(x) = e \cdot 1$ where $e = \bigoplus_{i=1}^{n} \mathbf{r}(x_i)$. It follows that $\mathrm{Idp}(S \rtimes G) = (\mathrm{Idp}\, S) \cdot 1$. Since $\mathrm{Idp}\, S$ is Boolean, so is $\mathrm{Idp}(S \rtimes 1)$. In particular, $\mathrm{Idp}(S \rtimes 1)$ is closed under the operation $(x, y) \mapsto x \smallsetminus y$. By Lemma 3.1.13, $S \rtimes G$ is a Boolean inverse semigroup.

Let $x \in S \rtimes G$, written as in (3.6.1), and let $y \in S$ such that $x \leq y \cdot 1$ within $S \rtimes G$. For each $i \in [n]$, $x_i \cdot g_i = \mathbf{r}(x_i \cdot g_i)(y \cdot 1) = \big(\mathbf{r}(x_i) \cdot 1\big)(y \cdot 1) = \mathbf{r}(x_i)y \cdot 1$, thus $x_i = \mathbf{r}(x_i)y$ (i.e., $x_i \leq y$) and either $g_i = 1$ or $x_i = 0$. In any case, $x_i \cdot g_i = x_i \cdot 1$. Then it follows from Lemma 3.6.1 that the x_i are pairwise orthogonal in S; whence $x = \big(\bigoplus_{i=1}^{n} x_i\big) \cdot 1$ belongs to the range of ε. Therefore, ε is a lower semigroup embedding. □

Our next encounter with crossed products of Boolean inverse semigroups, involving type monoids, will occur in Theorem 4.1.10.

3.7 Fundamental Boolean Inverse Semigroups and Boolean Inverse Meet-Semigroups

Two important subclasses of the class of all Boolean inverse semigroups will come up repeatedly in our work, namely *fundamental Boolean inverse semigroups* and *Boolean inverse meet-semigroups*. This will also motivate the introduction of a definition of *semisimplicity* for Boolean inverse semigroups, close in spirit to the eponymous ring-theoretical concept.

We start by recalling the following definition.

Definition 3.7.1 An inverse semigroup S is *fundamental* (cf. Howie [60], Munn [87])[5] if the identity is the only idempotent-separating congruence of S. Equivalently, every element of S, which commutes with all idempotent elements of S, is idempotent.

Denote by $\boldsymbol{\mu}$ the largest idempotent-separating congruence of an inverse semigroup S. By Howie [60, Theorem V.3.4], the quotient $S/\boldsymbol{\mu}$ is then fundamental and $\mathrm{Idp}(S/\boldsymbol{\mu}) \cong \mathrm{Idp}\, S$.

The following lemma records a useful basic property of fundamental inverse semigroups.

[5] In Wagner [112, 113], Zhitomirskiy [126, 127], such semigroups are called *antigroups*.

Lemma 3.7.2 *Let S be a fundamental inverse semigroup and let p and q be atoms of* $\operatorname{Idp} S$. *Then there is at most one element $x \in S$ such that $\mathbf{d}(x) = p$ and $\mathbf{r}(x) = q$.*

Proof We first deal with the case where $p = q$. Let $x \in S$ such that $\mathbf{d}(x) = \mathbf{r}(x) = p$. In particular, $x = pxp$. Since p is an atom, every $e \in \operatorname{Idp} S$ satisfies either $p \leq e$ or $pe = 0$. In the first case, $xe = ex = x$. In the second case, $xe = ex = 0$. In either case, $xe = ex$, so x commutes with every idempotent of S. Since S is fundamental, x is idempotent, so $x = p$.

Now we deal with the general case. Let $x, y \in S$ such that $\mathbf{d}(x) = \mathbf{d}(y) = p$ and $\mathbf{r}(x) = \mathbf{r}(y) = q$. It follows that $\mathbf{d}(x^{-1}y) = \mathbf{r}(x^{-1}y) = p$, thus, by the paragraph above, $x^{-1}y = p$. It follows that $y = qy = xx^{-1}y = xp = x$. □

Example 3.7.3 For any set X, the symmetric inverse monoid $\mathfrak{I}_X$ (cf. Example 3.1.8) is a fundamental Boolean inverse semigroup.

Example 3.7.4 For a group G, the monoid $G^{\sqcup 0}$ (cf. Definition 1.5.1) is a Boolean inverse semigroup, with the same unit as G. It is fundamental iff G is trivial.

Let $\tau_g : G \to G$, $x \mapsto gx$, for all $g \in G$. Then the assignment $\tau : G^{\sqcup 0} \to \mathfrak{I}_G$ (cf. Example 3.1.8), defined by $0 \mapsto \varnothing$, $g \mapsto \tau_g$, is a semigroup embedding from $G^{\sqcup 0}$ into $\mathfrak{I}_G$. The orthogonality relation on the range of τ is trivial, hence the range of τ is closed under finite orthogonal joins. Therefore, $G^{\sqcup 0}$ is isomorphic to an inverse subsemigroup of $\mathfrak{I}_G$, closed under finite orthogonal joins. More generally, recall from Corollary 3.3.2 that every Boolean inverse semigroup has an additive semigroup embedding into some symmetric inverse monoid, thus into some fundamental Boolean inverse semigroup. This shows that Lemma 3.1.13 does not extend to fundamental Boolean inverse semigroups.

Definition 3.7.5 The *pedestal* of a Boolean inverse semigroup S is defined as the set $\operatorname{Ped} S = \{x \in S \mid S \downarrow x \text{ is finite}\}$. We say that S is *semisimple* if $\operatorname{Ped} S = S$.

The terminology in Definition 3.7.5 is consistent with the one, introduced in Ara and Goodearl [10, Definition 2.3], for conical refinement monoids. With an eye on ring theory, it would seem reasonable to call the subset defined above the *socle* of S. However, the concept of the (left or right) socle, of a semigroup with zero (cf. Clifford and Preston [30, § 6.4]), is related, but not equivalent, to our concept of a pedestal, even in the particular case of Boolean inverse semigroups.

It is not hard to verify that an element x, in a Boolean inverse semigroup S, belongs to $\operatorname{Ped} S$ iff $\mathbf{d}(x)$ (equivalently, $\mathbf{r}(x)$) is a finite join of atoms of the Boolean ring $\operatorname{Idp} S$, iff x is a finite orthogonal join of atoms of S. Further, $\operatorname{Ped} S$ is an additive ideal of S. Observe also that every finite Boolean inverse semigroup is semisimple.

Proposition 3.7.6 *Every additive congruence $\boldsymbol{\theta}$ of a fundamental semisimple Boolean inverse semigroup S is ideal-induced, and $S/\boldsymbol{\theta}$ is a fundamental semisimple Boolean inverse semigroup.*

Proof By applying Proposition 3.4.9 to the canonical projection $\theta : S \twoheadrightarrow S/\boldsymbol{\theta}$, we obtain that the subset $I = 0/\boldsymbol{\theta}$ is an additive ideal of S. In order to prove that $\boldsymbol{\theta}$ is induced by that ideal, it suffices to prove that the additive semigroup homomorphism

$\overline{\theta}\colon S/I \twoheadrightarrow S/\boldsymbol{\theta}$ given by Proposition 3.4.9 is one-to-one, that is, $\theta(x) = \theta(y)$ implies that $x \equiv_I y$, for all $x, y \in S$.

We first settle the case where $x, y \in qSp$, for atoms p and q of B. Since S is fundamental and by Lemma 3.7.2, either $x = y$ or $0 \in \{x, y\}$. In the first case, $x \equiv_I y$ trivially. In the second case, say $x = 0$, then $\theta(y) = 0$, that is, $y \in I$, so $x \equiv_I y$.

Now we settle the general case. Since S is semisimple, there is a finite set P of atoms of $\operatorname{Idp} S$ whose (orthogonal) join contains $\mathbf{d}(x)$, $\mathbf{r}(x)$, $\mathbf{d}(y)$, $\mathbf{r}(y)$. For any $p, q \in P$, $\theta(qxp) = \theta(qyp)$, thus, by the paragraph above, $qxp \equiv_I qyp$. By evaluating the orthogonal join, over $p \in P$, of both sides of that equation, we obtain, since $\equiv_I$ is an additive congruence (cf. Proposition 3.4.6), the relation

$$\bigoplus_{p \in P} qxp \equiv_I \bigoplus_{p \in P} qyp\,,$$

thus, using Proposition 3.1.9, $qx \equiv_I qy$. By the same token, now summing up over q instead of p, we obtain $x \equiv_I y$. This completes the proof that $\overline{\theta}$ is one-to-one.

Observe that $x/\boldsymbol{\theta}$ is either zero or an atom, for every atom x of S, according to whether $x \in I$ or $x \notin I$, respectively. Since every element of S is a finite join of atoms, it follows that every element of $S/\boldsymbol{\theta}$ is a finite join of atoms, that is, $S/\boldsymbol{\theta}$ is semisimple.

Finally we prove that $S/\boldsymbol{\theta}$ is fundamental. This amounts to proving that for every $x \in S$, if $xe \equiv_{\boldsymbol{\theta}} ex$ for all $e \in \operatorname{Idp} S$, then $x/\boldsymbol{\theta}$ is idempotent in $S/\boldsymbol{\theta}$. Since $\overline{\theta}$ is one-to-one, we get $xe \equiv_I ex$, for all $e \in \operatorname{Idp} S$. The latter relation means that there is $z_e \in S$ such that $z_e \leq {xe \atop ex}$ and $\{xe \smallsetminus z_e, ex \smallsetminus z_e\} \subseteq I$. Set $v = \mathbf{d}(x) \vee \mathbf{r}(x)$ (any larger idempotent would do). Since S is semisimple, the set $P = (\operatorname{Idp} S) \downarrow v$ is finite. Since I is an additive ideal of S, the idempotent element

$$u = \bigvee_{e \in P} \bigl(\mathbf{d}(xe \smallsetminus z_e) \vee \mathbf{d}(ex \smallsetminus z_e)\bigr)$$

belongs to I; moreover, $u \leq v$. Observe that $xe \smallsetminus z_e = (xe \smallsetminus z_e)u$ for each $e \in P$. Hence, from $z_e \leq xe$ it follows that $xe(v \smallsetminus u) = z_e(v \smallsetminus u)$. Likewise, $ex(v \smallsetminus u) = z_e(v \smallsetminus u)$, so $xe(v \smallsetminus u) = ex(v \smallsetminus u)$. It follows that $x(v \smallsetminus u)$ commutes with every element of P, thus with every idempotent below v. Since it also commutes with every idempotent e orthogonal to v (for in that case, $xe = ex = 0$), it follows that $x(v \smallsetminus u)$ commutes with every idempotent of S. Since S is fundamental, $x \smallsetminus xu = x(v \smallsetminus u)$ is idempotent in S, thus $x/\boldsymbol{\theta} = x(v \smallsetminus u)/\boldsymbol{\theta}$ is idempotent in $S/\boldsymbol{\theta}$. □

Definition 3.7.7 An inverse semigroup S is an *inverse meet-semigroup* (cf. Leech [79], also Lawson [75]) if it is a meet-semilattice under $\leq$, that is, the meet $x \wedge y$ exists for all $x, y \in S$.

As witnessed by Example 3.1.7, not every finite inverse monoid with zero is an inverse meet-semigroup: in that example, $\frac{1}{2} \leq \frac{3}{4}$, but there is no x such that $\frac{1}{2} \leq x \leq \frac{3}{4}$. For Boolean inverse semigroups, this strange behavior does not occur.

Proposition 3.7.8 *Let S be a Boolean inverse semigroup, let $x \in \operatorname{Ped} S$, and let $y \in S$. Then $x \wedge y$ exists in S. In particular, every semisimple Boolean inverse semigroup S is an inverse meet-semigroup.*

Proof The set X, of all common lower bounds of x and y, is a compatible subset of the finite set $S \downarrow x$. Since S is Boolean, X has a join in S, which is necessarily the meet of $\{a, b\}$. □

The following example shows that not every fundamental Boolean inverse semigroup is an inverse meet-semigroup. By Proposition 3.7.8, any such example is infinite.

Example 3.7.9 Define S as the inverse subsemigroup of the symmetric inverse semigroup $\mathfrak{I}_{\mathbb{Z}^+}$ (cf. Example 3.1.8) consisting of all functions whose domain is either finite or cofinite. Then S is a fundamental Boolean inverse semigroup. However, for any permutation α of $\mathbb{Z}^+$ whose fixed point set consists of all even numbers, $\alpha \wedge \mathrm{id}_{\mathbb{Z}^+}$ does not exist in S. Hence S is not an inverse meet-semigroup.

It is well known that any compatible elements x and y in an inverse semigroup S have a meet, given (among many other expressions) by (3.1.3). In particular, *every semigroup homomorphism preserves compatible meets.* On the other hand, we will see shortly that additive semigroup homomorphisms between Boolean inverse meet-semigroups may not preserve meets (cf. Example 3.7.12). Nevertheless, the following result shows that under certain conditions, additive semigroup homomorphisms may preserve all meets.

Proposition 3.7.10 *Let S be a fundamental Boolean inverse semigroup, let T be a Boolean inverse semigroup, and let $f: S \to T$ be an additive semigroup homomorphism. Then $f(x \wedge y) = f(x) \wedge f(y)$, for all $x \in \operatorname{Ped} S$ and all $y \in S$.*

Note Although, by Proposition 3.7.8, the meet $x \wedge y$ exists in S, we are not assuming that T is an inverse meet-semigroup.

Proof Set $B = \operatorname{Idp} S$. The set P, of all atoms of B below $\mathbf{d}(x) \vee \mathbf{r}(x)$, is finite. Let $z \in T$ such that $z \leq \frac{f(x)}{f(y)}$. By multiplying those inequalities by $f(p)$ on the right side and $f(q)$ on the left side, we obtain

$$f(q)zf(p) \leq \frac{f(qxp)}{f(qyp)}, \quad \text{for any } p, q \in P. \tag{3.7.1}$$

It follows from Lawson [73, Proposition 1.4.19] that $(qxp) \wedge (qyp) = q(x \wedge y)p$. Further, by Lemma 3.7.2, either $qxp = qyp$ or $0 \in \{qxp, qyp\}$. Hence, in any case,

$$f(qxp) \wedge f(qyp) = f(q(x \wedge y)p) = f(q)f(x \wedge y)f(p),$$

and hence, by (3.7.1), we get $f(q)zf(p) \leq f(q)f(x \wedge y)f(p)$. This holds for all $p, q \in P$, thus, since $z \leq \bigoplus_{p \in P} f(p)$ and by using the additivity of f, we get $z \leq f(x \wedge y)$. □

The following two examples show that the assumption in Proposition 3.7.6, that $x \in \operatorname{Ped} S$, cannot be dropped. Moreover, Example 3.7.11 witnesses that Proposition 3.7.10 cannot be extended to arbitrary finite Boolean inverse semigroups S, and Example 3.7.12 witnesses that the finiteness assumption is necessary in Proposition 3.7.10, even for inverse meet-semigroups S.

Example 3.7.11 Finite Boolean inverse monoids S and T, together with a surjective, non one-to-one additive semigroup homomorphism $f: S \twoheadrightarrow T$ such that $\ker f = \{0\}$.

Proof Let G be any non-trivial group. Set $S = G^{\sqcup 0}$ and $T = \{0, \infty\}$ (the two-element join-semilattice), and let $f: S \twoheadrightarrow T$ the map that sends 0 to 0 and any element of G to ∞. Then f is an additive semigroup homomorphism and $\ker f = \{0\}$. Since G is non-trivial, f is not one-to-one. □

Example 3.7.12 Fundamental, unital, Boolean inverse meet-semigroups S and T, together with a surjective additive semigroup homomorphism $f: S \twoheadrightarrow T$, with an invertible element $\alpha \in S \setminus \{1\}$ such that $f(\alpha \wedge 1) < f(\alpha) = f(1)$ and $\alpha \not\equiv_{\ker f} 1$. In particular, f is not ideal-induced.

Proof Define S as the inverse submonoid of the symmetric inverse monoid $\mathfrak{I}_{\mathbb{Z}^+}$ (cf. Example 3.1.8) consisting of all bijections $x: A \to B$, where A and B are both either finite or cofinite subsets of $\mathbb{Z}^+$, and such that if A is cofinite, then there exists $n \in \mathbb{Z}$ such that $x(k) = n + k$ for all large enough $k \in A$ (*this condition is put there in order to ensure that S is an inverse meet-semigroup*). Further, define T as the two-element join-semilattice $\{0, \infty\}$, and define $f: S \twoheadrightarrow T$ by letting $f(x) = \infty$ iff the domain of x is cofinite, whenever $x \in S$. Then S and T are both fundamental unital Boolean inverse meet-semigroups and f is an additive semigroup homomorphism from S to T.

Now let α be any permutation of $\mathbb{Z}^+$ without fixed points (e.g., let α interchange $2n$ and $2n + 1$, for any $n \in \mathbb{Z}^+$). Then $f(\alpha) = f(\mathrm{id}) = \infty$ and $f(\alpha \wedge \mathrm{id}) = f(0) = 0$. If $\alpha \equiv_{\ker f} \mathrm{id}$, then α and id would need to agree on some cofinite subset of $\mathbb{Z}^+$, which is not the case. □

The following example shows that an additive homomorphic image of a fundamental unital Boolean inverse meet-semigroup may not be fundamental.

Example 3.7.13 A fundamental unital Boolean inverse meet-semigroup S, a unital Boolean inverse meet-semigroup T, and a surjective additive semigroup homomorphism $f: S \twoheadrightarrow T$, such that T is not fundamental.

Proof We use the same monoid S as in Example 3.7.12 together with the larger $T = \mathbb{Z}^{\sqcup 0}$. For every $x \in S$, we set $f(x) = 0$ in case the domain of x is finite. If x is infinite, we define $f(x)$ as the unique $n \in \mathbb{Z}$ such that $x(k) = n + k$ for all large enough k. Observe that T is not fundamental (cf. Example 3.7.4). □

We will need later the following preservation result for fundamental Boolean inverse semigroups and Boolean inverse meet-semigroups.

Proposition 3.7.14 *Let T be a Boolean inverse semigroup. If T is fundamental (resp., a Boolean inverse meet-semigroup), then so is any additive quasi-ideal of T, and so is $\mathrm{M}_{\Omega}^{\oplus}(T)$, for any set Ω.*

Proof Let S be any additive quasi-ideal of T. It follows from Proposition 3.1.18 that S is a lower inverse subsemigroup of T.

Set $\overline{T} = \mathrm{M}_{\Omega}^{\oplus}(T)$.

Suppose first that T is fundamental and let $x \in S$ commute with all idempotents of S. Since x commutes with both $\mathbf{d}(x)$ and $\mathbf{r}(x)$, we get $\mathbf{d}(x) = \mathbf{r}(x)$. Denote by a this element. Since S is a lower subset of T, $T \downarrow a$ is contained in $\mathrm{Idp}\, S$. By assumption, it follows that x commutes with all elements of $T \downarrow a$. On the other hand, $xe = ex = 0$ for any $e \in \mathrm{Idp}\, T$ orthogonal to a. Since $e = ea \oplus (e \smallsetminus a)$ for any $e \in \mathrm{Idp}\, T$ and by Proposition 3.1.9, it follows that x commutes with all idempotent elements of T. Since T is fundamental, x is idempotent. Therefore, S is fundamental.

Any element $x \in \overline{T}$ that commutes with all idempotents must commute with all $e_{(i,i)}$, where $e \in \mathrm{Idp}\, S$ and $i \in [n]$. It follows easily that x must be a diagonal matrix, each of whose diagonal entries commutes with all idempotents of T. Since T is fundamental, it follows that x is a diagonal matrix with idempotent entries; hence x is idempotent. Therefore, $\overline{T}$ is fundamental.

Finally, we only assume that T is an inverse meet-semigroup. Since S is a lower subset of T, it is a meet-subsemilattice of T, thus it is also a fundamental unital Boolean inverse meet-semigroup. Furthermore, since T is an inverse meet-semigroup and by Proposition 3.5.3(2), $\overline{T}$ is an inverse meet-semigroup and the meets in $\overline{T}$ are evaluated componentwise. □

3.8 Inner Endomorphisms and Automorphisms of a Boolean Inverse Semigroup

We set $\mathrm{ad}_g(x) = g\langle x\rangle = gxg^{-1}$, for all elements g and x in an inverse semigroup S. We call ad_g the *inner endomorphism* determined by g. If S is unital, then inner endomorphisms with respect to invertible elements are automorphisms, called *inner automorphisms* of S.

In order to extend this definition to the case where S is not unital, we add the assumption that S is Boolean, then we need to drop the assumption that g be invertible but we keep the assumption that $\mathbf{d}(g) = \mathbf{r}(g)$. Then we replace g

by $g \oplus e$, for large enough idempotents e ranging through the ideal $g^{\perp} = \{e \in \operatorname{Idp} S \mid e \perp g\} = \{e \in \operatorname{Idp} S \mid ge = eg = 0\}$ of the Boolean ring $\operatorname{Idp} S$. As the following lemma shows, for large enough $e \in g^{\perp}$, the value of $(g \oplus e)\langle x \rangle$ depends only on g and x.

Lemma 3.8.1 *Let S be a Boolean inverse semigroup and let $g, x \in S$. Then the value of $(g \oplus e)\langle x \rangle$, where $e \in g^{\perp}$ and $\mathbf{d}(x) \vee \mathbf{r}(x) \leq (\mathbf{d}(g) \vee \mathbf{r}(g)) \oplus e$, depends only on g and x.*

Proof Both elements $\overline{x} = \mathbf{d}(x) \vee \mathbf{r}(x)$ and $\overline{g} = \mathbf{d}(g) \vee \mathbf{r}(g)$ are idempotent. Let $e_i \in \operatorname{Idp} S$ such that $e_i \perp g$ and $\overline{x} \leq \overline{g} \oplus e_i$, for $i \in \{0, 1\}$. From $e_i = (\overline{g} \oplus e_i) \smallsetminus \overline{g}$ it follows, by multiplying on the left by x, that $xe_i = x \smallsetminus x\overline{g}$ is independent of i. Symmetrically, $e_i x = x \smallsetminus \overline{g}x$ is also independent of i. It follows that $e_i x e_i = xe_i \smallsetminus \overline{g}xe_i = (x \smallsetminus x\overline{g}) \smallsetminus \overline{g}(x \smallsetminus x\overline{g})$ is also independent of i. Therefore,

$$(g \oplus e_i)\langle x \rangle = (g \oplus e_i)x(g^{-1} \oplus e_i) = gxg^{-1} \oplus gxe_i \oplus e_i xg^{-1} \oplus e_i xe_i$$

is independent of i. □

Notation 3.8.2 We shall denote by $\operatorname{inn}_g(x)$ the constant value of $(g \oplus e)\langle x \rangle$, for large enough $e \in g^{\perp}$.

Hence, inn_g is the directed union, over all $e \in g^{\perp}$, of all maps $\operatorname{ad}_{g \oplus e}$.

We will be interested in situations where inn_g is an automorphism of S. We wish to identify those $g \in S$ such that inn_g defines an automorphism of aSa for any large idempotent a. Accordingly, we define a subset of S as follows.

Notation 3.8.3 We set $\operatorname{Self} S = \{g \in S \mid \mathbf{d}(g) = \mathbf{r}(g)\}$, for any Boolean inverse semigroup S.

Observe that $\operatorname{Self} S$ is usually not a subsemigroup of S.

Lemma 3.8.4 *The following statements hold, for any Boolean inverse semigroup S:*

(1) $\operatorname{inn}_{g \oplus e} = \operatorname{inn}_g$, *for any $g \in S$ and any $e \in g^{\perp}$. In particular, $\operatorname{inn}_e = \operatorname{id}_S$ whenever e is idempotent.*
(2) $\operatorname{inn}_{fg} = \operatorname{inn}_f \circ \operatorname{inn}_g$, *for any $f, g \in \operatorname{Self} S$ with $\mathbf{d}(f) = \mathbf{d}(g)$.*
(3) inn_g *is an automorphism of S, for any $g \in \operatorname{Self} S$. We call the automorphisms of that form the* inner automorphisms *of S.*
(4) $x \mathscr{D} \operatorname{inn}_g(x)$, *for any $x \in S$ and any $g \in \operatorname{Self} S$.*
(5) *The inner automorphisms of S form a subgroup of the automorphism group of S.*

Proof (1) is trivial.

(2) Set $a = \mathbf{d}(f) = \mathbf{d}(g)$. For all $x \in S$ and all large enough $e \in a^{\perp}$,

$$\begin{aligned}(\operatorname{inn}_f \circ \operatorname{inn}_g)(x) = (f \oplus e)\langle (g \oplus e)\langle x \rangle \rangle &= ((f \oplus e)(g \oplus e))\langle x \rangle \\ &= (fg \oplus e)\langle x \rangle \\ &= \operatorname{inn}_{fg}(x)\,.\end{aligned}$$

(3) follows trivially from (1) and (2).

(4) Since $x \mathscr{D} \mathbf{d}(x)$ for every x, a direct application of (3) reduces the problem to the case where x is idempotent. Set $a = \mathbf{d}(g) = \mathbf{r}(g)$ and let $e \in a^{\perp}$ such that $\mathbf{d}(x) \vee \mathbf{r}(x) \leq a \oplus e$. Setting $h = g \oplus e$, we get $\mathbf{d}(h) = \mathbf{r}(h) = a \oplus e$. It follows from Lemma 3.8.1 that $\mathrm{inn}_g(x) = h\,\langle x \rangle$; thus $\mathrm{inn}_g(x) = (hx)(hx)^{-1}$. Moreover,

$$(hx)^{-1}hx = x^{-1}(h^{-1}h)x = x^{-1}(a \oplus e)x = x^{-1}x = x\,.$$

Hence $x \mathscr{D} \mathrm{inn}_g(x)$.

(5) Let $f, g \in \mathrm{Self}\,S$, with respective domains a and b. We must prove that $\mathrm{inn}_f \circ \mathrm{inn}_g$ is an inner automorphism of S. By (1), we may replace f by $f \oplus (b \smallsetminus a)$ and g by $g \oplus (a \smallsetminus b)$, and thus suppose that $\mathbf{d}(f) = \mathbf{d}(g)$. The conclusion follows then immediately from (2). □

Observe that every inner automorphism of S fixes all elements in some $a^{\perp}$, where $a \in S$: if $g \in \mathrm{Self}\,S$ and $\mathbf{d}(g) = \mathbf{r}(g) = a$, then $\mathrm{inn}_g(x) = x$ for every $x \in a^{\perp}$.

Notation 3.8.5 We denote by $\mathrm{Inn}\,S$ the group of all inner automorphisms of S.

Of course, if S has a unit, then $\mathrm{Inn}\,S = \{\mathrm{ad}_g \mid g \text{ invertible element of } S\}$. However, $\mathrm{Inn}\,S$ is also defined if S has no unit. In fact, it can be proved that $\mathrm{Inn}\,S \cong \mathrm{Inn}\,\widetilde{S}$, where $\widetilde{S}$ is the Boolean inverse monoid, introduced in Sect. 6.6, which we will call the *Boolean unitization* of S.

Chapter 4
Type Monoids and V-Measures

The type monoid of a Boolean inverse semigroup is an abstraction of the concept of monoid of equidecomposability types of a Boolean ring under a group action. The latter concept has been studied in a wide array of works including Banach [17], Tarski [109]. Its relation with type monoids of Boolean inverse semigroups was recognized in Wallis' Ph.D. thesis [116], see also Kudryavtseva et al. [71], Lawson and Scott [77]. The type monoid is an analogue, for Boolean inverse semigroups, of the nonstable K-theory of a ring (cf. Sect. 1.3.4) or the dimension monoid of a lattice (cf. Wehrung [122]).

In Sect. 4.1, we introduce some basic material pertaining to type monoids of Boolean inverse semigroups, in particular calculating type monoids of direct products, directed colimits, and crossed products.

In Sect. 4.2, we express the concepts of quasi-ideal and additive enlargement, originating in Lawson [72], in terms of the type monoid, and we use this to calculate type monoids of Boolean inverse semigroups of generalized rook matrices ($\mathrm{M}_{\Omega}^{\oplus}(S)$) or local submonoids ($eSe$). We also establish a one-to-one correspondence between the additive ideals of a Boolean inverse semigroup and the o-ideals of its type monoid.

In Sect. 4.3, we describe the type monoid of a quotient of a Boolean inverse semigroup S by an additive ideal I, in monoid-theoretical terms involving the type monoids of S and I.

Section 4.4 will be mainly devoted to descriptions of examples, in terms of partial bijections (or their analogues for abstract Boolean rings), of fundamental Boolean inverse semigroups and Boolean inverse meet-semigroups. In that section, we will also formally relate type monoids of Boolean inverse semigroups with monoids of equidecomposability types of a Boolean ring with respect to a group action.

In Sect. 4.5 we will specialize the concept of a V-relation, introduced in Sect. 2.4, to the class of all Boolean rings, and recall some folklore results originating in Vaught's thesis [114]. We will also express the existence of a V-equivalence, between two Boolean algebras, in terms of elementary equivalence with respect to

F. Wehrung, *Refinement Monoids, Equidecomposability Types, and Boolean Inverse Semigroups*, Lecture Notes in Mathematics 2188,
DOI 10.1007/978-3-319-61599-8_4

infinitary sentences of first-order logic. Although most results of that section are known in some form, they are difficult to trace back to a definite bibliographical reference, thus we chose to state them here in some detail.

A *V-measure*, on a partial commutative monoid P, is a homomorphism of partial monoids, defined on P, whose graph is a V-relation. Section 4.6 will prepare the ground for relating that concept to the study of type monoids of Boolean inverse semigroups.

Section 4.7 sets the relation, initiated in Sect. 4.4, between type monoids of Boolean inverse semigroups and monoids of equidecomposability types of Boolean rings with respect to group actions (which will lead to *group-induced* measures) or inverse semigroup of partial transformations (which will lead to *groupoid-induced* measures). In particular, we will discuss sufficient conditions for "groupoid-induced" to be able to be turned to "group-induced".

Instead of focusing on the *measures* as Sect. 4.7, Sect. 4.8 will focus of the *monoids*, leading to the definitions of groupoid-measurable (resp., *group-measurable*) commutative monoids, as those which are representable as monoids of equidecomposability types with respect to partial (resp., full) automorphisms.

Section 4.9 will show that although type monoids of Boolean inverse meet-semigroups are the same as type monoids of Boolean inverse semigroups, most obvious extensions of this result that one may think of do not hold.

4.1 Type Monoids of Boolean Inverse Semigroups

The type monoid of a Boolean inverse semigroup is one of many constructs of the form $\mathrm{U_{mon}}(P/\Gamma)$, where Γ is an additive V-equivalence Γ on a partial refinement monoid P. The original form of this construction arises in Tarski [109], with the monoid of equidecomposability types of elements of a ring of subsets of a set Ω with respect to a collection of partial transformations of Ω (see Sect. 4.4 for more detail).

Let us first introduce some notation and terminology. For an inverse semigroup S and for $a, b, x \in S$ with a and b both idempotent, let $a \overset{x}{\longrightarrow} b$ hold if $\mathbf{d}(x) = a$ and $\mathbf{r}(x) = b$. Then for any idempotents $a, b \in S$, $a \mathrel{\mathscr{D}} b$ iff there exists $x \in S$ such that $a \overset{x}{\longrightarrow} b$ (cf. Howie [60, Proposition V.1.4]). The *trace product*[1] of elements $x, y \in S$ is defined as xy if $\mathbf{d}(x) = \mathbf{r}(y)$, undefined otherwise. The *trace product groupoid*[2] of S has objects the idempotents of S and arrows the elements of S, with

$$\mathrm{Hom}(a, b) = \left\{x \in S \mid a \overset{x}{\longrightarrow} b\right\}, \quad \text{for all } a, b \in \operatorname{Idp} S\,.$$

[1] A referee informed the author that the trace product was first studied by Charles Ehresmann.

[2] A *groupoid* is a category where every arrow is an isomorphism. Not to be confused with the groupoids in universal algebra, which are just sets endowed with a binary operation.

In particular, Green's relation $\mathscr{D}$ on Idp S is identical to the relation of isomorphy in the trace product groupoid of S.

For what follows, recall that V-relations are introduced in Definition 2.4.1.

Lemma 4.1.1 *Let S be a distributive inverse semigroup with zero. Then Green's relations $\mathscr{L}$, $\mathscr{R}$, and $\mathscr{D}$ are all additive and conical V-equivalences on S.*

Note The proofs of the statements about $\mathscr{L}$ and $\mathscr{R}$ require no further assumption than S being an inverse semigroup with zero. Only our proof of the additivity of $\mathscr{D}$ requires S being distributive (weaker assumptions would be sufficient, but we will not need such generalizations).

Proof Let $x = x_0 \oplus x_1$ and $y = y_0 \oplus y_1$ in S, with each $x_i \mathscr{L} y_i$. It follows from Lemma 3.1.4 that $\mathbf{d}(x) = \mathbf{d}(x_0) \oplus \mathbf{d}(x_1) = \mathbf{d}(y_0) \oplus \mathbf{d}(y_1) = \mathbf{d}(y)$, that is, $x \mathscr{L} y$. Hence $\mathscr{L}$ is additive.

Let $x, x_0, x_1, y \in S$ such that $x = x_0 \oplus x_1$ and $x \mathscr{L} y$. Using again Lemma 3.1.4, it follows that $\mathbf{d}(y) = \mathbf{d}(x_0) \oplus \mathbf{d}(x_1)$, thus, using Lemma 3.1.3 and setting $y_i = y\,\mathbf{d}(x_i)$, we get $y = y\,\mathbf{d}(y) = y\,\mathbf{d}(x_0) \oplus y\,\mathbf{d}(x_1) = y_0 \oplus y_1$, with each $\mathbf{d}(x_i) = \mathbf{d}(y_i)$. Hence $\mathscr{L}$ is a V-equivalence. It is trivially conical. Symmetrically, $\mathscr{R}$ is an additive, conical V-equivalence. Since $\mathscr{D}$ is the join of $\mathscr{L}$ and $\mathscr{R}$, it is also a conical V-equivalence.

Finally, we need to prove that $\mathscr{D}$ is additive. (This was already observed in Wallis [116, Lemma 4.2.1]; we include a proof for convenience.) Let $a = a_0 \oplus a_1$ and $b = b_0 \oplus b_1$ with $a_i \xrightarrow{u_i} b_i$ whenever $i \in \{0, 1\}$. Since $u_0 \perp u_1$ and S is a distributive inverse semigroup, the element $u = u_0 \oplus u_1$ exists. By Lemma 3.1.4, $\mathbf{d}(u) = a$ and $\mathbf{r}(u) = b$, so $a \xrightarrow{u} b$. □

Example 4.1.2 Green's relation $\mathscr{H}$, on a Boolean inverse semigroup, need not be a V-relation. Let $S = \mathfrak{I}_2$; define a as the identity on $\{1, 2\}$ and b as the transposition $(1\ 2)$. Then $a = a_1 \oplus a_2$, where each a_i is the identity function on $\{i\}$, and $a \mathscr{H} b$. Nevertheless, there is no decomposition $b = b_1 \oplus b_2$ where each $a_i \mathscr{H} b_i$.

Lemma 4.1.1 also fails to extend to Green's relation $\mathscr{J}$, on a Boolean inverse semigroup, with a far more involved counterexample (cf. Example 5.2.11).

Definition 4.1.3 Let S be a Boolean inverse semigroup. The *type interval* of S, denoted by Int S, is the quotient partial commutative monoid $S/\mathscr{D}$ (cf. Lemma 2.4.4). The *type monoid* of S, denoted by Typ S, is the enveloping monoid of Int S. That is, Typ $S = \mathrm{U_{mon}}(\mathrm{Int}\, S)$.

The type interval Int S was introduced in Wallis [116, § 4.2], where it was denoted by $A(S)$, under an additional assumption called "orthogonally separating", equivalent to saying that Int S is a (full) commutative monoid. The partial monoid Int S was studied further in Lawson and Scott [77], where it was denoted by $\mathsf{E}(S)$ and called the *partial algebra associated with S*. Since $E(S)$ often denotes, in many papers, the set of all idempotent elements of S, we gave preference to the notations Idp S (for the idempotents of S) and Int S (for the type interval of S).

The type monoid Typ S is introduced in Wallis [116] and Kudryavtseva et al. [71]; in the latter paper, it is denoted by $\mathsf{T}(S)$. In the present work, we define it in a different, though equivalent, fashion, taking advantage of the results of Chap. 2, on extending a partial refinement monoid to a full refinement monoid.

The symbolism used in Definition 4.1.3 can be decoded as follows. We emphasize (cf. Lemma 2.4.3) that $\mathscr{D}$ is a conical V-equivalence on S. The type interval of S is the partial commutative monoid consisting of all equivalence classes $x/\mathscr{D}$, with $x \in S$, endowed with the addition defined by

$$z/\mathscr{D} = x/\mathscr{D} \oplus y/\mathscr{D} \quad \text{whenever } x, y, z \in S \text{ and } z = x \oplus y \text{ in } S\,.$$

Owing to Lemma 3.1.3, it suffices to consider the case where x, y, and z are idempotent. In particular,

$$\operatorname{Int} S = \{x/\mathscr{D} \mid x \in S\} = \{x/\mathscr{D} \mid x \in \operatorname{Idp} S\}\,.$$

The type monoid of S can thus be defined as the commutative monoid freely generated by elements $\operatorname{typ}(x)$ (thought of as $x/\mathscr{D}$), where $x \in \operatorname{Idp} S$, subjected to the relations

$$\operatorname{typ}(0) = 0\,; \tag{4.1.1}$$

$$\operatorname{typ}(x) = \operatorname{typ}(y)\,, \quad \text{whenever } x, y \in \operatorname{Idp} S \text{ and } x \mathrel{\mathscr{D}} y\,; \tag{4.1.2}$$

$$\operatorname{typ}(z) = \operatorname{typ}(x) + \operatorname{typ}(y)\,, \quad \text{whenever } x, y, z \in \operatorname{Idp} S \text{ and } z = x \oplus y \text{ in } \operatorname{Idp} S\,. \tag{4.1.3}$$

Whenever convenient, we shall extend the notation $\operatorname{typ}(x)$ to the case where x is not necessarily idempotent, by setting $\operatorname{typ}(x) = \operatorname{typ}(\mathbf{d}(x)) = \operatorname{typ}(\mathbf{r}(x))$. The statements (4.1.1)–(4.1.3), now extended to $x, y, z \in S$, remain valid. We will also write $\operatorname{typ}_S(x)$ instead of $\operatorname{typ}(x)$ in case we feel that S should be specified.

By applying Theorem 2.2.3 together with Lemmas 2.4.4 and 4.1.1, we get immediately the following.

Corollary 4.1.4 *Let S be a Boolean inverse semigroup. Then* $\operatorname{Typ} S$ *is a conical refinement monoid, and* $\operatorname{Int} S$ *is a lower interval of* $\operatorname{Typ} S$.

Lemma 4.1.5 *The following statements hold, for any Boolean inverse semigroup S:*

(1) $x \leq y$ *(within S) implies that* $\operatorname{typ}(x) \leq^+ \operatorname{typ}(y)$ *(within* $\operatorname{Typ} S$*), for all $x, y \in S$.*

(2) $\operatorname{typ}(xy) \leq^+ \begin{matrix}\operatorname{typ}(x)\\ \operatorname{typ}(y)\end{matrix}$ *(within* $\operatorname{Typ} S$*), for all $x, y \in S$.*

Proof (1) From $y = x \oplus (y \smallsetminus x)$ (cf. Notation 3.1.11) it follows that $\operatorname{typ}(y) = \operatorname{typ}(x) + \operatorname{typ}(y \smallsetminus x)$, thus $\operatorname{typ}(x) \leq^+ \operatorname{typ}(y)$.

(2) From $\mathbf{d}(xy) = y^{-1}\mathbf{d}(x)y \leq y^{-1}y \leq \mathbf{d}(y)$ it follows that $\operatorname{typ}(xy) = \operatorname{typ}(\mathbf{d}(xy)) \leq^+ \operatorname{typ}(\mathbf{d}(y)) = \operatorname{typ}(y)$. Likewise, $\mathbf{r}(xy) \leq \mathbf{r}(x)$, thus $\operatorname{typ}(xy) = \operatorname{typ}(\mathbf{r}(xy)) \leq^+ \operatorname{typ}(\mathbf{r}(x)) = \operatorname{typ}(x)$. □

By Lemma 4.1.1, $\mathscr{D}$ is an additive, conical V-equivalence on S. Thus a direct application of Lemma 2.4.4(1) yields the following.

Lemma 4.1.6 *Let S be a Boolean inverse semigroup, let $z \in S$, and let $\boldsymbol{x}, \boldsymbol{y} \in \operatorname{Typ} S$ such that $\operatorname{typ}(z) = \boldsymbol{x} + \boldsymbol{y}$. Then there are $x, y \in S$ such that $z = x \oplus y$, $\boldsymbol{x} = \operatorname{typ}(x)$, and $\boldsymbol{y} = \operatorname{typ}(y)$.*

Definition 4.1.7 For a Boolean inverse semigroup S and a commutative monoid M, define an *M-valued dimension function* on S as a map $\mu\colon \operatorname{Idp} S \to M$ satisfying (4.1.1)–(4.1.3) above. That is, $\mu(0_S) = 0_M$, $x \mathscr{D} y$ implies that $\mu(x) = \mu(y)$, and $\mu(x \oplus y) = \mu(x) + \mu(y)$ whenever $x \oplus y$ exists, for all $x, y \in \operatorname{Idp} S$.

The map $(\operatorname{Idp} S \to \operatorname{Typ} S,\ x \mapsto \operatorname{typ}(x) = x/\mathscr{D})$ is then the initial object in the category of all dimension functions on S. That is, for every commutative monoid M and every dimension function $\mu\colon \operatorname{Idp} S \to M$, there exists a unique monoid homomorphism $\overline{\mu}\colon \operatorname{Typ} S \to M$ such that $\mu(x) = \overline{\mu}(\operatorname{typ}_S(x))$ for all $x \in \operatorname{Idp} S$. We will refer to this statement as the *universal property of the type monoid.*

The universality of the $S \mapsto \operatorname{Typ} S$ construction yields immediately the following result.

Lemma 4.1.8 *Let S and T be Boolean inverse semigroups and let $f\colon S \to T$ be an additive semigroup homomorphism. Then there are a unique homomorphism $\operatorname{Int} f\colon \operatorname{Int} S \to \operatorname{Int} T$ of partial monoids and a unique homomorphism $\operatorname{Typ} f\colon \operatorname{Typ} S \to \operatorname{Typ} T$ of monoids such that $(\operatorname{Int} f)(\operatorname{typ}_S(x)) = (\operatorname{Typ} f)(\operatorname{typ}_S(x)) = \operatorname{typ}_T(f(x))$ for all $x \in S$.*

In particular, the assignments $S \mapsto \operatorname{Typ} S$, $f \mapsto \operatorname{Typ} f$ define a functor from the category of all Boolean inverse semigroups with additive semigroup homomorphisms to the category of all conical refinement monoids with monoid homomorphisms.

The following result expresses basic preservation properties for the functor Typ.

Proposition 4.1.9

(1) *The functor* Typ *preserves finite direct products.*
(2) *The functor* Typ *preserves directed colimits.*

Proof (1) Let S and T be Boolean inverse semigroups. It is straightforward to verify that the inverse semigroup $S \times T$ is Boolean, and that the projection homomorphisms $p\colon S \times T \twoheadrightarrow S$ and $q\colon S \times T \twoheadrightarrow T$ are both additive. This yields monoid homomorphisms $\operatorname{Typ} p\colon \operatorname{Typ}(S \times T) \to \operatorname{Typ} S$ and $\operatorname{Typ} q\colon \operatorname{Typ}(S \times T) \to \operatorname{Typ} T$, thus a monoid homomorphism $\varphi\colon \operatorname{Typ}(S \times T) \to (\operatorname{Typ} S) \times (\operatorname{Typ} T)$, which sends any $z \in \operatorname{Typ}(S \times T)$ to the pair $((\operatorname{Typ} p)(z), (\operatorname{Typ} q)(z))$. Conversely, the assignments $e\colon S \hookrightarrow S \times T$, $x \mapsto (x, 0_T)$ and $f\colon T \hookrightarrow S \times T$, $y \mapsto (0_S, y)$ are both V-embeddings, which, by (1) above, give rise to monoid homomorphisms

$$\operatorname{Typ} e\colon \operatorname{Typ} S \to \operatorname{Typ}(S \times T) \text{ and } \operatorname{Typ} f\colon \operatorname{Typ} T \to \operatorname{Typ}(S \times T),$$

thus to a monoid homomorphism $\psi\colon (\operatorname{Typ} S) \times (\operatorname{Typ} T) \to \operatorname{Typ}(S \times T)$, $(x,y) \mapsto (\operatorname{Typ} e)(x) + (\operatorname{Typ} f)(y)$. It is straightforward to verify that φ and ψ are mutually inverse. In particular, φ is an isomorphism.

(2) Let $S = \varinjlim_{i\in I} S_i$, for a directed poset I, Boolean inverse semigroups S_i, additive transition maps $f_i^j\colon S_i \to S_j$ (for $i \leq j$ in I), limiting maps $f_i\colon S_i \to S$. Working in the variety of all biases, all f_i^j are bias homomorphisms (cf. Theorem 3.2.5), thus so are all f_i. In particular, $S = \bigcup (f_i[S_i] \mid i \in I)$, from which it follows trivially that

$$\operatorname{Idp} S = \bigcup (f_i[\operatorname{Idp} S_i] \mid i \in I)\,. \tag{4.1.4}$$

Set $M = \varinjlim_{i\in I} \operatorname{Typ} S_i$, with transition maps $\operatorname{Typ} f_i^j\colon \operatorname{Typ} S_i \to \operatorname{Typ} S_j$ (for $i \leq j$ in I) and limiting maps $\boldsymbol{f}_i\colon \operatorname{Typ} S_i \to M$ (for $i \in I$). By (4.1.4), every $x \in \operatorname{Idp} S$ can be written as $x = f_i(\overline{x})$, for some $i \in I$ and $\overline{x} \in \operatorname{Idp} S_i$. It can then be easily verified that the element $\boldsymbol{f}_i(\operatorname{typ}_{S_i}(\overline{x}))$ depends only on x. Denoting this element by $\mu(x)$, it is then straightforward to verify that μ is an M-valued dimension function on S. By the universal property of the type monoid, there is a unique monoid homomorphism $\varphi\colon \operatorname{Typ} S \to M$ such that $\varphi(\operatorname{typ}_S(x)) = \mu(x)$ for every $x \in \operatorname{Idp} S$.

Conversely, by the universal property of the colimit, there is a unique monoid homomorphism $\psi\colon M \to \operatorname{Typ} S$ such that $\operatorname{Typ} f_i = \psi \circ \boldsymbol{f}_i$ for all $i \in I$. It is straightforward to verify that φ and ψ are mutually inverse. In particular, they are both isomorphisms. □

The following result makes it possible to calculate the type monoid of a crossed product of a Boolean inverse semigroup by a group action (cf. Sect. 3.6), via the construction $M/\!\!/G$ introduced in Sect. 2.8. It is a Boolean inverse semigroup version of Proposition 2.8.4. Unlike Proposition 2.8.4, this result always yields an isomorphism.

Theorem 4.1.10 *Let a group G act by automorphisms on a Boolean inverse semigroup S. We endow the commutative monoid $\operatorname{Typ} S$ with the induced group action, that is, $g \cdot \operatorname{typ}_S(x) = \operatorname{typ}_S(g(x))$, for all $x \in S$ and all $g \in G$. Then there is a unique monoid isomorphism $\tau\colon \operatorname{Typ}(S)/\!\!/G \to \operatorname{Typ}(S \rtimes G)$ such that $\tau\big([\operatorname{typ}_S(x)]_G\big) = \operatorname{typ}_{S\rtimes G}(x \cdot 1)$ for every $x \in S$.*

Proof Denote by $\varepsilon\colon S \hookrightarrow S \rtimes G$, $x \mapsto x \cdot 1$ the canonical lower semigroup embedding (cf. Proposition 3.6.3). Since $\operatorname{Idp}(S \rtimes G) = \varepsilon[\operatorname{Idp} S]$ (cf. Proposition 3.6.3), the canonical map $\operatorname{Int} \varepsilon\colon \operatorname{Int}(S) \to \operatorname{Int}(S \rtimes G)$ is surjective.

Claim The kernel of $\operatorname{Int} \varepsilon$ is the restriction of $\simeq_G$ to $\operatorname{Int} S$.

Proof of Claim. An argument similar to the one of the proof of Proposition 2.8.4 yields that $g(x) \cdot 1 \mathrel{\mathscr{D}_{S\rtimes G}} x \cdot 1$, for any $x \in S$ and any $g \in G$ (it suffices to consider the case where x is idempotent). It follows that $\simeq_G$ is contained in the kernel of $\operatorname{Int} \varepsilon$.

Conversely, let $a, b \in \operatorname{Idp} S$ such that $a \cdot 1 \mathrel{\mathscr{D}_{S\rtimes G}} b \cdot 1$, that is, there is $x \in S \rtimes G$ such that $a \cdot 1 = \mathbf{d}(x)$ and $b \cdot 1 = \mathbf{r}(x)$. Writing x in the form (3.6.1), this means, using Lemma 3.6.1, that $a = \bigoplus_{i=1}^n g_i^{-1}\big(\mathbf{d}(x_i)\big)$ and $b = \bigoplus_{i=1}^n \mathbf{r}(x_i)$. Therefore, setting

$\boldsymbol{b}_i = \mathrm{typ}_S(\mathbf{d}(x_i)) = \mathrm{typ}_S(\mathbf{r}(x_i))$, we obtain that

$$\mathrm{typ}_S(a) = \sum_{i=1}^{n} g_i^{-1}\boldsymbol{b}_i \quad \text{and} \quad \mathrm{typ}_S(b) = \sum_{i=1}^{n} \boldsymbol{b}_i\,,$$

so $\mathrm{typ}_S(a) \simeq_G \mathrm{typ}_S(b)$. □ Claim.

Since $\simeq_G$ is a V-equivalence, it follows from Lemma 2.4.5 that $\mathrm{Int}\,\varepsilon$ induces an isomorphism $\tau_0\colon \mathrm{Int}(S)/\!\!/G \to \mathrm{Int}(S \rtimes G)$ of partial commutative monoids. By using the canonical isomorphism $\mathrm{Typ}(S)/\!\!/G \cong \mathrm{U}_{\mathrm{mon}}(\mathrm{Int}(S)/\!\!/G)$ (apply Theorem 2.4.6 to $\Gamma = \simeq_G$) and since $\mathrm{Typ}(S \rtimes G) = \mathrm{U}_{\mathrm{mon}}(\mathrm{Int}(S \rtimes G))$, τ_0 extends to a unique isomorphism τ as desired. □

4.2 Type Theory of Additive Quasi-Ideals and Additive Enlargements

In this section we shall introduce tools that will enable us to describe type monoids of further semigroups, such as ideals (of Boolean inverse semigroups), local submonoids (subsemigroups of the form eSe), semigroups of generalized rook matrices.

Definition 4.2.1 Let S and T be Boolean inverse semigroups. An additive map $f\colon S \to T$ is

- *type-expanding* if $\mathrm{Typ}\,f$ is a V-embedding of commutative monoids;
- *type-preserving* if $\mathrm{Typ}\,f$ is an isomorphism.

For an additive inverse subsemigroup S of T, we say that T is a *type-expanding extension* (resp., a *type-preserving extension*) if $\mathrm{Typ}\,\varepsilon$ is type-expanding (resp., type-preserving), where $\varepsilon\colon S \hookrightarrow T$ denotes the inclusion map.

Our next result relates the quasi-ideals (cf. Definition 3.1.16) and additive enlargements (cf. Definition 3.1.21) to the type theory of Boolean inverse semigroups.

Theorem 4.2.2 *Let S be an additive quasi-ideal in a Boolean inverse semigroup T. Then T is a type-expanding extension of S. Furthermore, T is a type-preserving extension of S iff T is an additive enlargement of S.*

Proof Denote by $\varepsilon\colon S \hookrightarrow T$ the inclusion map.

Let $a, b \in \mathrm{Idp}\,S$ such that $(\mathrm{Int}\,\varepsilon)(\mathrm{typ}_S(a)) = (\mathrm{Int}\,\varepsilon)(\mathrm{typ}_S(b))$, that is, $a \mathrel{\mathscr{D}_T} b$. There is $x \in T$ such that $a = \mathbf{d}(x)$ and $b = \mathbf{r}(x)$. Since S is a quasi-ideal of T and $x = bxa \in STS$, it follows that $x \in S$, so $a \mathrel{\mathscr{D}_S} b$, that is, $\mathrm{typ}_S(a) = \mathrm{typ}_S(b)$. Hence, $\mathrm{Int}\,\varepsilon$ is one-to-one.

Let $\boldsymbol{a}, \boldsymbol{b} \in \mathrm{Int}\,T$ and let $\boldsymbol{c} \in \mathrm{Int}\,S$ such that $(\mathrm{Int}\,\varepsilon)(\boldsymbol{c}) = \boldsymbol{a} \oplus \boldsymbol{b}$ within $\mathrm{Int}\,T$. Pick $c \in \boldsymbol{c}$. Since $c/\mathscr{D}_T = (\mathrm{Int}\,\varepsilon)(\boldsymbol{c}) = \boldsymbol{a} \oplus \boldsymbol{b}$ and by Lemma 2.4.4, there is $(a, b) \in \boldsymbol{a} \times \boldsymbol{b}$

such that $c = a \oplus b$ within T. Since S is a lower subset of T (cf. Proposition 3.1.18), a and b both belong to S and $c = a \oplus b$ within S. Observe that $\boldsymbol{a} = (\operatorname{Int}\varepsilon)(\operatorname{typ}_S(a))$ and $\boldsymbol{b} = (\operatorname{Int}\varepsilon)(\operatorname{typ}_S(b))$.

We have thus verified that $\operatorname{Int}\varepsilon$ is a V-embedding from $\operatorname{Int} S$ into $\operatorname{Int} T$. By Proposition 2.2.5, it follows that $\operatorname{Typ}\varepsilon$ is a V-embedding from $\operatorname{Typ} S$ into $\operatorname{Typ} T$.

Suppose now that T is an additive enlargement of S. We must prove that $\operatorname{Typ}\varepsilon$ is surjective. Since every element of $\operatorname{Typ} T$ is a finite sum of elements of $\operatorname{Int} T$, it suffices to prove that the range of $\operatorname{Typ}\varepsilon$ contains $\operatorname{Int} T$. Let $b \in \operatorname{Idp} T$. Since $T = (TST)^{\oplus}$, there is a decomposition of the form $b = \bigoplus_{i<n} x_i a_i y_i$, where all $x_i, y_i \in T$ and all $a_i \in S$. Observe that all elements $b_i = x_i a_i y_i$ are beneath b, thus they are idempotent. By Nambooripad [88, Theorem 1.6], for each $i < n$, there are $x'_i \leq x_i$, $a'_i \leq a_i$, and $y'_i \leq y_i$ such that $b_i = x'_i a'_i y'_i$ is a trace product.[3] In particular, $b_i \mathrel{\mathscr{D}_T} a'_i$. Since S is a lower subset of T (cf. Proposition 3.1.18), $a_i \in S$, and $a'_i \leq a_i$, we get $a'_i \in S$. Hence, $\operatorname{typ}_T(b_i) = \operatorname{typ}_T(a'_i)$ belongs to the range of $\operatorname{Typ}\varepsilon$. Therefore, $\operatorname{typ}_T(b) = \sum_{i<n} \operatorname{typ}_T(b_i)$ also belongs to the range of $\operatorname{Typ}\varepsilon$.

Suppose, finally, that S is an additive quasi-ideal of T and that $\operatorname{Typ}\varepsilon$ is an isomorphism. We must prove that every element b of T belongs to $(TST)^{\oplus}$. Since $\operatorname{Typ}\varepsilon$ is surjective, there is a decomposition of the form $\operatorname{typ}_T(\mathbf{d}(b)) = \sum_{i<n} \operatorname{typ}_T(a_i)$, where each $a_i \in \operatorname{Idp} S$. By Lemma 4.1.6, there is a decomposition $\mathbf{d}(b) = \bigoplus_{i<n} b_i$, where each $\operatorname{typ}_T(a_i) = \operatorname{typ}_T(b_i)$, that is, $a_i \mathrel{\mathscr{D}_T} b_i$. If $x_i \in T$ such that $\mathbf{d}(x_i) = a_i$ and $\mathbf{r}(x_i) = b_i$, then we get $b_i = x_i a_i x_i^{-1} \in TST$. Therefore, $b = b\,\mathbf{d}(b) = \bigoplus_{i<n} b x_i a_i x_i^{-1} \in (TST)^{\oplus}$. □

Example 3.1.25 shows that the assumption, of Theorem 4.2.2, that S be an additive quasi-ideal of T, cannot be relaxed to S be a lower inverse subsemigroup of T: for that example, $\big(\operatorname{Typ} S, \operatorname{typ}_S(1)\big) \cong \big(\mathbb{Z}^+ \times \mathbb{Z}^+, (1,1)\big)$ and $\big(\operatorname{Typ} T, \operatorname{typ}_T(1)\big) \cong (\mathbb{Z}^+, 2)$, so there is no monoid embedding from $\operatorname{Typ} S$ into $\operatorname{Typ} T$.

The following result is an analogue, for Boolean inverse semigroups, of a result stating that $\mathrm{V}(eRe) \cong \mathrm{V}(ReR)$ for any idempotent element e in a ring R (see the beginning of the proof of Lemma 7.3 in Ara and Facchini [8]). Observe that in the latter statement, ReR is defined as the two-sided ideal of R generated by e, as opposed to the set of all xey where $x, y \in R$.

Proposition 4.2.3 *Let e be an idempotent element in a Boolean inverse semigroup S. Then the additive ideal $(SeS)^{\oplus}$ is a type-preserving extension of the local submonoid eSe. In particular,* $\operatorname{Typ}(eSe) \cong \operatorname{Typ}\big((SeS)^{\oplus}\big)$.

Proof By Proposition 3.1.22, $(SeS)^{\oplus}$ is an additive enlargement of eSe. Apply Theorem 4.2.2. □

The following result relates the o-ideals of $\operatorname{Typ} S$ with the additive ideals of S, for any Boolean inverse semigroup S.

[3] A referee informed the author that this was long known before by Charles Ehresmann.

Proposition 4.2.4 *The following statements hold, for any Boolean inverse semigroup S:*

(1) *Let I be an additive ideal in S. Then S is a type-expanding extension of I, and the canonical image $\boldsymbol{I}$ of* $\operatorname{Typ} I$ *in* $\operatorname{Typ} S$ *is an o-ideal of* $\operatorname{Typ} S$.
(2) *Let $\boldsymbol{I}$ be an o-ideal of* $\operatorname{Typ} S$. *Then the set $I = \{x \in S \mid \operatorname{typ}(x) \in \boldsymbol{I}\}$ is an additive ideal of S.*
(3) *The assignments $I \mapsto \boldsymbol{I}$ from* (1) *and $\boldsymbol{I} \mapsto I$ from* (2) *are mutually inverse order-isomorphisms between the lattice of all additive ideals of S and the lattice of all o-ideals of* $\operatorname{Typ} S$.

Proof (1) Since I is an additive ideal of S, it is also an additive quasi-ideal. Denote by $\varepsilon\colon I \hookrightarrow S$ the inclusion map. By Theorem 4.2.2, $\operatorname{Typ}\varepsilon$ is a V-embedding, thus the range $\boldsymbol{I}$ of $\operatorname{Typ}\varepsilon$ is a lower subset of $\operatorname{Typ} S$. Since $\operatorname{Typ} I$ is a monoid and $\operatorname{Typ}\varepsilon$ is a monoid homomorphism, $\boldsymbol{I}$ is a submonoid of $\operatorname{Typ} S$.

(2) It follows immediately from Lemma 4.1.5 that $SIS \subseteq I$. Since $0 \in I$, it follows that I is an ideal of S. For all orthogonal elements $x, y \in I$, $\operatorname{typ}(x \oplus y) = \operatorname{typ}(x) + \operatorname{typ}(y) \in \boldsymbol{I}$, thus $x \oplus y \in I$. Hence I is an additive ideal of S.

(3) First, let I be an additive ideal of S. Denote by $\boldsymbol{I}$ the canonical image of $\operatorname{Typ} I$ in $\operatorname{Typ} S$ and set $I' = \{x \in S \mid \operatorname{typ}(x) \in \boldsymbol{I}\}$. We claim that $I = I'$. The containment $I \subseteq I'$ is trivial. Conversely, let $x \in I'$. There are a positive integer n and elements $y_0, \dots, y_{n-1} \in I$ such that $\operatorname{typ}(x) = \sum_{i<n} \operatorname{typ}(y_i)$. By Lemma 4.1.6, there is a decomposition $x = \bigoplus_{i<n} x_i$ such that $x_i \mathrel{\mathscr{D}} y_i$ whenever $i < n$. Since each $y_i \in I$ and I is an ideal, each $x_i \in I$. Since I is an additive ideal, $x \in I$, thus proving our claim.

Finally, let $\boldsymbol{I}$ be an o-ideal of $\operatorname{Typ} S$. Set $I = \{x \in S \mid \operatorname{typ}(x) \in \boldsymbol{I}\}$ and denote by $\boldsymbol{I}'$ the canonical image of $\operatorname{Typ} I$ in $\operatorname{Typ} S$. We claim that $\boldsymbol{I} = \boldsymbol{I}'$. First, every $\boldsymbol{x} \in \boldsymbol{I}'$ has the form $\sum_{i<n} \operatorname{typ}(x_i)$, where n is a positive integer and each $x_i \in I$. Since each $\operatorname{typ}(x_i) \in \boldsymbol{I}$, it follows that $\boldsymbol{x} \in \boldsymbol{I}$. Hence, $\boldsymbol{I}' \subseteq \boldsymbol{I}$. Conversely, let $\boldsymbol{x} \in \boldsymbol{I}$. Since $\boldsymbol{I} \subseteq \operatorname{Typ} S$, we can write $\boldsymbol{x} = \sum_{i<n} \operatorname{typ}(x_i)$ for a positive integer n and elements $x_i \in S$. From $\boldsymbol{x} \in \boldsymbol{I}$ it follows that $\operatorname{typ}(x_i) \in \boldsymbol{I}$, that is, $x_i \in I$, for each $i < n$. Hence, $\boldsymbol{x} \in \boldsymbol{I}'$, thus completing the proof of our claim. □

In view of Proposition 4.2.4, we will usually identify $\operatorname{Typ} I$ with the canonical image of $\operatorname{Typ} I$ in $\operatorname{Typ} S$, for any additive ideal I in a Boolean inverse semigroup S.

Another useful class of type-preserving embeddings is provided by the following result.

Proposition 4.2.5 *Let S be a Boolean inverse semigroup, let Ω be a set, and let $o \in \Omega$. Then the canonical additive semigroup embedding $\eta\colon S \hookrightarrow \mathrm{M}_{\Omega}^{\oplus}(S)$, $x \mapsto x_{(o,o)}$ is type-preserving. In particular,* $\operatorname{Typ} S \cong \operatorname{Typ} \mathrm{M}_{\Omega}^{\oplus}(S)$.

Proof By Corollary 3.5.4, η is a lower semigroup embedding and $\mathrm{M}_{\Omega}^{\oplus}(S)$ is an additive enlargement of $\eta[S]$. By Theorem 4.2.2, it follows that η is type-preserving. □

In the particular case where S is a monoid and $\Omega = [n]$, for a positive integer n, the result of Proposition 4.2.5 can be refined by observing the following sequence of equations (where we set $\overline{S} = \mathrm{M}_n^{\oplus}(S)$):

$$\begin{aligned}(\mathrm{Typ}\,\eta)\big(n \cdot \mathrm{typ}_S(1)\big) &= n \cdot \mathrm{typ}_{\overline{S}}(1_{(1,1)})\\ &= \mathrm{typ}_{\overline{S}}(1_{(1,1)}) + \cdots + \mathrm{typ}_{\overline{S}}(1_{(n,n)})\\ &= \mathrm{typ}_{\overline{S}}(1_{\overline{S}}),\end{aligned}$$

which yields the following.

Corollary 4.2.6 *Let S be a Boolean inverse monoid, let n be a positive integer, and set $\overline{S} = \mathrm{M}_n^{\oplus}(S)$. Then $\big(\mathrm{Typ}\,\overline{S}, \mathrm{typ}_{\overline{S}}(1_{\overline{S}})\big) \cong \big(\mathrm{Typ}\,S, n \cdot \mathrm{typ}_S(1_S)\big)$.*

The following result describes a "scaling" process of a Boolean inverse semigroup S, relatively to an element of $\mathrm{Typ}\,S$.

Theorem 4.2.7 *Let S be a Boolean inverse semigroup and let $\boldsymbol{e} \in \mathrm{Typ}\,S$. Then there are a positive integer n and an idempotent element e of $\mathrm{M}_n^{\oplus}(S)$ such that, setting $\overline{S} = e\,\mathrm{M}_n^{\oplus}(S)e$, the relation $\big((\mathrm{Typ}\,S){\restriction}\boldsymbol{e}, \boldsymbol{e}\big) \cong (\mathrm{Typ}\,\overline{S}, \mathrm{typ}_{\overline{S}}(1))$ holds.*

Note By Propositions 3.5.3 and 3.1.22, both $\mathrm{M}_n^{\oplus}(S)$ and $e\,\mathrm{M}_n^{\oplus}(S)e$ are Boolean inverse semigroups.

Proof There are a positive integer n and idempotents $e_1, \ldots, e_n$ of S such that $\boldsymbol{e} = \sum_{i=1}^{n} \mathrm{typ}_S(e_i)$. Set $S_1 = \mathrm{M}_n^{\oplus}(S)$ and denote by $\eta\colon S \hookrightarrow S_1$, $x \mapsto x_{(1,1)}$ the canonical additive semigroup embedding. By Proposition 4.2.5, $\mathrm{Typ}\,\eta$ is an isomorphism from $\mathrm{Typ}\,S$ onto $\mathrm{Typ}\,S_1$. For each $i \in [n]$, $(e_i)_{(1,1)}\ \mathscr{D}_{S_1}\ (e_i)_{(i,i)}$ (cf. (3.5.3)), thus

$$(\mathrm{Typ}\,\eta)\big(\mathrm{typ}_S(e_i)\big) = \mathrm{typ}_{S_1}\big((e_i)_{(1,1)}\big) = \mathrm{typ}_{S_1}\big((e_i)_{(i,i)}\big), \tag{4.2.1}$$

thus, denoting by e the diagonal matrix with diagonal entries $e_1, \ldots, e_n$ (so $e = \bigoplus_{i=1}^{n} (e_i)_{(i,i)}$) and adding together the equations in (4.2.1), we obtain

$$(\mathrm{Typ}\,\eta)(\boldsymbol{e}) = \sum_{i=1}^{n} \mathrm{typ}_{S_1}\big((e_i)_{(i,i)}\big) = \mathrm{typ}_{S_1}(e). \tag{4.2.2}$$

Now by Proposition 3.1.20, $S_2 = (S_1 e S_1)^{\oplus}$ is the additive ideal of S_1 generated by e. Denote by $\varepsilon\colon S_2 \hookrightarrow S_1$ the inclusion map. By Proposition 4.2.4, $\mathrm{Typ}\,\varepsilon$ is a V-embedding from $\mathrm{Typ}\,S_2$ into $\mathrm{Typ}\,S_1$, with range $(\mathrm{Typ}\,S_1){\restriction}\,\mathrm{typ}_{S_1}(e)$. By Proposition 4.2.3, the subset $\overline{S} = eS_1e$ is a quasi-ideal of S_1, and S_2 is a type-preserving extension of $\overline{S}$. Therefore,

$$\begin{aligned}\big(\mathrm{Typ}\,\overline{S}, \mathrm{typ}_{\overline{S}}(e)\big) &\cong \big(\mathrm{Typ}\,S_2, \mathrm{typ}_{S_2}(e)\big)\\ &\cong \big((\mathrm{Typ}\,S_1){\restriction}\,\mathrm{typ}_{S_1}(e), \mathrm{typ}_{S_1}(e)\big) \cong \big((\mathrm{Typ}\,S){\restriction}\boldsymbol{e}, \boldsymbol{e}\big).\end{aligned} \tag{4.2.3}$$

□

The particular case where $n = 1$ and $\boldsymbol{e} = \mathrm{typ}_S(e)$, where $e \in \mathrm{Idp}\,S$, in Theorem 4.2.7, yields the following instance of that theorem (which can also be easily deduced from Proposition 4.2.3).

Corollary 4.2.8 *Let e be an idempotent element in a Boolean inverse semigroup S, and denote by $\varepsilon\colon eSe \hookrightarrow S$ the inclusion map. Then the map $\mathrm{Typ}\,\varepsilon$ defines an isomorphism from $\big(\mathrm{Typ}(eSe), \mathrm{typ}_{eSe}(e)\big)$ onto $\big((\mathrm{Typ}\,S)\!\restriction \mathrm{typ}_S(e), \mathrm{typ}_S(e)\big)$.*

4.3 Type Monoids of Quotients

In this section we shall prove that type monoids of quotients S/I, for an additive ideal I of a Boolean inverse semigroup S, behave as one should expect.

Lemma 4.3.1 *Let S and T be Boolean inverse semigroups and let $f\colon S \twoheadrightarrow T$ be a surjective additive semigroup homomorphism. Then for all $a \in \mathrm{Idp}\,S$, all $n \in \mathbb{N}$, and all $b_1, \dots, b_n \in \mathrm{Idp}\,T$ such that $f(a) = \bigoplus_{i=1}^{n} b_i$, there is a decomposition $a = \bigoplus_{i=1}^{n} a_i$ in $\mathrm{Idp}\,S$ such that each $f(a_i) = b_i$.*

Proof An easy induction argument reduces the problem to the case where $n = 2$. Since f is surjective, there are $a_1, a_2 \in S$ such that $f(a_i) = b_i$ whenever $i \in \{1, 2\}$. From the idempotence of b_i it follows that $b_i = f(\mathbf{d}(a_i))$, thus we may assume that a_1 and a_2 are both idempotent. Since f is a semigroup homomorphism, $f(a_i a) = f(a_i)f(a) = b_i f(a) = b_i$, so we may assume that $a_i \leq a$, whenever $i \in \{1, 2\}$. Since f restricts, on idempotents, to a homomorphism of Boolean rings, and since $f(a_1 a_2) = b_1 b_2 = 0$, we may replace each a_i by $a_i \smallsetminus a_1 a_2$ and thus assume that $a_1 \perp a_2$. At this point, each $f(a_i) = b_i$ and $a_1 \oplus a_2 \leq a$. Keeping a_1 unchanged and replacing a_2 by $a_2 \oplus \big(a \smallsetminus (a_1 \oplus a_2)\big)$, we finally get a_1 and a_2 as desired. □

Theorem 4.3.2 *Let S and T be Boolean inverse semigroups and let $f\colon S \twoheadrightarrow T$ be a surjective additive semigroup homomorphism. Consider the additive ideal $I = \ker f$. Then the map $\mathrm{Typ}\,f\colon \mathrm{Typ}\,S \twoheadrightarrow \mathrm{Typ}\,T$ factors, through $\mathrm{Typ}\,S/\mathrm{Typ}\,I$, to an isomorphism $\mathrm{Typ}\,S/\mathrm{Typ}\,I \cong \mathrm{Typ}\,T$.*

Proof The set $\boldsymbol{I} = \mathrm{Typ}\,I$ is an o-ideal of $\mathrm{Typ}\,S$ (cf. Proposition 4.2.4), on which the map $\mathrm{Typ}\,f$ vanishes. We claim that for any $\boldsymbol{a}, \boldsymbol{b} \in \mathrm{Typ}\,S$, if $(\mathrm{Typ}\,f)(\boldsymbol{a}) = (\mathrm{Typ}\,f)(\boldsymbol{b})$, then $\boldsymbol{a}/\boldsymbol{I} = \boldsymbol{b}/\boldsymbol{I}$, that is, there are $\boldsymbol{x}, \boldsymbol{y} \in \boldsymbol{I}$ such that $\boldsymbol{a} + \boldsymbol{x} = \boldsymbol{b} + \boldsymbol{y}$. This claim clearly implies the desired result.

Write $\boldsymbol{a} = \sum_{i<m} \mathrm{typ}_S(a_i)$ and $\boldsymbol{b} = \sum_{j<n} \mathrm{typ}_S(b_j)$, for positive integers m and n together with idempotents $a_i, b_j \in S$, for $i < m$ and $j < n$. Since $\mathrm{Typ}\,T$ is a refinement monoid (cf. Corollary 4.1.4) and since

$$\sum_{i<m} \mathrm{typ}_T(f(a_i)) = (\mathrm{Typ}\,f)(\boldsymbol{a}) = (\mathrm{Typ}\,f)(\boldsymbol{b}) = \sum_{j<n} \mathrm{typ}_T(f(b_j)),$$

there is a refinement matrix as follows:

	$\mathrm{typ}_T f(b_j) (j < n)$
$\mathrm{typ}_T f(a_i) (i < n)$	$\boldsymbol{d}_{i,j}$

with all $\boldsymbol{d}_{i,j} \in \mathrm{Typ}\, T$.

For each $i < m$, it follows from the equality $\mathrm{typ}_T(f(a_i)) = \sum_{j<n} \boldsymbol{d}_{i,j}$ together with Lemma 4.1.6 that there is a decomposition of the form $f(a_i) = \bigoplus_{j<n} x_{i,j}$ within T, where each $\mathrm{typ}_T(x_{i,j}) = \boldsymbol{d}_{i,j}$. By Lemma 4.3.1, there is a decomposition $a_i = \bigoplus_{j<n} a_{i,j}$ within $\mathrm{Idp}\, S$, such that each $f(a_{i,j}) = x_{i,j}$. Hence, $\mathrm{typ}_T f(a_{i,j}) = \mathrm{typ}_T(x_{i,j}) = \boldsymbol{d}_{i,j}$. Likewise, for each $j < n$, there is a decomposition $b_i = \bigoplus_{i<m} b_{i,j}$ in $\mathrm{Idp}\, S$ such that each $\mathrm{typ}_T f(b_{i,j}) = \boldsymbol{d}_{i,j}$.

Let $i < m$ and $j < n$. From $\mathrm{typ}_T f(a_{i,j}) = \mathrm{typ}_T f(b_{i,j}) = \boldsymbol{d}_{i,j}$ it follows that $f(a_{i,j}) \mathscr{D}_T f(b_{i,j})$, thus $f(a_{i,j}) = \mathbf{d}(y)$ and $f(b_{i,j}) = \mathbf{r}(y)$ for some $y \in T$. Since f is surjective, $y = f(x)$ for some $x \in S$. Hence,

$$f(a_{i,j}) = f(\mathbf{d}(x)) \text{ and } f(b_{i,j}) = f(\mathbf{r}(x)). \tag{4.3.1}$$

Since f is an additive semigroup homomorphism, it restricts, on the idempotents, to a homomorphism of Boolean rings. Since $a_{i,j}$, $b_{i,j}$, $\mathbf{d}(x)$, $\mathbf{r}(x)$ are all idempotents of S, it follows that $a_{i,j} \equiv \mathbf{d}(x) \pmod I$ and $b_{i,j} \equiv \mathbf{r}(x) \pmod I$, where $x \equiv y \pmod I$ is shorthand for $\{x \smallsetminus y, y \smallsetminus x\} \subseteq I$. From the equalities $x \vee y = x \oplus (y \smallsetminus x) = y \oplus (x \smallsetminus y)$ it follows, in particular, that $x \equiv y \pmod I$ implies $\mathrm{typ}_S(x)/I = \mathrm{typ}_S(y)/I$. By applying this observation to (4.3.1), we obtain

$$\mathrm{typ}_S(a_{i,j})/\boldsymbol{I} = \mathrm{typ}_S(\mathbf{d}(x))/\boldsymbol{I} \text{ and } \mathrm{typ}_S(b_{i,j})/\boldsymbol{I} = \mathrm{typ}_S(\mathbf{r}(x))/\boldsymbol{I}.$$

Now $\mathbf{d}(x) \mathscr{D}_S \mathbf{r}(x)$, thus $\mathrm{typ}_S(\mathbf{d}(x)) = \mathrm{typ}_S(\mathbf{r}(x))$. Therefore, $\mathrm{typ}_S(a_{i,j})/\boldsymbol{I} = \mathrm{typ}_S(b_{i,j})/\boldsymbol{I}$, and therefore,

$$\boldsymbol{a}/\boldsymbol{I} = \sum_{i<m} \mathrm{typ}_S(a_i)/\boldsymbol{I} = \sum_{i<m,\, j<n} \mathrm{typ}_S(a_{i,j})/\boldsymbol{I} = \sum_{i<m,\, j<n} \mathrm{typ}_S(b_{i,j})/\boldsymbol{I} = \boldsymbol{b}/\boldsymbol{I},$$

as desired. □

By applying Theorem 4.3.2 to the canonical projection $S \twoheadrightarrow S/I$, for an additive ideal I of S, we obtain the following.

Corollary 4.3.3 *Let I be an additive ideal in a Boolean inverse semigroup S. Then* $\mathrm{Typ}(S/I) \cong \mathrm{Typ}\, S/\mathrm{Typ}\, I$.

4.4 Inverse Semigroups of Bi-measurable Partial Functions

In this section we will give a few constructions of Boolean inverse semigroups via partial bijections (on sets) or partial automorphisms (on Boolean rings). Further, we will give sufficient conditions for those constructions to yield fundamental Boolean inverse semigroups or Boolean inverse meet-semigroups.

Definition 4.4.1 Let $\mathcal{B}$ be a ring of subsets of a set Ω. A *bi-measurable partial function with respect to* $\mathcal{B}$ is a partial bijective function f on Ω, with domain and range both belonging to $\mathcal{B}$, such that both $f[X]$ and $f^{-1}[X]$ belong to $\mathcal{B}$ whenever $X \in \mathcal{B}$.

We leave to the reader the straightforward proof of the following result.

Proposition 4.4.2 *The set* $\mathbf{pMeas}(\mathcal{B})$*, of all bi-measurable partial functions with respect to* $\mathcal{B}$*, is an additive Boolean inverse subsemigroup of the symmetric inverse semigroup* $\mathfrak{I}_\Omega$*. Furthermore, the idempotent elements of* $\mathbf{pMeas}(\mathcal{B})$ *are exactly the identity functions on the elements of* $\mathcal{B}$*, and the orthogonal join in* $\mathbf{pMeas}(\mathcal{B})$ *is given by disjoint union of functions.*

The proof of the following lemma is equally straightforward.

Lemma 4.4.3 *A subset* $\mathcal{F}$ *of* $\mathbf{pMeas}(\mathcal{B})$ *has a meet, in* $\mathbf{pMeas}(\mathcal{B})$*, iff there is a largest element of* $\mathcal{B}$ *contained in the set* $\{x \in \Omega \mid f(x) = g(x) \text{ for all } f, g \in \mathcal{F}\}$.

Corollary 4.4.4 *If the set* $\|f = g\| = \{x \in (\operatorname{dom} f) \cap (\operatorname{dom} g) \mid f(x) = g(x)\}$ *belongs to* $\mathcal{B}$ *whenever* $f, g \in \mathbf{pMeas}(\mathcal{B})$*, then* $\mathbf{pMeas}(\mathcal{B})$ *is an inverse meet-semigroup.*

Lemma 4.4.5 *The following are equivalent, for any* $f \in \mathbf{pMeas}(\mathcal{B})$:

(i) f *commutes with every idempotent element of* $\mathbf{pMeas}(\mathcal{B})$.
(ii) $f[X] \cap X \neq \varnothing$ *for any* $X \in \mathcal{B}$ *such that* $\varnothing \subsetneq X \subseteq \operatorname{dom} f$.
(iii) $f[X] \subseteq X$ *for any* $X \in \mathcal{B} \downarrow \operatorname{dom} f$.
(iv) $f[X] \subseteq X$ *for any* $X \in \mathcal{B}$.
(v) $f[X] = X$ *for any* $X \in \mathcal{B} \downarrow \operatorname{dom} f$.

Proof (i)⇒(iv) Let $X \in \mathcal{B}$. From $f \restriction_X = f \circ \mathrm{id}_X = \mathrm{id}_X \circ f = f \restriction_{f^{-1}[X]}$ it follows that $X \cap \operatorname{dom} f = f^{-1}[X] \cap \operatorname{dom} f$, thus $f[X] \subseteq X$.

(iv)⇒(iii) and (iii)⇒(ii) are both trivial.

(ii)⇒(iii) Let $X \in \mathcal{B} \downarrow \operatorname{dom} f$ and suppose that $f[X] \nsubseteq X$. This means that the subset $Y = X \setminus f^{-1}[X]$ is nonempty. Since $Y \in \mathcal{B}$ and $\varnothing \subsetneq Y \subseteq \operatorname{dom} f$, it follows that $f[Y] \cap Y \neq \varnothing$, so $Y \cap f^{-1}[Y] \neq \varnothing$, a contradiction since $f^{-1}[Y] \subseteq f^{-1}[X]$ and $Y \cap f^{-1}[X] = \varnothing$.

(iii)⇒(v) Let $X \in \mathcal{B} \downarrow \operatorname{dom} f$. It follows from our assumption that $f[X] \subseteq X$. Further, the subset $Y = X \setminus f[X]$ belongs to $\mathcal{B} \downarrow \operatorname{dom} f$, thus $f[Y] \subseteq Y$. Since $f[Y] \subseteq f[X]$, it follows that $f[Y] = \varnothing$, thus, since $Y \subseteq \operatorname{dom} f$, we get $Y = \varnothing$, that is, $f[X] = X$.

(v)⇒(i) Both sets $Y = X \cap \operatorname{dom} f$ and $Z = f^{-1}[X] \cap \operatorname{dom} f$ belong to $\mathcal{B} \downarrow \operatorname{dom} f$, for any $X \in \mathcal{B}$. From $Y \subseteq \operatorname{dom} f$ and our assumption it follows that $Y \subseteq f^{-1}[Y]$, thus $Y \subseteq Z$. From $Z \subseteq \operatorname{dom} f$ and our assumption it follows that $Z \subseteq f[Z]$, thus $Z \subseteq Y$. Therefore, $Y = Z$, that is, $f \circ \mathrm{id}_X = \mathrm{id}_X \circ f$. □

Corollary 4.4.6 *The inverse semigroup* **pMeas**$(\mathcal{B})$ *is fundamental iff for every* $f \in$ **pMeas**$(\mathcal{B})$, *if* f *is not an identity, then there exists a nonempty subset* $X \in \mathcal{B} \downarrow \operatorname{dom} f$ *such that* $f[X] \cap X = \varnothing$.

Example 4.4.7 Define a *partial homeomorphism* on a Hausdorff topological space Ω as a homeomorphism between compact open[4] subsets of Ω. The set **pHomeo**(Ω) of all partial homeomorphisms of Ω is an inverse subsemigroup of the inverse semigroup of all partial one-to-one functions on Ω. In fact, **pHomeo**(Ω) = **pMeas**$(\mathcal{B})$, where $\mathcal{B}$ denotes the Boolean ring of all compact open subsets of Ω. An immediate application of Corollary 4.4.6 yields the following result.

Proposition 4.4.8 *Let* Ω *be a Hausdorff topological space. Then the inverse semigroup* **pHomeo**(Ω) *is a Boolean inverse semigroup. Its idempotent elements are the identity functions on the compact open subsets of* Ω. *Furthermore, if* Ω *is Hausdorff zero-dimensional, then* **pHomeo**(Ω) *is fundamental.*

Example 4.4.9 Let B be a Boolean ring. A *partial automorphism of* B is an automorphism $f: B \downarrow a \to B \downarrow b$, where $a, b \in B$. The inverse of a partial automorphism is a partial automorphism. Partial automorphisms $f: B \downarrow a \to B \downarrow b$ and $g: B \downarrow c \to B \downarrow d$ can be composed, by setting

$$gf: B \downarrow f^{-1}(bc) \to B \downarrow g(bc)\,, \quad x \mapsto g(f(x))\,.$$

The composition map defined above endows the collection $\operatorname{Inv}(B)$ of all partial automorphisms of B with a structure of a fundamental Boolean inverse semigroup, called the *Munn semigroup* of the Boolean algebra. Two partial automorphisms $f: B \downarrow a \to B \downarrow b$ and $g: B \downarrow c \to B \downarrow d$ are orthogonal iff $ac = bd = 0$, and then $f \oplus g: B \downarrow (a \oplus b) \to B \downarrow (c \oplus d)$, $x \mapsto f(x \wedge a) \oplus g(x \wedge b)$. The proofs are straightforward abstractions of those in Example 4.4.7 (cf. Proposition 4.4.8) and we omit them.

By the following result, Examples 4.4.7 (for Ω Hausdorff zero-dimensional) and 4.4.9 are essentially the same object.

Proposition 4.4.10 *Let* Ω *be the set of all prime filters of a Boolean ring* B. *There are mutually inverse isomorphisms* $\varepsilon: \operatorname{Inv}(B) \to$ **pHomeo**(Ω) *and* $\eta:$ **pHomeo**$(\Omega) \to \operatorname{Inv}(B)$, *given by the following rules:*

[4]We stray away from the usual definition of a partial homeomorphism, which involves *open* subsets as opposed to the *compact open* used here.

(1) *Let* $f : B \downarrow a \to B \downarrow b$ *in* $\mathrm{Inv}(B)$. *Then* $\varepsilon(f) : \Omega(a) \to \Omega(b)$ *sends any prime filter* $\mathfrak{p} \in \Omega(a)$ *to the filter generated by* $f[\mathfrak{p}]$.
(2) *Let* $g : \Omega(a) \to \Omega(b)$ *in* $\mathbf{pHomeo}(\Omega)$. *Then* $g[\Omega(x)] = \Omega_{\eta(g)(x)}$ *whenever* $x \in B \downarrow a$.

The proof of Proposition 4.4.10, although somewhat tedious, is a routine application of the results of Sect. 1.4, and we leave it to the reader.

As an immediate consequence of Propositions 4.4.8 and 4.4.10, $\mathrm{Inv}(B)$ *is a fundamental Boolean inverse monoid.*

Let us now move to a refinement of the definition of $\mathbf{pMeas}(\mathcal{B})$.

Definition 4.4.11 Let $\mathcal{B}$ be a ring of subsets of a set Ω. An action η of a group G on Ω is $\mathcal{B}$-*measurable* if $g[X] \in \mathcal{B}$ whenever $g \in G$ and $X \in \mathcal{B}$. For such an action, a bijection $f : A \to B$, where $A, B \in \mathcal{B}$, is *piecewise in* G *relatively to* $\mathcal{B}$ if there are $m \in \mathbb{Z}^+$, elements $f_0, \dots, f_{m-1} \in G$, and decompositions of the form

$$A = \bigsqcup_{i<m} A_i \text{ and } B = \bigsqcup_{i<m} B_i\,,$$

with all $A_i, B_i \in \mathcal{B}$, such that

$$f(x) = \eta_{f_i}(x)\,, \quad \text{whenever } i < m \text{ and } x \in A_i\,. \tag{4.4.1}$$

Of course, those conditions imply that f is bi-measurable with respect to $\mathcal{B}$, that is, $f \in \mathbf{pMeas}(\mathcal{B})$. We denote by $\mathbf{pMeas}(\mathcal{B}, \eta)$, or $\mathbf{pMeas}(\mathcal{B}, G)$ if there is no ambiguity on the group action η, the set of all elements of $\mathbf{pMeas}(\mathcal{B})$ that are piecewise in G with respect to $\mathcal{B}$.

In the context of Definition 4.4.11, define a *support* of f as any subset of G containing $\{f_i \mid i < m\}$. If X is a support of f, then $X^{-1} = \{x^{-1} \mid x \in X\}$ is a support of f^{-1}, and if, moreover, Y is a support of g, then $XY = \{xy \mid (x, y) \in X \times Y\}$ is a support of $f \circ g$. Say that an inverse subsemigroup S of an inverse semigroup T is *wide* if S and T have the same idempotents. As a consequence, we obtain easily the following result.

Proposition 4.4.12 *Let* $\mathcal{B}$ *be a ring of subsets of a set* Ω *and let* η *be a* $\mathcal{B}$-*measurable group action on* Ω. *Then* $\mathbf{pMeas}(\mathcal{B}, \eta)$ *is a wide additive inverse subsemigroup of* $\mathbf{pMeas}(\mathcal{B})$.

The following result gives a convenient sufficient condition for $\mathbf{pMeas}(\mathcal{B}, \eta)$ to be an inverse meet-semigroup.

Proposition 4.4.13 *Let* $\mathcal{B}$ *be a ring of subsets of a set* Ω, *and let* η *be a* $\mathcal{B}$-*measurable action of a group* G *on* Ω. *If* η *is fixed point free, then* $\mathbf{pMeas}(\mathcal{B}, \eta)$ *is a Boolean inverse meet-semigroup.*

Proof For any partial bijections $f : A \to B$ and $g : C \to D$ in $\mathbf{pMeas}(\mathcal{B}, \eta)$, there are decompositions $A = \bigsqcup_{i=1}^{m} A_i$, $B = \bigsqcup_{i=1}^{m} B_i$, $C = \bigsqcup_{i=1}^{n} C_i$, and $D = \bigsqcup_{i=1}^{n} D_i$ in $\mathcal{B}$, together with group elements $f_1, \dots, f_m, g_1, \dots, g_n$ such that each $f_i A_i = B_i$, each

$g_jC_j = D_j$, and

$$f(x) = f_ix \text{ whenever } i \in [m] \text{ and } x \in A_i\,,$$
$$g(x) = g_jx \text{ whenever } j \in [n] \text{ and } x \in C_j\,,$$

for all $x \in \Omega$. Since each $f_i^{-1}g_j$ is either the unit element of G or has no fixed point, we get, setting $\Delta = \{(i,j) \in [m] \times [n] \mid f_i = g_j\}$,

$$\|f = g\| = \bigsqcup_{(i,j)\in\Delta} (A_i \cap C_j)\,.$$

In particular, $\|f = g\|$ belongs to $\mathcal{B}$. By Lemma 4.4.3, the meet $f \wedge g$ exists in **pMeas**($\mathcal{B}$). The domain of $f \wedge g$ is $\|f = g\|$, which is the disjoint union of the sets $A_i \cap C_j$ where $(i,j) \in \Delta$. For each such (i,j), the restriction of $f \wedge g$ to $A_i \cap C_j$ is left multiplication by f_i (or, equivalently, g_j). Therefore, $f \wedge g$ belongs to **pMeas**($\mathcal{B}, \eta$). □

Example 4.4.14 Let us be given a continuous action η of a discrete group G on a Hausdorff topological space Ω. Since G is discrete, the continuity of the action only means that η_g is a self-homeomorphism of Ω whenever $g \in G$. Defining $\mathcal{B}$ as the Boolean ring of all compact open subsets of Ω, it follows from Proposition 4.4.12 that the structure **pHomeo**(Ω, η) = **pMeas**($\mathcal{B}, \eta$) is a wide additive Boolean inverse subsemigroup of **pHomeo**(Ω).

Now suppose that Ω is Hausdorff zero-dimensional. Then **pHomeo**(Ω) is fundamental (cf. Proposition 4.4.8). Since **pHomeo**(Ω, η) is wide in **pHomeo**(Ω), it follows that **pHomeo**(Ω, η) *is a fundamental Boolean inverse semigroup.*

Following an earlier convention, we will write **pHomeo**(Ω, G) instead of **pHomeo**(Ω, η) in case there is no ambiguity on the action of the group G.

In parallel to the step from Examples 4.4.7 to 4.4.14, we can construct the following variant of Example 4.4.9.

Example 4.4.15 For an action η of a group G on a Boolean ring B, denote by $\mathrm{Inv}(B,\eta)$, or $\mathrm{Inv}(B,G)$ if there is no ambiguity on η, the set of all partial isomorphisms $f\colon B \downarrow a \to B \downarrow b$ for which there are decompositions $a = \bigoplus_{i<m} a_i$ and $b = \bigoplus_{i<m} b_i$ in B, together with group elements $f_0, \dots, f_{m-1} \in G$, such that $f_ia_i = b_i$ for each $i < m$, and

$$f(x) = \bigoplus_{i<m} f_i(xa_i)\,, \quad \text{for all } x \in B \downarrow a\,. \tag{4.4.2}$$

Of course the formula (4.4.2) is the abstract analogue of (4.4.1). Define a *support* of f as any subset of G containing $\{f_i \mid i < m\}$. If X is a support of f, then $X^{-1} = \{x^{-1} \mid x \in X\}$ is a support of f^{-1}, and if, moreover, Y is a support of g, then $XY = \{xy \mid (x,y) \in X \times Y\}$ is a support of fg. Hence, $\mathrm{Inv}(B,\eta)$ is an inverse subsemigroup of $\mathrm{Inv}(B)$. This inverse subsemigroup is closed under finite orthogonal joins. Therefore, $\mathrm{Inv}(B, \eta)$ *is also a Boolean inverse semigroup*. Since it is wide in $\mathrm{Inv}(B)$, it follows that $\mathrm{Inv}(B, \eta)$ *is a fundamental Boolean inverse semigroup*.

As usual, we will write $\mathrm{Inv}(B, G)$ instead of $\mathrm{Inv}(B, \eta)$ in case there is no ambiguity on the action of G.

The following analogue of Proposition 4.4.10 shows that Examples 4.4.14 (for Ω Hausdorff zero-dimensional) and 4.4.15 describe essentially the same object. The proof is somewhat tedious, but straightforward, and we leave it to the reader.

Proposition 4.4.16 *Let a group G act by automorphisms on a Boolean ring B. The assignment $(g, \mathfrak{p}) \mapsto g[\mathfrak{p}]$ defines a continuous action $\overline{\eta}$ of G on Ω. Furthermore, the mutually inverse isomorphisms $\varepsilon\colon \mathrm{Inv}(B) \to \mathbf{pHomeo}(\Omega)$ and $\eta\colon \mathbf{pHomeo}(\Omega) \to \mathrm{Inv}(B)$ of Proposition* 4.4.10 *define, by restriction, mutually inverse isomorphisms between* $\mathrm{Inv}(B, G)$ *and* $\mathbf{pHomeo}(\Omega, G)$.

The following result shows a universality property of the fundamental Boolean inverse semigroups introduced in Example 4.4.7. Its first two statements are particular cases of a result proved in Zhitomirskiy [126] about arbitrary (not necessarily Boolean) fundamental inverse semigroups. We include an outline of a proof for convenience. We use the notation of Sect. 1.4 for Boolean rings. Recall that by Proposition 3.4.5, $S/\boldsymbol{\mu}$ is a fundamental Boolean inverse semigroup.

Theorem 4.4.17 *Let S be a Boolean inverse semigroup. Set $B = \mathrm{Idp}\, S$ and denote by Ω the set of all prime filters of B. For any $x \in S$, with $a = \mathbf{d}(x)$ and $b = \mathbf{r}(x)$, the assignment*

$$x^{\mu}\colon \mathfrak{p} \mapsto \text{filter of } B \text{ generated by } \left\{xux^{-1} \mid u \in \mathfrak{p}\right\} \tag{4.4.3}$$

defines a partial homeomorphism $x^{\mu}\colon \Omega(a) \to \Omega(b)$. Furthermore, the assignment $x \mapsto x^{\mu}$ defines an additive semigroup homomorphism from S to $\mathbf{pHomeo}(\Omega)$, with kernel the largest idempotent-separating congruence $\boldsymbol{\mu}$ of S.

Outline of Proof An easy application of Lemma 1.4.4 shows that $\mathfrak{p} \in \Omega(a)$ implies that $x^{\mu}(\mathfrak{p}) \in \Omega(b)$, for each $\mathfrak{p} \in \Omega$. Straightforward calculations show that x^{μ} and $(x^{-1})^{\mu}$ are mutually inverse functions between $\Omega(a)$ and $\Omega(b)$.

Further straightforward calculations yield that $x^{\mu}(\mathfrak{p}) \in \Omega(c)$ iff $\mathfrak{p} \in \Omega(x^{-1}cx)$, whenever $c \leq b$ and $\mathfrak{p} \in \Omega(a)$. In particular, x^{μ} is continuous. By applying a similar argument to x^{-1}, we obtain that x^{μ} is a partial homeomorphism from $\Omega(a)$ onto $\Omega(b)$. The verification of the equality $(xy)^{\mu} = x^{\mu}y^{\mu}$, whenever $x, y \in S$, is routine.

Further, we must verify that $(x, y) \in \boldsymbol{\mu}$ iff $x^{\mu} = y^{\mu}$, whenever $x, y \in S$. The implication from the left to the right follows trivially from (3.4.3) together with the definition of x^{μ}. Suppose, conversely, that $x^{\mu} = y^{\mu}$. Set $a = \mathbf{d}(x)$, $b = \mathbf{r}(x)$, $a' = \mathbf{d}(y)$, $b' = \mathbf{r}(y)$. The maps x^{μ} and y^{μ} are identical, thus they have the same domain, that is, $\Omega(a) = \Omega(a')$, so $a = a'$. Likewise, $b = b'$. Let $e \in \mathrm{Idp}\, S$, we must prove that $xex^{-1} = yey^{-1}$. Since $x = xa$ and $y = ya$, we may replace e by ae and thus assume that $e \leq a$. Suppose that $xex^{-1} \neq yey^{-1}$, say $xex^{-1} \nleq yey^{-1}$.

Since xex^{-1} and yey^{-1} are both idempotents below b and S is Boolean, there exists a nonzero $d \leq xex^{-1}$ such that $dyey^{-1} = 0$. The idempotent $c = x^{-1}dx$ is nonzero and beneath e, with $(xcx^{-1})(ycy^{-1}) \leq dyey^{-1} = 0$, so

$$(xcx^{-1})(ycy^{-1}) = 0\,. \tag{4.4.4}$$

Let $\mathfrak{p} \in \Omega(c)$. Then $xcx^{-1} \in x^{\mu}(\mathfrak{p})$ and $ycy^{-1} \in y^{\mu}(\mathfrak{p})$, with $x^{\mu}(\mathfrak{p}) = y^{\mu}(\mathfrak{p})$, in contradiction with (4.4.4). Therefore, $\boldsymbol{\mu}$ is the kernel of the map $x \mapsto x^{\mu}$, so the image $S^{\mu} = \{x^{\mu} \mid x \in S\}$ is isomorphic to $S/\boldsymbol{\mu}$.

By Proposition 3.4.5, it follows that the map $x \mapsto x^{\mu}$ is additive. □

We shall hence from now on identify $S/\boldsymbol{\mu}$ with S^{μ}.

Corollary 4.4.18 *Every Boolean inverse semigroup embeds additively into* **pHomeo**(Ω), *for some compact, Hausdorff, zero-dimensional topological space* Ω.

Proof By Corollary 3.3.2, every Boolean inverse semigroup S has an additive semigroup embedding into $\mathfrak{I}_X$ for some set X. Since $\mathfrak{I}_X$ is a fundamental Boolean inverse semigroup, it follows from Theorem 4.4.17 that $\mathfrak{I}_X$ has an additive semigroup embedding into **pHomeo**(Ω), where $\Omega = \beta X$ is the Čech-Stone compactification of X. □

While Corollary 3.3.2 gives an additive *embedding* of any Boolean inverse semigroup into some $\mathfrak{I}_\Omega$, Theorem 4.4.17 yields only an additive *homomorphism*. On the other hand, the apparently weaker construction offers some advantages:

- The set Ω of all ultrafilters of S, encountered in Corollary 3.3.2 (cf. Remark 3.3.3), can be endowed with several interesting topologies (much more about this can be found in Lawson and Lenz [76]). Among those, there is the topology induced by the product topology on the powerset of S (via the bijection $\operatorname{Pow} S \cong \{0,1\}^S$). Hence this topology is Hausdorff and zero-dimensional. However, unlike what happens for the construction in Theorem 4.4.17, the set $\Omega(x) = \{\mathfrak{p} \in \Omega \mid x \in \mathfrak{p}\}$ may not be compact for all $x \in S$: in fact, it is not hard to prove that $\Omega(x)$ is compact iff $x \wedge y$ exists for all $y \in S$. For example, if S is the Boolean inverse monoid of Example 3.3.5, one can compute that this topology makes the set Ω of all ultrafilters infinite discrete (indexed by $\mathbb{Z}^+ \sqcup G$), with each subset $\Omega([\mathbb{Z}^+, g])$ infinite (indexed by $\mathbb{Z}^+ \sqcup \{g\}$); thus none of them is compact.
  The compactness of all $\Omega(x)$, for the space Ω of Theorem 4.4.17, will be put to use in the proof of Theorem 5.3.8 (about type monoids with respect to supramenable groups).
- We will see shortly (cf. Theorem 4.4.19) that the canonical projection map $S \twoheadrightarrow S/\boldsymbol{\mu}$ is type-preserving (with $S/\boldsymbol{\mu}$ a fundamental Boolean inverse semigroup). By contrast, there is Example 4.9.3, showing that nothing of that sort can be done with inverse meet-semigroups.

Theorem 4.4.17 can be completed by the following observation, also completing Proposition 3.4.5, that S and $S/\boldsymbol{\mu}$ have the same type monoids. Since the map $x \mapsto x^{\mu}$ is additive, there is a well defined homomorphism $\tilde{\mu}\colon \operatorname{Int} S \to \operatorname{Int}(S/\boldsymbol{\mu})$, $\operatorname{typ}_S(x) \mapsto \operatorname{typ}_{S/\boldsymbol{\mu}}(x^{\mu})$ of partial monoids.

Theorem 4.4.19 *Let S be a Boolean inverse semigroup, with largest idempotent-separating congruence $\boldsymbol{\mu}$. Then the canonical homomorphism $\tilde{\mu}\colon \operatorname{Int} S \to \operatorname{Int}(S/\boldsymbol{\mu})$ is an isomorphism of partial monoids. In particular, we get* $\operatorname{Int} S \cong \operatorname{Int}(S/\boldsymbol{\mu})$ *and* $\operatorname{Typ} S \cong \operatorname{Typ}(S/\boldsymbol{\mu})$.

Proof Observe that $\operatorname{Idp}(S/\boldsymbol{\mu}) = \{x^{\mu} \mid x \in \operatorname{Idp} S\}$. We claim that $a \mathscr{D} b$ (within S) iff $a^{\mu} \mathscr{D} b^{\mu}$ (within $S/\boldsymbol{\mu}$), for all $a, b \in \operatorname{Idp} S$. We prove the non-trivial direction only. If $a^{\mu} \mathscr{D} b^{\mu}$, then there exists $x \in S$ such that $a \equiv_{\mu} \mathbf{d}(x)$ and $b \equiv_{\mu} \mathbf{r}(x)$. Since $\boldsymbol{\mu}$ is idempotent-separating, it follows that $a = \mathbf{d}(x)$ and $b = \mathbf{r}(x)$, so $a \mathscr{D} b$. It follows that $\tilde{\mu}$ is a bijection from $\operatorname{Int} S$ onto $\operatorname{Int}(S/\boldsymbol{\mu})$. Since $x \mapsto x^{\mu}$ preserves finite orthogonal joins and $\boldsymbol{\mu}$ is idempotent-separating, it follows that $\tilde{\mu}$ is an isomorphism of partial monoids. □

An alternate proof of Theorem 4.4.19 follows from a direct application of Theorem 4.3.2 to the canonical projection $\mu\colon S \twoheadrightarrow S/\boldsymbol{\mu}$. Indeed, since $\boldsymbol{\mu}$ separates idempotents, $0/\boldsymbol{\mu} = \{0\}$, so Theorem 4.4.19 yields that $\operatorname{Typ} \mu$ is an isomorphism. However, the proof of Theorem 4.4.19 given above is direct.

Theorem 4.4.19 implies that the canonical projection $S \twoheadrightarrow S/\boldsymbol{\mu}$ is type-preserving (cf. Definition 4.2.1). The given result is in fact slightly stronger, as it states more than an isomorphism $\operatorname{Typ} S \cong \operatorname{Typ}(S/\boldsymbol{\mu})$ of monoids, namely an isomorphism $\operatorname{Int} S \cong \operatorname{Int}(S/\boldsymbol{\mu})$ of partial monoids.

As the following easy result shows, the type theory of fundamental Boolean inverse semigroups contains the theory of equidecomposability types monoids $\mathbb{Z}^+\langle B\rangle /\!\!/ G$. It will turn out later (cf. Proposition 4.8.5) that the two theories are actually the same.

Proposition 4.4.20 *Let a group G act by automorphisms on a Boolean ring B. Then* $\operatorname{Typ}(\operatorname{Inv}(B, G)) \cong \mathbb{Z}^+\langle B\rangle /\!\!/ G$, *via an isomorphism that sends* $\operatorname{typ}(\operatorname{id}_{B\downarrow a})$ *to* $[a]_G$ *for every $a \in B$.*

Proof The idempotent elements of the fundamental Boolean inverse semigroup $S = \operatorname{Inv}(B, G)$ are exactly the partial automorphisms $\operatorname{id}_{B\downarrow a}$, where $a \in B$. Furthermore, for all $a, b \in B$,

$$
\begin{aligned}
\operatorname{id}_{B\downarrow a} \mathscr{D} \operatorname{id}_{B\downarrow b} &\Leftrightarrow (\exists f \in S)(\mathbf{d}(f) = \operatorname{id}_{B\downarrow a} \text{ and } \mathbf{r}(f) = \operatorname{id}_{B\downarrow b}) \\
&\Leftrightarrow \Big(\exists \text{ decompositions } a = \bigoplus_{i<m} a_i \text{ and } b = \bigoplus_{i<m} f_i a_i \\
&\qquad\qquad \text{with each } a_i \in B \text{ and } f_i \in G\Big) \\
&\qquad\qquad \text{(cf. Example 4.4.15)} \\
&\Leftrightarrow a \sim_G^{+} b \qquad \text{(cf. Example 2.8.9)}\,.
\end{aligned}
$$

Therefore, $\operatorname{Typ} S = \operatorname{U_{mon}}\big((\operatorname{Idp} S)/\mathscr{D}\big) \cong \operatorname{U_{mon}}(B/{\sim_G^{+}}) \cong \mathbb{Z}^+\langle B\rangle /\!\!/ G$. □

It is sometimes interesting to deal with more general kinds of Boolean inverse semigroups than the fundamental ones. Hence we state the following result.

Proposition 4.4.21 *Let $\mathcal{B}$ be a ring of subsets of a set Ω, and let η be a $\mathcal{B}$-measurable action of a group G on Ω. Then* $\operatorname{Typ}(\mathbf{pMeas}(\mathcal{B}, \eta)) \cong \mathbb{Z}^+\langle\mathcal{B}\rangle /\!\!/ G$, *via an isomorphism that sends* $\operatorname{typ}(\mathrm{id}_A)$ *to* $[A]_G$ *for every* $A \in \mathcal{B}$.

The proof of Proposition 4.4.21 is similar to the one of Proposition 4.4.20, and we omit it.

4.5 V-Relations and $\mathcal{L}_{\infty,\omega}$-Equivalence on Boolean Algebras

Every Boolean algebra is a conical partial refinement monoid under disjoint join (cf. Example 2.2.7), which makes it possible to apply some of the results of Sects. 2.2 and 2.4. Since Boolean algebras are very special kinds of partial refinement monoids, it is not surprising that much more can be said.

In various references, the concepts handled (and, mostly, surveyed) in the present section are often credited to Vaught's thesis [114], which is not available to me. Further relevant references include Hanf [58] and Pierce [94].

Lemma 4.5.1 *Let A and B be Boolean algebras and let A_0 be a subalgebra of A, with A countable and A_0 finite. Let $\Gamma \subseteq A \times B$ be an additive, right conical, left V-relation. Then every embedding $f_0 \colon A_0 \hookrightarrow B$ of Boolean algebras, with graph contained in Γ, extends to some embedding $f \colon A \hookrightarrow B$, with graph contained in Γ.*

Proof Suppose first that A is generated by $A_0 \cup \{a\}$, for some a. For every atom u of A_0, it follows from the equation $u = (u \wedge a) \oplus (u \wedge \neg a)$, the relation $u \,\Gamma\, f_0(u)$, and the assumption that Γ is a right conical left V-relation that there are $f'(u), f''(u) \in B$ such that $f_0(u) = f'(u) \oplus f''(u)$, $(u \wedge a)\,\Gamma\, f'(u)$, $(u \wedge \neg a)\,\Gamma\, f''(u)$, and the following conditions hold:

$$u \wedge a = 0_A \Leftrightarrow f'(u) = 0_B\,, \tag{4.5.1}$$

$$u \wedge \neg a = 0_A \Leftrightarrow f''(u) = 0_B\,. \tag{4.5.2}$$

Since 1_B is the disjoint join, in B, of all elements $f_0(u)$, it is also the disjoint join of all $f'(u)$ and $f''(u)$, so it follows from (4.5.1) and (4.5.2) that there exists a unique embedding $f \colon A \to B$ of Boolean algebras such that $f(u \wedge a) = f'(u)$ and $f(u \wedge \neg a) = f''(u)$ for every atom u of A_0. Then $f(u) = f(u \wedge a) \oplus f(u \wedge \neg a) = f'(u) \oplus f''(u) = f_0(u)$, for every atom u of A_0; whence f extends f_0. Furthermore, since Γ is additive, the graph of f is contained in Γ.

In the general case, we can write $A = \bigcup_{n<\omega} A_n$, for finite subalgebras A_n such that each A_{n+1} is generated by $A_n \cup \{a_n\}$ for some a_n. By the paragraph above, every embedding $f_n \colon A_n \to B$ with graph contained in Γ extends to an embedding $f_{n+1} \colon A_{n+1} \to B$ with graph contained in Γ. The union of all f_n is an embedding from A into B with graph contained in Γ. □

Lemma 4.5.2 (Vaught's Theorem) *Let A and B be countable Boolean algebras, let A_0 be a finite subalgebra of A, let B_0 be a finite subalgebra of B, and let $\Gamma \subseteq A \times B$*

be an additive conical V-relation. Then every isomorphism $f_0: A_0 \to B_0$ of Boolean algebras, with graph contained in Γ, extends to an isomorphism $f: A \to B$ with graph contained in Γ.

Note If A is trivial (i.e., $0_A = 1_A$), then, since $0_A = 1_A \,\Gamma\, f_0(1_A) = 1_B$ and Γ is conical, we get $0_B = 1_B$. Hence, *A is trivial iff B is trivial.* It follows that if $1_A \,\Gamma\, 1_B$, then the subalgebras $A_0 = \{0_A, 1_A\}$ and $B_0 = \{0_B, 1_B\}$ are as above; in particular, Γ contains the graph of an isomorphism from A onto B.

Proof Let $A = \{a_n \mid n < \omega\}$ and $B = \{b_n \mid n < \omega\}$. Suppose having defined finite subalgebras A_n of A and B_n of B, with an isomorphism $f_n: A_n \to B_n$ with graph contained in Γ. If $n = 2m$ for some m, define A_{n+1} as the subalgebra of A generated by $A_n \cup \{a_m\}$. It follows from Lemma 4.5.1 that f_n extends to an embedding $f_{n+1}: A_{n+1} \hookrightarrow B$ with graph contained in Γ. Hence f_{n+1} is an isomorphism from A_{n+1} onto $B_{n+1} = f_{n+1}[A_{n+1}]$. Similarly, if $n = 2m + 1$ for some m, define B_{n+1} as the subalgebra of B generated by $B_n \cup \{b_m\}$, then extend f_n to an isomorphism $f_{n+1}: A_{n+1} \to B_{n+1}$ for some finite subalgebra A_{n+1} of A. The union of all f_n is an isomorphism from A onto B, with graph contained in Γ. □

The following concepts can be found in Barwise [19, Chap. VII], where they are stated for arbitrary models of first-order theories (not only Boolean algebras).

Definition 4.5.3 Let A and B be Boolean algebras. A *partial isomorphism* from A to B is an isomorphism from a subalgebra of A onto a subalgebra of B. A *back-and-forth system* between A and B is a nonempty set $\boldsymbol{I}$ of partial isomorphisms from A to B such that for every $x \in A$ (resp., $y \in B$), every $f \in \boldsymbol{I}$ extends to some $g \in \boldsymbol{I}$ such that x belongs to the domain of f (resp., y belongs to the range of f); then we write $\boldsymbol{I}: A \cong_{\mathrm{p}} B$. Let $A \cong_{\mathrm{p}} B$ hold if there exists $\boldsymbol{I}$ such that $\boldsymbol{I}: A \cong_{\mathrm{p}} B$.

The infinitary language $\mathcal{L}_{\infty,\omega}$ is built on a first-order language $\mathcal{L}$ by allowing conjunctions and disjunctions of arbitrary (transfinite) length (all strings of quantifiers $\forall$ and $\exists$ are finite). For first-order structures A and B, say that $A \equiv_{\infty,\omega} B$ if A and B satisfy the same $\mathcal{L}_{\infty,\omega}$ sentences.

For Boolean algebras, we obtain an alternate description of the relation $\equiv_{\infty,\omega}$.

Proposition 4.5.4 *The following statements are equivalent, for any Boolean algebras A and B:*

(i) $A \cong_{\mathrm{p}} B$.
(ii) $A \equiv_{\infty,\omega} B$.
(iii) *There exists a conical V-relation Γ on $A \times B$ such that $1_A \,\Gamma\, 1_B$.*
(iv) *There exists an additive conical V-relation Γ on $A \times B$ such that $1_A \,\Gamma\, 1_B$.*

Proof The equivalence between (i) and (ii) is established in Karp [67], see also Barwise [19, Theorem VII.5.3].

(i)⇒(iii) Let $\boldsymbol{I}: A \cong_{\mathrm{p}} B$ and define Γ as the union of (the graphs of) all members of $\boldsymbol{I}$. Trivially, $1_A \,\Gamma\, 1_B$. Let $y \in B$ such that $0_A \,\Gamma\, y$. This means that there exists $f \in \boldsymbol{I}$ such that $f(0_A) = y$, that is, $y = 0_B$. Now let $x = x_0 \oplus x_1$ in A and let $y \in B$ such that $x \,\Gamma\, y$. There exists $f \in \boldsymbol{I}$ such that $f(x) = y$. Since $\boldsymbol{I}$ is a back-and-forth system, f extends to some $g \in \boldsymbol{I}$ with x_0 and x_1 both in the domain of g. Set $y_i = g(x_i)$, for each $i < 2$. Then $y = y_0 \oplus y_1$ and $x_i \,\Gamma\, y_i$ for each $i < 2$. By interchanging the roles of A and B, it follows that Γ is a V-relation on $A \times B$.

(iii)⇒(iv) follows trivially from Lemma 2.4.3.

(iv)⇒(i) Let Γ be an additive conical V-relation on $A \times B$ such that $1_A \,\Gamma\, 1_B$. Define $\boldsymbol{I}$ as the set of all finite partial isomorphisms from A to B with graph contained in Γ. It follows immediately from the finite case of Lemma 4.5.2 that $\boldsymbol{I}$ is a back-and-forth system between A and B. □

4.6 Vaught Measures

Vaught measures, or *V-measures*, are a large source of interesting V-equivalences on a given Boolean ring. They originate in Vaught's thesis [114] and have been much studied by Dobbertin; see, in particular, Dobbertin [33, § 3]. In particular, we will recall, in this section, a bunch of sufficient conditions, due to Dobbertin, for a conical refinement monoid to be the range of a V-measure (Theorem 4.6.7). This result is a precursor of various representability results as type monoids of Boolean inverse semigroups.

Definition 4.6.1 Let B be a Boolean ring and let M be a commutative monoid. A map $\mu: B \to M$ is a *premeasure* if it is a homomorphism of partial monoids, that is, $\mu(0) = 0$ and $\mu(x \oplus y) = \mu(x) + \mu(y)$ for all disjoint $x, y \in B$. If, in addition, μ is conical (i.e., $\mu^{-1}\{0\} = \{0\}$), we say that μ is a *measure*. If μ is a premeasure satisfying the following *V-condition*:

Whenever $\boldsymbol{a}, \boldsymbol{b} \in M$ and $c \in B$ with $\mu(c) = \boldsymbol{a} + \boldsymbol{b}$, there is a decomposition

$$c = a \oplus b \text{ in } B \text{ such that } \mu(a) = \boldsymbol{a} \text{ and } \mu(b) = \boldsymbol{b}\,, \tag{4.6.1}$$

then we say that μ is a *V-premeasure*. If, in addition, μ is a measure, we say that it is a *V-measure*.

A pointed commutative monoid $(M, \boldsymbol{c})$ (cf. Sect. 1.5) is *V-measurable* if there are a Boolean ring A and a V-measure $\mu: A \to M$ such that $\boldsymbol{c}$ belongs to the range of μ. We also say that $\boldsymbol{c}$ is *V-measurable in M*.

We say that M is *V-measurable* if every element of M is V-measurable in M.

The following are easy examples of V-measures.

Example 4.6.2 Let B be the powerset algebra of a set Ω and let M be any lower subset of the cardinals containing $\{n \cdot \operatorname{card}\Omega \mid n \in \mathbb{Z}^+\}$, endowed with cardinal addition. Set $\mu(x) = \operatorname{card} x$, for any $x \in B$. Then μ is an M-valued V-measure on B.

Example 4.6.3 Let a group G act on a Boolean ring B. Then the commutative monoid $M = \mathbb{Z}^+\langle B\rangle$ is the positive cone of a lattice-ordered group (cf. Example 2.2.7). By Lemma 2.8.8, the map $\mu_G: x \mapsto [x]_G$ defines an M-valued V-measure on B.

Example 4.6.4 Let S be a Boolean inverse semigroup. By Lemma 4.1.6, the assignment $x \mapsto \mathrm{typ}_S(x)$ defines a V-measure on the Boolean ring $\mathrm{Idp}\,S$. We will thus call this map the *canonical V-measure* on S.

For a family $(B_i \mid i \in I)$ of Boolean rings, the direct sum $\bigoplus (B_i \mid i \in I)$ is also a Boolean ring. Its elements are the families $(x_i \mid i \in I)$ where all $x_i \in B_i$ and the support $\{i \in I \mid x_i \neq 0\}$ is finite.

An easy application of refinement yields immediately the following property.

Lemma 4.6.5 *Let M be a refinement monoid and let $(\mu_i : B_i \to M \mid i \in I)$ be a family of V-measures. Then the assignment*

$$\bigoplus (B_i \mid i \in I) \to M\,, \quad (x_i \mid i \in I) \mapsto \sum_{i \in I} \mu_i(x_i)$$

defines a V-measure on $\bigoplus (B_i \mid i \in I)$.

The following observation gathers a few basic consequences of V-measurability.

Proposition 4.6.6 *The following statements hold, for every commutative monoid M and $P = \{\boldsymbol{c} \in M \mid (M, \boldsymbol{c}) \text{ is V-measurable}\}$:*

(1) *For every Boolean ring B, the range of a V-measure $\mu : B \to M$ is a lower subset of M (with respect to the algebraic preordering $\leq^+$ of M).*
(2) *The set P is a lower subset of M.*
(3) *An element $\boldsymbol{c} \in M$ is V-measurable in M iff there are a unital Boolean ring B and a V-measure $\mu : B \to M$ with $\mu(1) = \boldsymbol{c}$ (we will say that μ is a* normalized V-measure *from B to $(M, \boldsymbol{c})$).*
(4) *If P is nonempty, then M is conical, and the lower interval $M \downarrow \boldsymbol{c}$ satisfies refinement whenever $\boldsymbol{c} \in P$.*
(5) *If M is a conical refinement monoid, then P is an o-ideal of M.*
(6) *Suppose that M is a refinement monoid and let $\boldsymbol{e}$ be an order-unit of M. Then M is V-measurable iff $(M, \boldsymbol{e})$ is V-measurable.*
(7) *M is V-measurable iff there are a Boolean ring B and a V-measure with range M.*

Proof

(1) Let $\boldsymbol{a} \leq^+ \boldsymbol{b}$ in M. There exists $\boldsymbol{c} \in M$ such that $\boldsymbol{b} = \boldsymbol{a} + \boldsymbol{c}$. If $\boldsymbol{b} = \mu(b)$ with $b \in B$, then, by the V-condition, there is a decomposition $b = a \oplus c$ with $\mu(a) = \boldsymbol{a}$ and $\mu(c) = \boldsymbol{c}$. Hence $\boldsymbol{a}$ belongs to the range of μ.
(2) follows trivially from (1).
(3) Let B be a Boolean ring with $c \in B$ and a V-measure $\mu : B \to M$ such that $\mu(c) = \boldsymbol{c}$. Then $A = B \downarrow c$ is a unital Boolean ring, and the restriction of μ to A is a normalized V-measure from A to $(M, \boldsymbol{c})$.
(4) Let $\boldsymbol{c} \in P$. There are a Boolean ring B with $c \in B$ and a V-measure $\mu : B \to M$ such that $\mu(c) = \boldsymbol{c}$. For any $\boldsymbol{x}, \boldsymbol{y} \in M$, if $\boldsymbol{x} + \boldsymbol{y} = 0$, then $\boldsymbol{x} + \boldsymbol{y} + \boldsymbol{c} = \boldsymbol{c}$, thus, by the V-condition, there is a decomposition $c = x \oplus y \oplus z$ such that $\mu(x) = \boldsymbol{x}$, $\mu(y) = \boldsymbol{y}$, and $\mu(z) = \boldsymbol{c}$. Since $\mu(x \oplus y) = \mu(x) + \mu(y) = \boldsymbol{x} + \boldsymbol{y} = 0$ and μ

is a measure, we get $x \oplus y = 0$, thus $x = y = 0$, so $\boldsymbol{x} = \boldsymbol{y} = 0$, thus proving that M is conical.
Now let $\boldsymbol{z} = \boldsymbol{x}_0 + \boldsymbol{x}_1 = \boldsymbol{y}_0 + \boldsymbol{y}_1$ in $M \downarrow \boldsymbol{c}$. By (1) above, there is $z \leq c$ such that $\mu(z) = \boldsymbol{z}$. Since μ is a V-measure, there are decompositions $z = x_0 \oplus x_1 = y_0 \oplus y_1$ such that $\mu(x_i) = \boldsymbol{x}_i$ and $\mu(y_i) = \boldsymbol{y}_i$ for each $i \in \{0, 1\}$. The following is a refinement matrix in M:

	$\boldsymbol{y}_0$	$\boldsymbol{y}_1$
$\boldsymbol{x}_0$	$\mu(x_0 \wedge y_0)$	$\mu(x_0 \wedge y_1)$
$\boldsymbol{x}_1$	$\mu(x_1 \wedge y_0)$	$\mu(x_1 \wedge y_1)$

Thus $M \downarrow \boldsymbol{c}$ has refinement.

(5) Since M is conical, $0 \in P$. Let $\boldsymbol{a}, \boldsymbol{b} \in P$. By (3) above, there are unital Boolean rings A and B together with V-measures $\alpha: A \to M$ and $\beta: B \to M$ such that $\alpha(1_A) = \boldsymbol{a}$ and $\beta(1_B) = \boldsymbol{b}$. Since M is a refinement monoid, the assignment $A \times B \to M$, $(x, y) \mapsto \alpha(x) + \beta(y)$ is a V-measure (cf. Lemma 4.6.5). It sends $1_{A \times B} = (1_A, 1_B)$ to $\boldsymbol{a} + \boldsymbol{b}$. Hence, $\boldsymbol{a} + \boldsymbol{b} \in P$.

(6) By definition, M is V-measurable iff $P = M$, while $\boldsymbol{e}$ is V-measurable in M iff $\boldsymbol{e} \in P$. However, by (5) above and since $\boldsymbol{e}$ is an order-unit, $\boldsymbol{e} \in P$ iff $P = M$.

(7) Suppose first that M is V-measurable. There is a family $(\mu_{\boldsymbol{x}} \mid \boldsymbol{x} \in M)$ of V-measures, each $\mu_{\boldsymbol{x}}: B_{\boldsymbol{x}} \to M$ with $\mu_{\boldsymbol{x}}(1_{B_{\boldsymbol{x}}}) = \boldsymbol{x}$. Since M is a refinement monoid, the assignment

$$\mu: \bigoplus (B_{\boldsymbol{x}} \mid \boldsymbol{x} \in M) \to M\,, \quad (b_{\boldsymbol{x}} \mid \boldsymbol{x} \in M) \mapsto \sum_{\boldsymbol{x} \in M} \mu_{\boldsymbol{x}}(b_{\boldsymbol{x}}) \tag{4.6.2}$$

defines a V-measure (cf. Lemma 4.6.5), which is obviously surjective. Conversely, if M is the range of a V-measure, then, by (1) above, every element of M is V-measurable in M. □

The following result sums up many known representation results by V-measures. Recall that regular commutative monoids are introduced in Definition 1.5.1.

Theorem 4.6.7 (Dobbertin) *Every monoid in each of the following classes is V-measurable:*

(1) *All conical refinement monoids M such that* $\operatorname{card}(M \downarrow \boldsymbol{e}) \leq \aleph_1$ *for each $\boldsymbol{e} \in M$ (cf. [33, Theorem 3.4]).*
(2) *All distributive lattices with zero under join (cf. [34, Corollary 1.3]).*
(3) *All conical simple regular commutative monoids (cf. [34, Corollary 1.5]).*
(4) *All positive cones of Abelian lattice-ordered groups (cf. [35, Theorem 13]).*
(5) *All* $(\vee, 0)$*-semilattices of all finitely generated lower subsets of arbitrary posets (cf. [35, Theorem 14]).*

An important additional uniqueness statement was obtained by Dobbertin for the case of *countable* conical refinement monoids. The essence of the following result is contained in Dobbertin [32, Lemma 5.1].

Theorem 4.6.8 *The following statements hold, for any conical refinement monoids $(M, \boldsymbol{a})$ and $(N, \boldsymbol{b})$ with order-unit, any Boolean algebras A and B, any normalized measure $\alpha: A \to (M, \boldsymbol{a})$, any normalized V-measure $\beta: B \to (N, \boldsymbol{b})$, and any homomorphism $\boldsymbol{f}: (M, \boldsymbol{a}) \to (N, \boldsymbol{b})$ of pointed monoids:*

(1) *If A is countable, then there is an embedding $f: A \hookrightarrow B$ of Boolean algebras such that $\boldsymbol{f} \circ \alpha = \beta \circ f$.*
(2) *If A and B are both countable, α is a V-measure, and $\boldsymbol{f}$ is an isomorphism, then there is an isomorphism $f: A \to B$ of Boolean algebras such that $\boldsymbol{f} \circ \alpha = \beta \circ f$.*

Proof Since α and β are both measures, the following binary relation

$$\Gamma = \{(x, y) \in A \times B \mid \alpha(x) = \beta(y)\}$$

is both additive and conical on $A \times B$. Furthermore, since β is a V-measure, Γ is a left V-relation.

(1) By Lemma 4.5.1, Γ contains the graph of an embedding $f: A \hookrightarrow B$ of Boolean algebras.

(2) Since α and β are both V-measures, Γ is a V-relation. By Lemma 4.5.2, Γ contains the graph of an isomorphism $f: A \to B$ of Boolean algebras. □

The problem whether every conical refinement monoid is measurable was stated in Dobbertin [33, Problem 4]. A negative answer was established in Wehrung [124] (Theorem 2.8 of that paper for positive cones of dimension groups, Theorem 2.15 for distributive semilattices with zero).

Theorem 4.6.9 *For every cardinal $\kappa \geq \aleph_2$, there are a dimension group G with order-unit and a bounded distributive semilattice S, both of cardinality κ, such that neither G^+ nor S is V-measurable.*

While the dimension group G constructed in Wehrung [124] is a vector space over the rationals, it is also possible to construct G and its order-unit $\boldsymbol{u}$ in such a way that $\boldsymbol{u}$ has index 2 in G^+ (cf. Wehrung [125]). For that example, G^+ is neither isomorphic to the dimension monoid of any lattice (as defined in Wehrung [122]) nor to $\mathrm{V}(R)$ for any regular ring R.

In view of the uniqueness result given by Theorem 4.6.8(2), this looks somehow counter-intuitive. The point is that the isomorphism $\boldsymbol{f}$ given by Theorem 4.6.8(2) may not be unique: in short, *uniqueness up to isomorphism does not imply uniqueness up to unique isomorphism.*

4.7 Measures and Inverse Semigroups

We shall consider those V-measures that can be associated in a natural way to Boolean inverse semigroups (resp., to actions of groups on Boolean rings). These measures will be called *groupoid-induced* (resp., *group-induced*).

4.7.1 Groupoid-Induced and Group-Induced V-Measures

We recall that the fundamental Boolean inverse semigroup $\mathrm{Inv}(B)$ is introduced in Example 4.4.9.

Definition 4.7.1 Let M be a commutative monoid. An *M-valued premeasure space* is a pair (B, μ), where B is a Boolean ring and $\mu: B \to M$ is a premeasure. If μ is a measure then we will say that (B, μ) is a *measure space*. For M-valued premeasure spaces (A, α) and (B, β), a partial function f from a subset of A to B is *measure-preserving* if $\beta(f(x)) = \alpha(x)$ for every x in the domain of f.

Notation 4.7.2 Let B be a Boolean ring, let M be a commutative monoid, and let $\mu: B \to M$ be a premeasure. We introduce the following sets:

(1) The set $\mathrm{Inv}(B, \mu)$ of all measure-preserving $f \in \mathrm{Inv}(B)$.
(2) The set $\mathrm{Aut}(B, \mu)$ of all measure-preserving automorphisms f of the Boolean ring B.

Proposition 4.7.3 *The following statements hold, for any Boolean ring B, any commutative monoid M, and any premeasure $\mu: B \to M$:*

(1) $\mathrm{Aut}(B, \mu)$ *is a subgroup of the automorphism group* $\mathrm{Aut}(B)$ *of B.*
(2) $\mathrm{Inv}(B, \mu)$ *is an inverse subsemigroup of* $\mathrm{Inv}(B)$*, closed under finite orthogonal joins. It is also a fundamental Boolean inverse semigroup.*

Proof The statement (1) is trivial. It is also straightforward to verify that $\mathrm{Inv}(B, \mu)$ is an inverse subsemigroup of $\mathrm{Inv}(B)$, closed under finite orthogonal joins. Since the two semigroups have the same idempotents and $\mathrm{Inv}(B)$ is a fundamental Boolean inverse semigroup, it follows that $\mathrm{Inv}(B, \mu)$ is also a fundamental Boolean inverse semigroup. □

We are interested in the case where a V-measure $\mu: B \to M$ determines an isomorphism between M and $\mathrm{Typ}(\mathrm{Inv}(B, \mu))$. Accordingly, we introduce the following binary relations $\sim_\mu^{\mathrm{gpd}}$, $\sim_\mu^{\mathrm{gp}}$, and $\simeq_\mu^{\mathrm{gp}}$ on B.

Notation 4.7.4 Let B be a Boolean ring, let M be a commutative monoid, and let $\mu: B \to M$ be a premeasure. For $a, b \in B$, we define

- $a \sim_\mu^{\mathrm{gpd}} b$ if there exists $f \in \mathrm{Inv}(B, \mu)$ with $b = f(a)$.
- $a \sim_\mu^{\mathrm{gp}} b$ if there exists $f \in \mathrm{Aut}(B, \mu)$ with $b = f(a)$.
- $\simeq_\mu^{\mathrm{gp}}$ is the additive closure of $\sim_\mu^{\mathrm{gp}}$: that is, $a \simeq_\mu^{\mathrm{gp}} b$ if there are decompositions $a = \bigoplus_{i<n} a_i$, $b = \bigoplus_{i<n} b_i$, with all $b_i = f_i(a_i)$ where $f_i \in \mathrm{Aut}(B, \mu)$.

Observe, in particular, the obvious implications

$$a \sim_{\mu}^{\mathrm{gp}} b \Longrightarrow a \simeq_{\mu}^{\mathrm{gp}} b \Longrightarrow a \sim_{\mu}^{\mathrm{gpd}} b \Longrightarrow B \downarrow a \cong B \downarrow b\,,$$

for any $a, b \in B$.

The reason why we do not need to introduce the additive closure of $\sim_{\mu}^{\mathrm{gpd}}$ is contained in (3) of the following basic result.

Proposition 4.7.5 *The following statements hold, for any Boolean ring B, any commutative monoid M, and any premeasure $\mu\colon B \to M$:*

(1) *$\sim_{\mu}^{\mathrm{gp}}$ is contained in $\sim_{\mu}^{\mathrm{gpd}}$, which is contained in* $\operatorname{Ker} \mu$.
(2) *The binary relations $\sim_{\mu}^{\mathrm{gpd}}$ and $\sim_{\mu}^{\mathrm{gp}}$ are both refining equivalence relations.*
(3) *The binary relation $\sim_{\mu}^{\mathrm{gpd}}$ is additive.*

Proof The proofs of (1) and (2) are straightforward. For (3), let $a = a_0 \oplus a_1$ and $b = b_0 \oplus b_1$ in B, with $a_i \sim_{\mu}^{\mathrm{gpd}} b_i$ for each $i \in \{0, 1\}$. Pick $f_i \in \operatorname{Inv}(B, \mu)$ such that $f_i(a_i) = b_i$. By restricting f_i to $B \downarrow a_i$ we may assume that $f_i\colon B \downarrow a_i \to B \downarrow b_i$. The function $f = f_0 \oplus f_1$ (cf. Example 4.4.9) belongs to $\operatorname{Inv}(B, \mu)$, and $f(a) = b$. □

In certain cases it is possible to reduce $\sim_{\mu}^{\mathrm{gpd}}$ to $\sim_{\mu}^{\mathrm{gp}}$:

Proposition 4.7.6 *The following statements hold, for any Boolean ring B, any commutative monoid M, any $a, b, c \in B$, and any premeasure $\mu\colon B \to M$:*

(1) *Suppose that $a, b \leq c$. If $a \sim_{\mu}^{\mathrm{gpd}} b$ and $c \smallsetminus a \sim_{\mu}^{\mathrm{gpd}} c \smallsetminus b$, then $a \sim_{\mu}^{\mathrm{gp}} b$.*
(2) *Let B be unital. Then $a \sim_{\mu}^{\mathrm{gp}} b$ iff $a \sim_{\mu}^{\mathrm{gpd}} b$ and $\neg a \sim_{\mu}^{\mathrm{gpd}} \neg b$.*
(3) *If $a \wedge b = 0$, then $a \sim_{\mu}^{\mathrm{gp}} b$ iff $a \sim_{\mu}^{\mathrm{gpd}} b$.*
(4) *If $(a \vee b) \wedge c = 0$ and $a \sim_{\mu}^{\mathrm{gpd}} b \sim_{\mu}^{\mathrm{gpd}} c$, then $a \sim_{\mu}^{\mathrm{gp}} b$.*

Proof

(1) There are $f, g \in \operatorname{Inv}(B, \mu)$ such that $f(a) = b$ and $g(c \smallsetminus a) = c \smallsetminus b$. Trimming down f and g, we may assume that f and g have domain $B \downarrow a$ and $B \downarrow (c \smallsetminus a)$, respectively. Then $h = f \oplus g$ belongs to $\operatorname{Inv}(B, \mu)$, $h(a) = b$, and $h(c) = c$. Extend h to an automorphism $\overline{h}$ of B the standard way, that is, $\overline{h}(x \oplus y) = h(x) \oplus y$ whenever $x \leq c$ and $y \wedge c = 0$. Then $\overline{h} \in \operatorname{Aut}(B, \mu)$ and $\overline{h}(a) = b$.
(2) follows immediately from (1), by letting c be the unit of B.
(3) The non-trivial implication follows immediately from (1), by taking $c = a \oplus b$.
(4) Since $a \sim_{\mu}^{\mathrm{gpd}} c$ and $a \wedge c = 0$, it follows from (3) above that $a \sim_{\mu}^{\mathrm{gp}} c$. Likewise, $b \sim_{\mu}^{\mathrm{gp}} c$. The desired conclusion follows from the transitivity of $\sim_{\mu}^{\mathrm{gp}}$. □

Definition 4.7.7 Let B be a Boolean ring and let M be a conical commutative monoid. We say that a V-measure $\mu\colon B \to M$ is *groupoid-induced*[5] (resp., *group-induced*) if the kernel of μ is $\sim_{\mu}^{\mathrm{gpd}}$ (resp., $\simeq_{\mu}^{\mathrm{gp}}$). In particular, μ is groupoid-induced iff for all $a, b \in B$, $\mu(a) = \mu(b)$ iff there exists $f \in \operatorname{Inv}(B, \mu)$ such that $f(a) = b$.

---

[5] By reference to the trace product groupoid of the inverse semigroup $\operatorname{Inv}(B, \mu)$ (cf. Sect. 4.1).

Example 4.7.8 A non groupoid-induced V-measure on a Boolean algebra of cardinality $\aleph_1$.

Proof Fix an atomless Boolean algebra B_k of cardinality $\aleph_k$, whenever $k \in \{0, 1\}$, set $B = B_0 \times B_1$, and denote by $\mathbf{2} = \{0, \infty\}$ the two-element semilattice. Then the unique zero-separating map $\mu: B \to \mathbf{2}$ is a V-measure. In particular, $\mu(1, 0) = \mu(0, 1) = \infty$. However, the Boolean algebras $B \downarrow (1, 0) \cong B_0$ and $B \downarrow (0, 1) \cong B_1$ are not isomorphic; whence μ is not groupoid-induced. □

For an example involving a complete atomic Boolean algebra, see Example 5.4.3.

If a measure μ is group-induced, then it is groupoid-induced. As the following example shows, the converse does not hold as a rule.

Example 4.7.9 Denote by P the commutative monoid defined by generators ε and 1, subjected to the relation $\varepsilon + 1 = 1$. Hence $P = \mathbb{Z}^+\varepsilon \sqcup \mathbb{N}$, and P is a conical refinement monoid (it is a so-called *primitive monoid*). Denote by $\mathcal{B}$ the Boolean ring of subsets X of $\mathbb{Z}^+$ that are either finite or cofinite. Set $\mu(X) = (\operatorname{card} X)\varepsilon$ in the first case, $\mu(X) = 1$ in the second case. Then μ is a P-valued V-measure on $\mathcal{B}$. It is easy to see that μ is groupoid-induced and that the automorphisms of (B, μ) are the automorphisms induced by the permutations of Ω. Furthermore, $\mu(\mathbb{Z}^+) = \mu(\mathbb{N}) = 1$.

Suppose that $\mathbb{Z}^+ \simeq_\mu^{\mathrm{gp}} \mathbb{N}$. This means that there are $n \in \mathbb{Z}^+$, permutations $g_0, \dots, g_n$ of $\mathbb{Z}^+$, and decompositions of the form $\mathbb{Z}^+ = \bigsqcup_{i=0}^n X_i$, $\mathbb{N} = \bigsqcup_{i=0}^n Y_i$ such that each $Y_i = g_i X_i$. Necessarily, there is exactly one i such that X_i is infinite. The finite sets $\mathbb{Z}^+ \setminus X_i$ and $\mathbb{N} \setminus Y_i$ have the same cardinality, a contradiction since $Y_i = g_i X_i$ and g_i is a permutation of $\mathbb{Z}^+$. Therefore, μ is not group-induced.

The following result shows that for many types of pointed monoids, groupoid-induced is equivalent to group-induced.

Proposition 4.7.10 *Let B be a unital Boolean ring, let $(M, \boldsymbol{e})$ be a pointed conical refinement monoid, and let $\mu: B \to (M, \boldsymbol{e})$ be a normalized groupoid-induced V-measure. Then each of the following conditions implies that μ is group-induced:*

(1) *There are $\boldsymbol{a} \in M$ and a positive integer m such that $2\boldsymbol{a} \leq^+ \boldsymbol{e} \leq^+ m\boldsymbol{a}$.*
(2) *For all $\boldsymbol{a}, \boldsymbol{b}, \boldsymbol{c} \in M$, if $\boldsymbol{a} + \boldsymbol{c} = \boldsymbol{b} + \boldsymbol{c} \leq^+ \boldsymbol{e}$, then there exists $\boldsymbol{d} \in M$ such that $2\boldsymbol{d} \leq^+ \boldsymbol{c}$ and $\boldsymbol{a} + \boldsymbol{d} = \boldsymbol{b} + \boldsymbol{d}$.*
(3) *M is cancellative.*

Proof (1) Let $\boldsymbol{a}' \in M$ such that $\boldsymbol{e} = 2\boldsymbol{a} + \boldsymbol{a}'$. Since M is a refinement monoid, $\boldsymbol{a}' \leq^+ m\boldsymbol{a}$, and by Wehrung [118, Lemma 1.9], there are decompositions of the form

$$\boldsymbol{a} = \sum_{k=0}^{m} \boldsymbol{a}_k \text{ and } \boldsymbol{a}' = \sum_{k=0}^{m} k\boldsymbol{a}_k \quad \text{in } M\,.$$

It follows that $\boldsymbol{e} = \sum_{k=0}^{m} (k+2)\boldsymbol{a}_k$.

Since each $k+2 \geq 2$, we obtain, by reindexing the $\boldsymbol{a}_k$, an integer $n \geq 2$ and a decomposition of the form $\boldsymbol{e} = \sum_{k<n} \boldsymbol{e}_k$ where for each $i < n$ there is $j < n$ such that $i \neq j$ and $\boldsymbol{e}_i = \boldsymbol{e}_j$. Since $\mu(1) = \boldsymbol{e} = \sum_{k<n} \boldsymbol{e}_k$ and μ is a V-measure, there is a decomposition of the form $e = \bigoplus_{k<n} e_k$ such that each $\mu(e_k) = \boldsymbol{e}_k$.

Now let $x, y \in B$ such that $\mu(x) = \mu(y)$, in other words, $\sum_{i<n} \mu(x \wedge e_i) = \sum_{j<n} \mu(y \wedge e_j)$. Since M is a refinement monoid, there is a refinement matrix of the form

	$\mu(y \wedge e_j)(j<n)$
$\mu(x \wedge e_i)(i<n)$	$z_{i,j}$

with all $z_{i,j} \in M$.

Since μ is a V-measure, for each $i < n$ there is a decomposition $x \wedge e_i = \bigoplus_{j<n} x_{i,j}$ with each $\mu(x_{i,j}) = z_{i,j}$. Similarly, for each $j < n$ there is a decomposition $y \wedge e_j = \bigoplus_{i<n} y_{i,j}$ with each $\mu(y_{i,j}) = z_{i,j}$. Since μ is groupoid-induced, $x_{i,j} \sim_{\mu}^{\mathrm{gpd}} y_{i,j}$ for all $i, j < n$.

We claim that $x_{i,j} \sim_{\mu}^{\mathrm{gp}} y_{i,j}$. If $i \neq j$, then this follows from the relation $x_{i,j} \sim_{\mu}^{\mathrm{gpd}} y_{i,j}$, together with $x_{i,j} \wedge y_{i,j} = 0$ (because $x_{i,j} \leq e_i$ and $y_{i,j} \leq e_j$) and Proposition 4.7.6(3). Now suppose that $i = j$ and pick $k < n$ with $i \neq k$ and $\boldsymbol{e}_i = \boldsymbol{e}_k$. Since $\mu(x_{i,i}) \leq^+ \mu(x \wedge e_i) \leq^+ \mu(e_i) = \mu(e_k)$ and μ is a V-measure, there is $z \leq e_k$ such that $\mu(x_{i,j}) = \mu(z)$. From $x_{i,i} \vee y_{i,i} \leq e_i$ and $z \leq e_k$ it follows that $(x_{i,i} \vee y_{i,i}) \wedge z = 0$. By Proposition 4.7.6(4), we get $x_{i,i} \sim_{\mu}^{\mathrm{gp}} y_{i,i}$. Therefore, the relation $x_{i,j} \sim_{\mu}^{\mathrm{gp}} y_{i,j}$ holds in every case. Since $x = \bigoplus_{i,j<n} x_{i,j}$ and $y = \bigoplus_{i,j<n} y_{i,j}$, it follows that $x \simeq_{\mu}^{\mathrm{gp}} y$.

(2) Let $x, y \in B$ such that $\mu(x) = \mu(y)$. Setting $\boldsymbol{a} = \mu(x \setminus y)$, $\boldsymbol{b} = \mu(y \setminus x)$, and $\boldsymbol{c} = \mu(x \wedge y)$, we get $\boldsymbol{a} + \boldsymbol{c} = \boldsymbol{b} + \boldsymbol{c}$, thus, by assumption, there is $\boldsymbol{d} \in M$ such that $2\boldsymbol{d} \leq^+ \boldsymbol{c}$ and $\boldsymbol{a} + \boldsymbol{d} = \boldsymbol{b} + \boldsymbol{d}$. Since μ is a V-measure, there is a decomposition $x \wedge y = z_0 \oplus z_1 \oplus z_2$ such that $\mu(z_0) = \mu(z_1) = \boldsymbol{d}$. Set $x' = (x \setminus y) \oplus z_0$ and $y' = (y \setminus x) \oplus z_1$. Then $x' \wedge y' = 0$ and $\mu(x') = \boldsymbol{a} + \boldsymbol{d} = \boldsymbol{b} + \boldsymbol{d} = \mu(y')$. By Proposition 4.7.6(3), it follows that $x' \sim_{\mu}^{\mathrm{gp}} y'$. By the same token, $z_0 \sim_{\mu}^{\mathrm{gp}} z_1$. Since $x = x' \oplus z_1 \oplus z_2$ and $y = y' \oplus z_0 \oplus z_2$, it follows that $x \simeq_{\mu}^{\mathrm{gp}} y$.

The sufficiency of (3) is a trivial consequence of the sufficiency of (2). □

Proposition 4.7.11 *The following statements hold, for any Boolean ring B, any conical refinement monoid M, and any groupoid-induced V-measure* $\mu\colon B \to M$:

(1) *The restriction of* μ *to any ideal of B is also a groupoid-induced V-measure.*
(2) *Let I be a set with at least two elements and denote by* $B^{(I)}$ *the direct sum of I copies of B. Then* $\mu^{(I)}\colon B^{(I)} \to M$, $(x_i \mid i \in I) \mapsto \sum_{i \in I} \mu(x_i)$ *is a group-induced V-measure.*

Proof Since (1) is trivial, we only give a proof of (2). It follows from Lemma 4.6.5 that $\mu^{(I)}$ is a V-measure. Now let $\vec{a} = (a_i \mid i \in I)$ and $\vec{b} = (b_i \mid i \in I)$ be elements of $B^{(I)}$ with $\mu^{(I)}(\vec{a}) = \mu^{(I)}(\vec{b})$. We need to prove that $\vec{a} \simeq_{\mu^{(I)}}^{\mathrm{gp}} \vec{b}$. Fix a finite subset K of I, with at least two elements, such that $\vec{a}, \vec{b} \in B^K$. Since $\sum_{i \in K} \mu(a_i) =$

$\sum_{j \in K} \mu(b_j)$ and M is a refinement monoid, there exists a refinement matrix of the form

	$\mu(b_j)(j \in K)$
$\mu(a_i)(i \in K)$	$\boldsymbol{c}_{i,j}$

with all $\boldsymbol{c}_{i,j} \in M$.

Since μ is a V-measure, for each $i \in K$ there is a decomposition $a_i = \bigoplus_{j \in K} a_{i,j}$ with each $\mu(a_{i,j}) = \boldsymbol{c}_{i,j}$. Likewise, for each $j \in K$ there is a decomposition $b_j = \bigoplus_{i \in K} b_{i,j}$ with each $\mu(b_{i,j}) = \boldsymbol{c}_{i,j}$. Denoting by $\varepsilon_i : B \hookrightarrow B^{(I)}$ the ith canonical embedding whenever $i \in I$, we get

$$\vec{a} = \bigoplus_{i \in K} \varepsilon_i(a_i) = \bigoplus_{(i,j) \in K \times K} \varepsilon_i(a_{i,j}),$$

and, likewise,

$$\vec{b} = \bigoplus_{j \in K} \varepsilon_j(b_j) = \bigoplus_{(i,j) \in K \times K} \varepsilon_j(b_{i,j}).$$

Hence it suffices to prove that $\varepsilon_i(a_{i,j}) \sim^{\mathrm{gp}}_{\mu^{(I)}} \varepsilon_j(b_{i,j})$ for each $(i,j) \in K \times K$. First observe that $\mu(a_{i,j}) = \boldsymbol{c}_{i,j} = \mu(b_{i,j})$, thus, since μ is groupoid-induced, there is a measure-preserving isomorphism $B{\downarrow}a_{i,j} \to B{\downarrow}b_{i,j}$. Since ε_i and ε_j are both measure-preserving V-embeddings, we get measure-preserving isomorphisms $B{\downarrow}a_{i,j} \cong B^{(I)}{\downarrow}\varepsilon_i(a_{i,j})$ and $B \downarrow b_{i,j} \cong B^{(I)} \downarrow \varepsilon_j(b_{i,j})$, thus also a measure-preserving isomorphism $B^{(I)}{\downarrow}\varepsilon_i(a_{i,j}) \to B^{(I)}{\downarrow}\varepsilon_j(b_{i,j})$. This means that $\varepsilon_i(a_{i,j}) \sim^{\mathrm{gpd}}_{\mu^{(I)}} \varepsilon_j(b_{i,j})$. Now we separate cases.

Case 1. $i \neq j$. Then $\varepsilon_i(a_{i,j}) \wedge \varepsilon_j(b_{i,j}) = 0$, thus, by Proposition 4.7.6(3), we get $\varepsilon_i(a_{i,j}) \sim^{\mathrm{gp}}_{\mu^{(I)}} \varepsilon_j(b_{i,j})$.

Case 2. $i = j$. Pick $k \in K \setminus \{i\}$. The argument used for Case 1 above shows that $\varepsilon_i(a_{i,i}) \sim^{\mathrm{gp}}_{\mu^{(I)}} \varepsilon_k(a_{i,i})$ and $\varepsilon_k(a_{i,i}) \sim^{\mathrm{gp}}_{\mu^{(I)}} \varepsilon_j(b_{i,i})$. Hence, we get again $\varepsilon_i(a_{i,j}) \sim^{\mathrm{gp}}_{\mu^{(I)}} \varepsilon_j(b_{i,j})$.

This completes the proof that $\vec{a} \simeq^{\mathrm{gp}}_{\mu^{(I)}} \vec{b}$. □

4.7.2 Exhaustive Sets on Boolean Rings

In view of further applications, it might be interesting to explore the degree of freedom on the Boolean inverse semigroups representing, via the type monoid, a given conical refinement monoid. This is the purpose of exhaustive sets.

Definition 4.7.12 Let B be a Boolean ring, let M be a commutative monoid, and let $\mu\colon B \to M$ be a premeasure.

(1) A set S of measure-preserving partial functions between subsets of B is μ-*exhaustive* if the following condition holds:

For all $a, b \in B$ with $\mu(a) = \mu(b)$, there are finite decompositions

$$a = \bigoplus_{i<n} a_i\,,\ b = \bigoplus_{i<n} b_i\,,\ \text{and elements } f_i \in S \text{ such that each } b_i = f_i(a_i)\,. \tag{4.7.1}$$

(2) An action of a group G on B by measure-preserving automorphisms is μ-*exhaustive* if the subgroup of $\operatorname{Aut}(B, \mu)$ induced by the action of G is μ-exhaustive, that is,

For all $a, b \in B$ with $\mu(a) = \mu(b)$, there are finite decompositions

$$a = \bigoplus_{i<n} a_i\,,\ b = \bigoplus_{i<n} b_i\,,\ \text{and elements } g_i \in G \text{ such that each } b_i = g_i a_i\,. \tag{4.7.2}$$

The following result relates exhaustive sets of partial functions and type monoids of Boolean inverse semigroups.

Theorem 4.7.13 *Let M be a conical refinement monoid, let B be a Boolean ring, and let $\mu\colon B \to M$ be a groupoid-induced V-measure, with range generating M as a submonoid. Then the following statements hold:*

(1) *There exists a μ-exhaustive Boolean wide inverse subsemigroup S of* $\operatorname{Inv}(B, \mu)$.
(2) *Any such inverse semigroup S is a fundamental Boolean inverse semigroup, and there exists a unique monoid homomorphism $\tau\colon \operatorname{Typ} S \to M$ such that $\tau(\operatorname{typ}(\overline{a})) = \mu(a)$ for all $a \in B$, where we denote by $\overline{a} = \mathrm{id}_{B\downarrow a}$ the natural image of a in S. Furthermore, τ is a monoid isomorphism.*

Proof (1) Take $S = \operatorname{Inv}(B, \mu)$.

(2) It follows from our assumptions on S, together with $\operatorname{Inv}(B, \mu)$ being a fundamental Boolean inverse semigroup, that S is a fundamental Boolean inverse semigroup.

Let $a, b \in B$ such that $\operatorname{typ}(\overline{a}) = \operatorname{typ}(\overline{b})$, that is, $\overline{a} \mathrel{\mathscr{D}} \overline{b}$. There is $f \in S$ such that $\mathbf{d}(f) = \overline{a}$ and $\mathbf{r}(f) = \overline{b}$, that is, f is an isomorphism $B \downarrow a \to B \downarrow b$. In particular, $f(a) = b$. From $S \subseteq \operatorname{Inv}(B, \mu)$ it follows that $\mu(a) = \mu(b)$.

Conversely, let $a, b \in B$ such that $\mu(a) = \mu(b)$. Since S is μ-exhaustive and closed under finite orthogonal joins, there exists $f \in S$ such that $f(a) = b$. Since $\overline{a}$ is an idempotent of S, we may replace f by $f\overline{a}$ and thus assume that $f\colon B \downarrow a \to B \downarrow b$. It follows that $\mathbf{d}(f) = \overline{a}$ and $\mathbf{r}(f) = \overline{b}$, so $\overline{a} \mathrel{\mathscr{D}} \overline{b}$, that is, $\operatorname{typ}(\overline{a}) = \operatorname{typ}(\overline{b})$.

We have thus proved that the kernel of the canonical V-homomorphism $\overline{\mu}\colon \operatorname{Idp} S \to M$, $\overline{a} \mapsto \mu(a)$ is equal to $\mathscr{D}$. By applying Lemma 2.4.5 to $\overline{\mu}$, it follows that $\overline{\mu}$ induces a V-embedding $\tau_0\colon \operatorname{Int} S \hookrightarrow M$, $\operatorname{typ}(\overline{a}) \mapsto \mu(a)$.

By Proposition 2.2.4, τ_0 extends to a unique V-embedding $\tau\colon \operatorname{Typ} S \hookrightarrow M$. Since the range of τ_0 is the range of μ, it generates M, thus τ is an isomorphism. □

The following result relates exhaustive group actions and monoids of equidecomposability types.

Theorem 4.7.14 *Let M be a conical refinement monoid, let B be a Boolean ring, and let $\mu\colon B \to M$ be a group-induced V-measure with range generating M as a submonoid. Then the following statements hold:*

(1) *There exists a μ-exhaustive action of a group G on B.*
(2) *For any such group action, there exists a unique monoid homomorphism $\tau\colon \mathbb{Z}^+\langle B\rangle /\!\!/ G \to M$ such that $\tau([a]_G) = \mu(a)$ for every $a \in B$. Furthermore, τ is a monoid isomorphism.*

Proof (1) Consider the natural action of $G = \operatorname{Aut}(B, \mu)$ on B.

(2) The assumption that the action of G is μ-exhaustive means that the kernel of μ is the restriction of $\simeq_G$ to B. By Lemma 2.4.5, μ induces the V-embedding $\tau_0\colon B/\!\!/G \hookrightarrow M$, $[a]_G \mapsto \mu(a)$. By Proposition 2.2.4, τ_0 extends to a unique V-embedding $\tau\colon \mathbb{Z}^+\langle B\rangle /\!\!/ G \hookrightarrow M$. Since the range of τ_0 is equal to the range of μ, it generates M, thus τ is surjective. □

We can paraphrase Theorems 4.7.13 and 4.7.14 as follows: for a conical refinement monoid M,

- any groupoid-induced V-measure $\mu\colon B \to M$ induces an isomorphism $\operatorname{Typ} S \cong M$, for any large enough Boolean inverse subsemigroup S of $\operatorname{Inv}(B, \mu)$;
- any group-induced V-measure $\mu\colon B \to M$ induces an isomorphism $\mathbb{Z}^+\langle B\rangle /\!\!/ G \cong M$, for any large enough subgroup G of $\operatorname{Aut}(B, \mu)$.

Although the trivial implication from group-induced to groupoid-induced may seem lost there, we recover it quickly once we remember the isomorphism between $\operatorname{Typ}(\operatorname{Inv}(B, G))$ and $\mathbb{Z}^+\langle B\rangle /\!\!/ G$ given by Proposition 4.4.20.

4.8 Groupoid- and Group-Measurable Monoids

Following the lead given by Theorems 4.7.13 and 4.7.14, we shall strengthen the definition of V-measurability for pointed monoids (cf. Definition 4.6.1) as follows.

Definition 4.8.1 A pointed commutative monoid $(M, \boldsymbol{e})$ is *groupoid-measurable* (resp., *group-measurable*) if there are a unital Boolean ring B and a groupoid-induced (resp., group-induced) V-measure $\mu\colon B \to M$ such that $\mu(1) = \boldsymbol{e}$.

Proposition 4.8.2 *The following statements hold, for any conical refinement monoid M with order-unit $\boldsymbol{e}$:*

(1) *$(M, \boldsymbol{e})$ is groupoid-measurable iff there is a Boolean inverse monoid S such that $(M, \boldsymbol{e}) \cong (\operatorname{Typ} S, \operatorname{typ}(1))$. Moreover, S can be taken fundamental.*
(2) *$(M, \boldsymbol{e})$ is group-measurable iff there is a group G, acting by automorphisms on a unital Boolean ring B, such that $(M, \boldsymbol{e}) \cong (\mathbb{Z}^+\langle B\rangle /\!\!/ G, [1]_G)$.*

Proof (1) If $(M, \boldsymbol{e})$ is groupoid-measurable via a groupoid-induced V-measure $\mu: B \to M$, then, by Theorem 4.7.13, the fundamental Boolean inverse monoid $S = \operatorname{Inv}(B, \mu)$ satisfies $(M, \boldsymbol{e}) \cong (\operatorname{Typ} S, \operatorname{typ}(1))$. Let, conversely, S be a unital Boolean inverse semigroup such that $(M, \boldsymbol{e}) = (\operatorname{Typ} S, \operatorname{typ}(1))$. Set $B = \operatorname{Idp} S$. It follows from Lemma 4.1.6 that the assignment $x \mapsto \operatorname{typ}(x)$ defines a V-measure from B to M, sending 1 to $\boldsymbol{e}$. For any $a, b \in B$ such that $\operatorname{typ}(a) = \operatorname{typ}(b)$, there exists $f \in S$ such that $\mathbf{d}(f) = a$ and $\mathbf{r}(f) = b$. Hence the assignment $x \mapsto fxf^{-1}$ is a measure-preserving isomorphism from $B \downarrow a$ onto $B \downarrow b$.

(2) If $(M, \boldsymbol{e})$ is group-measurable via a group-induced V-measure $\mu: B \to M$, then, by Theorem 4.7.14, $(M, \boldsymbol{e}) \cong (\mathbb{Z}^+\langle B\rangle /\!\!/ G, [1]_G)$ where $G = \operatorname{Aut}(B, \mu)$. Let, conversely, $(M, \boldsymbol{e}) = (\mathbb{Z}^+\langle B\rangle /\!\!/ G, [1]_G)$ for some action η of a group G on a unital Boolean ring B. We claim that the canonical V-measure $\mu_G: B \to M$, $x \mapsto [x]_G$ (cf. Lemma 2.8.8) is group-induced. Let $a, b \in B$ such that $[a]_G = [b]_G$. This means that $a \simeq_G b$, that is, there are decompositions $a = \bigoplus_{i<n} a_i$ and $b = \bigoplus_{i<n} b_i$ where each $b_i = g_i a_i$ where $g_i \in G$. It follows that each η_{g_i} belongs to $\operatorname{Aut}(B, \mu)$ and $b_i = \eta_{g_i}(a_i)$. Therefore, $a \simeq_\mu^{\mathrm{gp}} b$. □

The following result shows that for non-pointed commutative monoids, the separation between "group" and "groupoid" becomes immaterial.

Proposition 4.8.3 *The following statements are equivalent, for every conical refinement monoid M:*

(i) *There are a Boolean ring B and a groupoid-induced V-measure $\mu: B \to M$ with generating range.*
(ii) *There are a Boolean ring B and a surjective group-induced V-measure $\mu: B \to M$.*

Proof (ii)⇒(i) is trivial.

(i)⇒(ii). By Proposition 4.7.11, $\mu^{(\mathbb{Z}^+)}: B^{(\mathbb{Z}^+)} \to M$ is a group-induced V-measure. Since the range of μ generates M, $\mu^{(\mathbb{Z}^+)}$ is surjective. □

Definition 4.8.4 A conical refinement monoid M is *group-measurable* if one of the equivalent conditions of Proposition 4.8.3 holds.

The analogue of Proposition 4.8.2 for non-pointed monoids is the following. The proof is similar to the one of Proposition 4.8.2 and we omit it.

Proposition 4.8.5 *The following are equivalent, for any conical refinement monoid M:*

(1) *M is group-measurable.*
(2) *There is a group G, acting by automorphisms on a Boolean ring B, such that $M \cong \mathbb{Z}^+\langle B\rangle /\!\!/ G$.*
(3) *There is a Boolean inverse semigroup S such that $M \cong \operatorname{Typ} S$. Moreover, S can be taken fundamental.*

For an extension of Proposition 4.8.5 to Boolean inverse meet-semigroups, see Theorem 4.9.1.

Theorem 4.8.6 *Let M be a conical refinement monoid and let B be a countable Boolean ring. Then any V-measure $\mu: B \to M$ is groupoid-induced.*

Proof Let $a, b \in B$ such that $\mu(a) = \mu(b)$. Since μ is a V-measure, the binary relation

$$\Gamma = \{(x, y) \in (B \downarrow a) \times (B \downarrow b) \mid \mu(x) = \mu(y)\}$$

is an additive, conical V-relation on $(B \downarrow a) \times (B \downarrow b)$ (cf. Definition 2.4.1). By Vaught's Theorem (cf. Lemma 4.5.2), the graph of Γ contains an isomorphism $\varphi: B\downarrow a \to B \downarrow b$. Then $\varphi \in \mathrm{Inv}(B, \mu)$ and $\varphi(a) = b$. □

Theorem 4.8.7 *Every countable conical refinement monoid $(M, \boldsymbol{e})$ with order-unit is groupoid-measurable. In fact, there exists a countable fundamental Boolean inverse monoid S such that $(M, \boldsymbol{e}) \cong (\mathrm{Typ}\, S, \mathrm{typ}_S(1))$.*

Proof By the countable case of Theorem 4.6.7, $(M, \boldsymbol{e})$ is V-measurable, via a normalized V-measure $\mu: B \to (M, \boldsymbol{e})$. By a standard Löwenheim-Skolem type argument, B can be taken countable. By Theorem 4.8.6, μ is groupoid-induced. For each $a, b \in B$ with $\mu(a) = \mu(b)$, pick $f_{a,b} \in \mathrm{Inv}(B, \mu)$ sending a to b. By the Löwenheim-Skolem Theorem, $\mathrm{Inv}(B, \mu)$ contains a countable fundamental Boolean inverse semigroup S such that all $f_{a,b}$ belong to S. By Theorem 4.7.13(2), it follows that $(\mathrm{Typ}\, S, \mathrm{typ}_S(1)) \cong (M, \boldsymbol{e})$. □

Example 4.7.9, giving a groupoid-induced measure that is not group-induced, yields immediately the following example.

Example 4.8.8 A groupoid-measurable, non group-measurable, countable conical refinement monoid with order-unit.

Proof As in Example 4.7.9, denote by P the commutative monoid defined by generators ε, 1 and the relation $\varepsilon + 1 = 1$. We have seen in Example 4.7.9 that $(P, 1)$ is groupoid-measurable.

Suppose that there is a group-induced normalized V-measure $\nu: B \to (P, 1)$. A standard Löwenheim-Skolem type argument shows that B can be taken countable. But then, by Theorem 4.6.8, (B, ν) is isomorphic to the pair $(\mathcal{B}, \mu)$ of Example 4.7.9, which we proved there is not group-induced. □

The non-unital version of Theorem 4.8.7 runs as follows.

Theorem 4.8.9 *Every countable conical refinement monoid is group-measurable; thus it is isomorphic to the type monoid of a fundamental Boolean inverse semigroup.*

Proof Let M be a countable conical refinement monoid and embed M as an ideal into a countable conical refinement monoid $(N, \boldsymbol{e})$ with order-unit. For example, $N = M \sqcup \{\infty\}$ with $x + \infty = \infty$ for all $x \in N$, and $\boldsymbol{e} = \infty$. By Theorem 4.8.7, there are a countable fundamental Boolean inverse semigroup T and an isomorphism

$\varepsilon\colon \operatorname{Typ} T \to N$. It follows from Proposition 4.2.4 that $S = \{x \in T \mid \varepsilon(\operatorname{typ}_T(x)) \in M\}$ is an additive ideal of T and $\operatorname{Typ} S \cong M$. By Proposition 4.8.3, M is group-measurable. □

4.9 Type Monoids of Boolean Inverse Meet-Semigroups

Recall from Proposition 4.8.5 that a conical refinement monoid is group-measurable iff it is isomorphic to the type monoid of a Boolean inverse semigroup (which can then be taken fundamental). The following result enables Boolean inverse meet-semigroups to enter that picture as well.

Theorem 4.9.1 *Every group-measurable conical refinement monoid is isomorphic to the type monoid of a Boolean inverse meet-semigroup.*

Proof By Proposition 4.8.5, it suffices to prove that for every group G, acting by automorphisms on a Boolean ring $\mathcal{B}$, there is a Boolean inverse meet-semigroup S such that $\mathbb{Z}^+\langle\mathcal{B}\rangle /\!\!/ G \cong \operatorname{Typ} S$. We may assume without loss of generality that $\mathcal{B}$ is a ring of subsets of a set Ω, on which G acts $\mathcal{B}$-measurably (define Ω as the prime filter space of the original $\mathcal{B}$, and then replace $\mathcal{B}$ by the ring of all compact open subsets of Ω; then G acts continuously on Ω). Now let G act on $\Omega' = \Omega \times G$ via $g\cdot(\mathfrak{p}, h) = (g\mathfrak{p}, gh)$ whenever $g \in G$ and $(\mathfrak{p}, h) \in \Omega'$. The set $\mathcal{B}' = \{X \times G \mid X \in \mathcal{B}\}$ is a ring of subsets of Ω', isomorphic to $\mathcal{B}$ (as a Boolean ring), and the action of G on Ω' is $\mathcal{B}'$-measurable. Further, two elements $X, Y \in \mathcal{B}$ are G-equidecomposable with pieces from $\mathcal{B}$ iff $X{\times}G$ and $Y{\times}G$ are G-equidecomposable with pieces from $\mathcal{B}'$. It follows that $\mathbb{Z}^+\langle\mathcal{B}'\rangle /\!\!/ G \cong \mathbb{Z}^+\langle\mathcal{B}\rangle /\!\!/ G \cong M$. By Proposition 4.4.21, it follows that $\operatorname{Typ}(\mathbf{pMeas}(\mathcal{B}', G)) \cong M$.

Now the action of G on $\Omega' = \Omega \times G$ is fixed point free, thus, by Proposition 4.4.13, $\mathbf{pMeas}(\mathcal{B}', G)$ is a Boolean inverse meet-semigroup. □

By combining that result with Theorem 4.8.9 and a final Löwenheim-Skolem type argument, we get the following.

Theorem 4.9.2 *Every countable conical refinement monoid is isomorphic to the type monoid of some countable Boolean inverse meet-semigroup.*

The proof of Theorem 4.9.1 contains a touch of ad-hoc-ness that does not appear in the proof of the "fundamental semigroup" part of Proposition 4.8.5: while the canonical projection from a Boolean inverse semigroup S onto its maximal fundamental quotient S/μ is type-preserving (cf. Theorem 4.4.19), there is no convenient "canonical Boolean inverse meet-semigroup image" that we could use to that end. (However, remember that every Boolean inverse semigroup has an additive semigroup embedding into some symmetric inverse monoid—thus into some unital Boolean inverse meet-semigroup, see Corollary 3.3.2.) More dramatically, the following example shows that embedding a given Boolean inverse semigroup into any Boolean inverse *meet-semigroup* may introduce some collapsing at the type monoid level.

Example 4.9.3 A countable, fundamental Boolean inverse monoid S without any type-expanding map into any Boolean inverse meet-semigroup. That is, for every Boolean inverse meet-semigroup T and every additive semigroup homomorphism $\varphi: S \to T$, the monoid homomorphism $\operatorname{Typ} \varphi: \operatorname{Typ} S \to \operatorname{Typ} T$ is not one-to-one.

Proof Let $\tau: \mathbb{Z}/3\mathbb{Z} \to \mathbb{Z}/3\mathbb{Z}$, $t \mapsto 2-t$. We set $\Omega = \mathbb{Z}^+ \times (\mathbb{Z}/3\mathbb{Z})$, and we define S as the set of all partial functions $x \in \mathfrak{I}_\Omega$ such that the domain of x is either finite or cofinite in Ω, and, in the latter case, there are $m \in \mathbb{Z}^+$, $h \in \mathbb{Z}$, and $\varepsilon \in \{0, 1\}$ such that

$$\text{for all } n \in [m, \infty) \text{ and all } t \in \mathbb{Z}/3\mathbb{Z}\,,\ x(n, t) = \big(n + h, \tau^\varepsilon(t)\big)\,. \tag{4.9.1}$$

It is straightforward to verify that S is a fundamental, additive Boolean inverse submonoid of $\mathfrak{I}_\Omega$. The idempotent elements of S are the identity functions id_X on the subsets X of Ω that are either finite or cofinite, and the pedestal of S (cf. Definition 3.7.5) consists exactly of the elements of $\mathfrak{I}_\Omega$ with finite domain. The zero element of S is the empty function.

We set $\Omega' = \Omega \setminus \{(0,0), (0,2)\}$. The functions $a = \mathrm{id}_\Omega$ and $b = \mathrm{id}_{\Omega'}$ are idempotent elements of S.

Claim $(a, b) \notin \mathscr{D}_S$.

Proof of Claim. By way of contradiction, suppose that there exists a bijection $x: \Omega \to \Omega'$ in S. Since the domain of x is cofinite, there are $m \in \mathbb{N}$, $h \in \mathbb{Z}$, and $\varepsilon \in \{0, 1\}$ such that (4.9.1) holds. We may choose $m > -h$. It follows from (4.9.1) that $x\big[[m, \infty) \times (\mathbb{Z}/3\mathbb{Z})\big] = [m + h, \infty) \times (\mathbb{Z}/3\mathbb{Z})$. Since $x[\Omega] = \Omega'$ and since x is one-to-one, it follows, by evaluating set-theoretical differences, that

$$x\big[[0, m) \times (\mathbb{Z}/3\mathbb{Z})\big] = \big([0, m + h) \times (\mathbb{Z}/3\mathbb{Z})\big) \setminus \{(0,0), (0,2)\}\,,$$

thus, evaluating cardinalities, $3m = 3(m + h) - 2$, that is, $3h = 2$, a contradiction. □ Claim.

Hence, in order to conclude the proof, it suffices to prove that $\varphi(a)\ \mathscr{D}_T\ \varphi(b)$ whenever T is a Boolean inverse meet-semigroup and $\varphi: S \to T$ is an additive semigroup homomorphism. To this end, let s and g the self-maps of Ω defined by the rules $s(n, t) = (n + 1, t)$ and $g(n, t) = (n, 2 - t)$ whenever $(n, t) \in \Omega$. Observe that s and g both belong to S, $g^2 = \mathrm{id}_\Omega$, and $g \circ s = s \circ g$. Set $X_0 = \{0\} \times (\mathbb{Z}/3\mathbb{Z})$ and $\Omega_1 = \Omega \setminus X_0$. Then $\mathrm{id}_{X_0} \wedge g \restriction_{\Omega_1} = \mathrm{id}_{\Omega_1} \wedge g \restriction_{X_0} = 0$, thus, since id_{X_0} and $g \restriction_{X_0}$ both belong to the pedestal of S and by Proposition 3.7.10, we obtain

$$\varphi(\mathrm{id}_{X_0}) \wedge \varphi(g \restriction_{\Omega_1}) = \varphi(\mathrm{id}_{\Omega_1}) \wedge \varphi(g \restriction_{X_0}) = 0\,. \tag{4.9.2}$$

Further, $\mathrm{id}_{X_0} \wedge g \restriction_{X_0} = \mathrm{id}_{\{(0,1)\}}$. Since id_{X_0} belongs to the pedestal of S and by Proposition 3.7.10, we obtain

$$\varphi(\mathrm{id}_{X_0}) \wedge \varphi(g \restriction_{X_0}) = \varphi(\mathrm{id}_{\{(0,1)\}})\,. \tag{4.9.3}$$

Further, it follows from the additivity of φ that

$$\varphi(\mathrm{id}_{\Omega}) = \varphi(\mathrm{id}_{X_0}) \oplus \varphi(\mathrm{id}_{\Omega_1}) \text{ and } \varphi(g) = \varphi(g \restriction_{X_0}) \oplus \varphi(g \restriction_{\Omega_1}) . \tag{4.9.4}$$

Therefore, putting together (4.9.2)–(4.9.4) and using the distributivity of $\wedge$ on $\oplus$ (cf. Proposition 3.1.9), we obtain

$$\varphi(\mathrm{id}_{\Omega}) \wedge \varphi(g) = \varphi(\mathrm{id}_{\{(0,1)\}}) \oplus \big(\varphi(\mathrm{id}_{\Omega_1}) \wedge \varphi(g \restriction_{\Omega_1})\big) . \tag{4.9.5}$$

We further compute, using the abbreviation $x\langle y\rangle = xyx^{-1}$ (cf. (3.4.2)),

$$\begin{aligned} \varphi(s)\,\langle\varphi(\mathrm{id}_{\Omega})\rangle &= \varphi(s\,\langle\mathrm{id}_{\Omega}\rangle) = \varphi(\mathrm{id}_{\Omega_1}) , \\ \varphi(s)\,\langle\varphi(g)\rangle &= \varphi(s\,\langle g\rangle) = \varphi(g \restriction_{\Omega_1}) , \end{aligned}$$

thus

$$\begin{aligned} \varphi(s)\,\langle\varphi(\mathrm{id}_{\Omega}) \wedge \varphi(g)\rangle &= \varphi(s)\,\langle\varphi(\mathrm{id}_{\Omega})\rangle \wedge \varphi(s)\,\langle\varphi(g)\rangle \\ &= \varphi(\mathrm{id}_{\Omega_1}) \wedge \varphi(g \restriction_{\Omega_1}) , \end{aligned}$$

and thus, using the abbreviation $x \smallsetminus y = x \smallsetminus (x \wedge y)$ (cf. Notation 3.1.11),

$$\begin{aligned} \varphi(s)\,\langle\varphi(\mathrm{id}_{\Omega}) \smallsetminus \varphi(g)\rangle &= \varphi(s)\,\big\langle\varphi(\mathrm{id}_{\Omega}) \smallsetminus \big(\varphi(\mathrm{id}_{\Omega}) \wedge \varphi(g)\big)\big\rangle \\ &= \varphi(s)\,\langle\varphi(\mathrm{id}_{\Omega})\rangle \smallsetminus \varphi(s)\,\langle\varphi(\mathrm{id}_{\Omega}) \wedge \varphi(g)\rangle \\ &= \varphi(\mathrm{id}_{\Omega_1}) \smallsetminus \big(\varphi(\mathrm{id}_{\Omega_1}) \wedge \varphi(g \restriction_{\Omega_1})\big) \\ &= \varphi(\mathrm{id}_{\Omega_1}) \smallsetminus \varphi(g \restriction_{\Omega_1}) . \end{aligned}$$

It follows that

$$\begin{aligned} \varphi(b) = \varphi(\mathrm{id}_{\Omega'}) &= \varphi(\mathrm{id}_{\{(0,1)\}}) \oplus \varphi(\mathrm{id}_{\Omega_1}) \\ &= \varphi(\mathrm{id}_{\{(0,1)\}}) \oplus \big(\varphi(\mathrm{id}_{\Omega_1}) \wedge \varphi(g \restriction_{\Omega_1})\big) \oplus \big(\varphi(\mathrm{id}_{\Omega_1}) \smallsetminus \varphi(g \restriction_{\Omega_1})\big) \\ &= \varphi(\mathrm{id}_{\{(0,1)\}}) \oplus \big(\varphi(\mathrm{id}_{\Omega_1}) \wedge \varphi(g \restriction_{\Omega_1})\big) \oplus \varphi(s)\,\langle\varphi(\mathrm{id}_{\Omega}) \smallsetminus \varphi(g)\rangle , \end{aligned}$$

thus, by (4.9.5),

$$\varphi(b) = \big(\varphi(\mathrm{id}_{\Omega}) \wedge \varphi(g)\big) \oplus \varphi(s)\,\langle\varphi(\mathrm{id}_{\Omega}) \smallsetminus \varphi(g)\rangle . \tag{4.9.6}$$

From $\mathbf{d}(\varphi(s)) = \varphi(\mathbf{d}(s)) = \varphi(\mathrm{id}_{\Omega}) \geq \varphi(\mathrm{id}_{\Omega}) \smallsetminus \varphi(g)$ it follows that

$$\varphi(s)\,\langle\varphi(\mathrm{id}_{\Omega}) \smallsetminus \varphi(g)\rangle \mathrel{\mathscr{D}_T} \varphi(\mathrm{id}_{\Omega}) \smallsetminus \varphi(g) . \tag{4.9.7}$$

Since

$$\varphi(a) = \varphi(\mathrm{id}_\Omega) = \big(\varphi(\mathrm{id}_\Omega) \wedge \varphi(g)\big) \oplus \big(\varphi(\mathrm{id}_\Omega) \smallsetminus \varphi(g)\big),$$

and $\mathscr{D}_T$ is additive, it follows, by using (4.9.6) and (4.9.7), that $\varphi(a)\ \mathscr{D}_T\ \varphi(b)$. □

On the other hand, using the notation of the proof of Theorem 4.9.1, it is not hard to verify that $\mathbf{pMeas}(\mathcal{B}, G)$ is a type-preserving quotient of $\mathbf{pMeas}(\mathcal{B}', G)$.

The following example shows that there is no "best of two worlds" improving both Theorems 4.8.9 and 4.9.2: in that example, "fundamental Boolean inverse semigroup" and "Boolean inverse meet-semigroup" cannot be reached simultaneously.

Example 4.9.4 A countable, conical refinement monoid M with order-unit, such that there is no fundamental Boolean inverse meet-semigroup S with $\mathrm{Typ}\, S \cong M$.

Proof We consider again the conical refinement monoid P, introduced in Example 4.7.9, defined by the generators ε, 1 and the relation $\varepsilon + 1 = 1$. We denote by $\pi : P \twoheadrightarrow \mathbb{Z}^+$ the unique monoid homomorphism such that $\pi(1) = 1$ (and thus $\pi(\varepsilon) = 0$).

The additive group G of all eventually constant sequences $\big(x_n \mid n \in \mathbb{Z}^+\big)$ of integers, ordered componentwise, is an Abelian lattice-ordered group. For $x \in G$, we denote by $x(\infty)$ the constant value of $x(n)$ for large enough n. We also denote by χ_X the characteristic function of any subset X of $\mathbb{Z}^+$. The set

$$M = \big\{(n, x) \in P \times G^+ \mid \pi(n) = x(\infty)\big\}$$

is a conical submonoid of $P \times G^+$, with order-unit $\boldsymbol{e} = (1, \chi_{\mathbb{Z}^+})$.

For every $n \in \mathbb{Z}^+$, the set $M_n = \big\{(n, x) \in P \times G^+ \mid \pi(n) = x(k) \text{ for all } k \geq m\big\}$ is a submonoid of M, isomorphic to $P \times (\mathbb{Z}^+)^m$ via $(n, x) \mapsto \big(n, x + \pi(n)\chi_{[m,\infty)}\big)$. In particular, M_n is a refinement monoid. Since M is the directed union of all M_n, it follows that M is also a refinement monoid.

Now suppose that $M \cong \mathrm{Typ}\, T$ for some fundamental Boolean inverse meet-semigroup T. Since $\boldsymbol{e}$ is an order-unit of M and by Theorem 4.2.7, there are a positive integer n and an idempotent e of $\mathrm{M}_n^{\oplus}(T)$ such that, setting $S = e\,\mathrm{M}_n^{\oplus}(T)e$, the following relation holds:

$$(M, \boldsymbol{e}) \cong \big(\mathrm{Typ}\, S, \mathrm{typ}_S(1)\big).$$

Since T is a fundamental Boolean inverse meet-semigroup, so are $\mathrm{M}_n^{\oplus}(T)$ and S (apply Propositions 3.1.22 and 3.7.14). Pick an isomorphism $\iota : (\mathrm{Typ}\, S, \mathrm{typ}_S(1)) \to (M, \boldsymbol{e})$. Set $\mu(x) = \iota(\mathrm{typ}_S(x))$, for each $x \in S$. Observe that μ defines a V-measure from $\mathrm{Idp}\, S$ to M (cf. Example 4.6.4). The elements $\boldsymbol{a} = (\varepsilon, 0)$ and $\boldsymbol{b}_n = (0, \chi_{\{n\}})$, for $n \in \mathbb{Z}^+$ all belong to M.

Claim 1 For each $n \in \mathbb{Z}^+$, there is a unique $b_n \in \mathrm{Idp}\, S$ such that $\boldsymbol{b}_n = \mu(b_n)$.

Proof of Claim. Let $x, y \in \operatorname{Idp} S$ such that $\mu(x) = \mu(y) = \boldsymbol{b}_n$ and suppose that $x \neq y$. Since μ is a V-measure and since $\boldsymbol{b}_n$ is a minimal element of $M \setminus \{0\}$, the elements x and y are both atoms of $\operatorname{Idp} S$. Since $x \neq y$, it follows that $x \perp y$, thus $x \oplus y \leq 1$, and thus $2\boldsymbol{b}_n = \mu(x) + \mu(y) = \mu(x \oplus y) \leq^+ \mu(1) = \boldsymbol{e}$, a contradiction. This deals with the uniqueness part. The existence part follows from μ being a V-measure together with the inequalities $\boldsymbol{b}_n \leq^+ \boldsymbol{e} = \mu(1)$. □ Claim 1.

Since $\boldsymbol{a} + \boldsymbol{e} = \boldsymbol{e} = \mu(1)$ and μ is a V-measure, there are $a, e \in \operatorname{Idp} S$ such that $1 = a \oplus e$, $\mu(a) = \boldsymbol{a}$, and $\mu(e) = \boldsymbol{e}$. From $\mu(1) = \mu(e)$ it follows that $1 \mathrel{\mathscr{D}_S} e$, that is, there is $x \in S$ such that $\mathbf{d}(x) = 1$ and $\mathbf{r}(x) = e$. Set $a_k = x^k \langle a \rangle$, for each $k \in \mathbb{Z}^+$. From $1 = a \oplus x \langle 1 \rangle$ and an easy induction argument, we get

$$1 = \bigoplus_{k<n} a_k \oplus x^n \langle 1 \rangle\,, \quad \text{for each } n \in \mathbb{Z}^+. \tag{4.9.8}$$

Since S is an inverse meet-semigroup, the element $b = 1 \wedge x$ exists in S. From $b \leq 1$ it follows that b is idempotent. If $b \leq x^n$, then $b = b^2 \leq xb \leq x^{n+1}$; whence $b \leq x^n$ for each $n \in \mathbb{Z}^+$. It follows that $b \leq \mathbf{r}(x^n) = x^n \langle 1 \rangle$, for each $n \in \mathbb{Z}^+$, and thus, by (4.9.8),

$$\bigoplus_{k<n} a_k \leq 1 \smallsetminus b\,, \quad \text{for each } n \in \mathbb{Z}^+.$$

Since $a \mathrel{\mathscr{D}} a_k$ for all k, it thus follows that $(n\varepsilon, 0) = n\boldsymbol{a} \leq^+ \mu(1 \smallsetminus b)$, for all $n \in \mathbb{Z}^+$. Since $\mu(1 \smallsetminus b) \leq^+ \mu(1) = (1, \chi_{\mathbb{Z}^+})$, it follows from the definition of M that

$$\mu(1 \smallsetminus b) = (1, \chi_V)\,, \quad \text{for some cofinite subset } V \subseteq \mathbb{Z}^+. \tag{4.9.9}$$

Claim 2 The inequality $b_n \leq b$ holds, for each $n \in \mathbb{Z}^+$.

Proof of Claim. Since $b_n \leq 1 = \mathbf{d}(x)$, we get $b_n \mathrel{\mathscr{D}} x \langle b_n \rangle$. By Claim 1, it follows that $x \langle b_n \rangle = b_n$, thus $xb_n = b_n x = b_n x b_n$. Moreover, from $ex = x$ and $x \langle b_n \rangle = b_n$ it follows that $eb_n = b_n$, that is, $b_n \leq e$.

If $b_n x = 0$, then $b_n = b_n e = b_n x x^{-1} = 0$, a contradiction. Hence $b_n x \neq 0$. Since b_n is an atom of $\operatorname{Idp} S$ and by Lemma 3.7.2 together with the assumption that S is fundamental, it follows that $b_n x b_n \in \{0, b_n\}$. Since $b_n x = b_n x b_n$ and $b_n x \neq 0$, it follows that $b_n x = b_n$, that is (since b_n is idempotent), $b_n \leq x$. Since $b_n \leq 1$, the desired conclusion follows. □ Claim 2.

Since the $\boldsymbol{b}_n$ are pairwise meet-orthogonal in M and all $\boldsymbol{b}_n = \mu(b_n)$, the b_n are pairwise orthogonal in M. By Claim 2, it follows that $\bigoplus_{i<n} b_i \leq b$, for each $n \in \mathbb{Z}^+$, and thus, evaluating the two sides under μ, it follows that $\sum_{i<n} \boldsymbol{b}_i \leq^+ \mu(b)$, that is, $(0, \chi_{[0,n)}) \leq^+ \mu(b)$. This holds for every $n \in \mathbb{Z}^+$, thus, since $\mu(b) \leq^+ \mu(1) = (1, \chi_{\mathbb{Z}^+})$, it follows that $\mu(b) = (1, \chi_{\mathbb{Z}^+})$. Consequently, $(1, \chi_{\mathbb{Z}^+}) = \mu(1) = \mu(b) + \mu(1 \smallsetminus b) = (1, \chi_{\mathbb{Z}^+}) + \mu(1 \smallsetminus b)$, in contradiction with (4.9.9). □

Chapter 5
Type Theory of Special Classes of Boolean Inverse Semigroups

While Theorem 4.8.9 implies that the type monoid of a Boolean inverse semigroup S can be any countable conical refinement monoid, there are situations in which the structure of S impacts greatly the one of $\operatorname{Typ} S$. A basic illustration of this is given by the class of *AF inverse semigroups*, introduced in Lawson and Scott [77], which is the Boolean inverse semigroup version of the class of AF C*-algebras. Another Boolean inverse semigroup version of a class of C*-algebras, which we will not consider here, is given by the *Cuntz inverse monoids* studied in Lawson and Scott [77, § 3].

Section 5.1 introduces the class of *locally matricial inverse semigroups* (named by analogy with locally matricial rings), which are just the directed colimits of finite products of finite symmetric inverse semigroups, and the countable members of that class, the *AF inverse semigroups*. Mimicking the classical ring-theoretical proofs, we describe the type monoids of those Boolean inverse semigroups.

In Sect. 5.2, we describe a different class of fundamental Boolean inverse semigroups, whose type monoids are exactly the positive cones of Abelian lattice-ordered groups. Moreover, we show that the projectable Abelian lattice-ordered groups are exactly those arising from Boolean inverse meet-semigroups from that class.

It follows from Tarski [109] that monoids of equidecomposability types, with respect to exponentially bounded groups, are strongly separative (i.e., they satisfy the implication $\mathsf{x} + \mathsf{z} = \mathsf{y} + 2\mathsf{z} \Rightarrow \mathsf{x} = \mathsf{y} + \mathsf{z}$). In Sect. 5.3, we state a Boolean inverse semigroup version of that result. This result enables us, in particular, to extend Tarski's result to all supramenable groups.

In Sect. 5.4, we partly survey the impact of various amounts of completeness, of a Boolean inverse semigroup S, on the type monoid of S, and we illustrate the sharpness of those results by various examples and counterexamples.

F. Wehrung, *Refinement Monoids, Equidecomposability Types, and Boolean Inverse Semigroups*, Lecture Notes in Mathematics 2188,
DOI 10.1007/978-3-319-61599-8_5

5.1 Type Monoids of AF Inverse Semigroups

Most of the results presented in this section are analogues, for Boolean inverse semigroups, of classical results for nonstable K-theory of regular rings; see for example Goodearl [50, Chap. 15]. Some of the ideas that we will state here can also be found, with different formulations, in Lawson and Scott [77, § 2]. The inverse semigroups proofs are straightforward translations of the ring proofs, and we present only that part of them that we believe will help the reader to get familiar with the context.

Recall that symmetric inverse semigroups are introduced in Example 3.1.8.

Definition 5.1.1 An inverse semigroup S is

- *matricial* if it is a finite direct product of finite symmetric inverse monoids (equivalently, if it is a finite fundamental Boolean inverse semigroup);
- *locally matricial* if it is a directed colimit of finite fundamental Boolean inverse semigroups and additive semigroup homomorphism;
- *approximately finite*, or *AF*, if it is both countable and locally matricial.

It is fairly easy, although not completely trivial, to see that the definition of AF given in Definition 5.1.1 is equivalent to the one of Lawson and Scott [77]. That this is indeed the case will follow from Proposition 5.1.2.

It follows from Lawson [75, Theorem 4.18] that the finite fundamental Boolean inverse semigroups are exactly[1] the isomorphic copies of the $\prod_{i=1}^{m} \mathfrak{I}_{n_i}$, where $m \in \mathbb{Z}^+$ and all $n_i \in \mathbb{N}$.

This yields the following alternative description of locally matricial inverse semigroups

Proposition 5.1.2 *An inverse semigroup S is locally matricial iff it is a directed union of finite fundamental Boolean inverse semigroups that are additive in S. If those conditions are satisfied, then S is a fundamental Boolean inverse meet-semigroup.*

Proof If S is a directed union of finite fundamental Boolean inverse semigroups S_i that are additive in S, then it is also the directed colimit of the S_i, the transition maps being defined as the inclusion maps $S_i \hookrightarrow S_j$ whenever $S_i \subseteq S_j$.

Suppose, conversely, that $S = \varinjlim_{i \in I} S_i$, where I is a directed poset, with all the S_i finite fundamental Boolean inverse semigroups and all transition maps $f_i^j : S_i \to S_j$ additive. Since all f_i^j are bias homomorphisms (cf. Theorem 3.2.5), so are all limiting maps $f_i : S_i \to S$, for $i \in I$; that is, all f_i are additive. (See the comments following Theorem 3.2.5.) The subset $K_i = \ker f_i$ is an additive ideal in S_i, for each $i \in I$ (cf. Proposition 3.4.9). Denote by $p_i : S_i \twoheadrightarrow S_i / K_i$ the canonical projection.

[1]Those inverse semigroups are called *semisimple* in the first version of Lawson and Scott [77], which conflicts with the usual meaning of that word in ring theory. We chose instead to introduce semisimplicity through Definition 3.7.5.

By Proposition 3.7.6, S_i/K_i is a fundamental Boolean inverse semigroup and there is a unique additive semigroup embedding $\overline{f}_i: S_i/K_i \to S$ such that $f_i = \overline{f}_i \circ p_i$. Since S is the directed union of all $f_i[S_i]$, it is also the directed union of all (finite fundamental Boolean inverse semigroups) $\overline{f}_i[S_i/K_i]$.

Hence, the S_i, f_i^j, and f_i can be modified in such a way that each S_i is an additive inverse subsemigroup of S, and the f_i^j and f_i are all inclusion maps. By Proposition 3.7.10, all f_i^j are meet-homomorphisms. It follows that all f_i are meet-homomorphisms and S is a meet-semilattice, thus a Boolean inverse meet-semigroup.

Let $x \in S$ commute with all idempotents of S. Then $x \in S_i$ for some $i \in I$, and x commutes with every idempotent of S_i. Since S_i is a fundamental Boolean inverse semigroup, x is idempotent. Hence S is a fundamental Boolean inverse semigroup. □

Corollary 5.1.3 *The following statements hold, for any additive ideal S in an inverse semigroup T:*

(1) *If T is a finite symmetric inverse semigroup, then either $S = \{0\}$ or $S = T$.*
(2) *If T is a finite fundamental Boolean inverse semigroup, then so is S.*
(3) *If T is locally matricial, then so is S.*

Proof (1) All the atoms of $\operatorname{Idp} T$ are pairwise $\mathscr{D}$-equivalent, thus if one of them belongs to S, then all of them do. In the first case, $S = \{0\}$. In the second case, S contains all the idempotents of T, thus all its matrix units (because it is an ideal), thus $S = T$ (because S is closed under finite orthogonal joins).

(2) Let $T = \prod_{i=1}^{m} \mathfrak{I}_{n_i}$, where $m \in \mathbb{Z}^+$ and all $n_i \in \mathbb{N}$. Then each $\mathfrak{I}_{n_i}$ embeds, as an additive ideal, in T. Since S is an additive ideal, $S = \prod_{i=1}^{m} S_i$ where each $S_i = S \cap \mathfrak{I}_{n_i}$. By (1) above, either $S_i = \{0\}$ or $S_i = \mathfrak{I}_{n_i}$.

(3) By Proposition 5.1.2, T is an additive directed union of finite fundamental Boolean inverse semigroups T_i. Each $S \cap T_i$ is an additive ideal of T_i, thus, by (2) above, $S \cap T_i$ is a finite fundamental Boolean inverse semigroup. Since S is the additive directed union of all $S \cap T_i$, it is locally matricial. □

It is trivial that every locally matricial fundamental Boolean inverse semigroup S is locally finite. An easy application of Lemma 3.8.4 yields then that the group $\operatorname{Inn} S$ of all inner automorphisms of S is locally finite. As the following example shows, the converse does not hold.

Example 5.1.4 A fundamental unital Boolean inverse meet-semigroup S such that $\operatorname{Inn} S$ is locally finite but S has elements of infinite order (thus it is not locally finite).

Proof We consider the fundamental unital Boolean inverse meet-semigroup S introduced in Example 3.7.12 and the additive semigroup homomorphism $f: S \twoheadrightarrow \mathbb{Z}^{\sqcup 0}$ introduced in Example 3.7.13. The shift mapping $\sigma: \mathbb{Z}^+ \to \mathbb{Z}^+$, $n \mapsto n + 1$ is an element of S generating an infinite subsemigroup, and $f(\sigma) = 1$.

Now let u be an invertible element of S and set $m = f(u)$. Since u is a permutation of $\mathbb{Z}^+$ and $u \in S$, there exists $n \in \mathbb{Z}^+$ such that $u(k) = m + k$ for every $k \geq n$. In particular, u maps $[n, \infty)$ onto $[m + n, \infty)$. Since u is a bijection, it follows that $m + n = n$, that is, $m = 0_{\mathbb{Z}}$. Therefore, the invertible elements of S are exactly the permutations u of $\mathbb{Z}^+$ such that $u(k) = k$ for all large enough k. It follows easily that the invertible elements of S form a locally finite subgroup of S, thus that $\operatorname{Inn} S$ is locally finite. □

For the following result, we remind the reader that simplicial monoids are introduced in Sect. 1.5 and exhaustive group actions are introduced in Definition 4.7.12.

Proposition 5.1.5 *The following statements hold, for any Boolean inverse semigroup S:*

(1) *If S is a finite symmetric inverse semigroup, then either* $\operatorname{Typ} S = \{0\}$ *or* $\operatorname{Typ} S \cong \mathbb{Z}^+$.
(2) *If S is a finite fundamental Boolean inverse semigroup, then* $\operatorname{Typ} S$ *is a simplicial monoid.*
(3) *If S is locally matricial, then* $\operatorname{Typ} S$ *is the positive cone of a dimension group. Moreover, the action of* $\operatorname{Inn} S$ *on* $\operatorname{Idp} S$ *by restriction is exhaustive with respect to the dimension function* typ_S*; furthermore,* $\operatorname{Typ} S \cong \mathbb{Z}^+ \langle \operatorname{Idp} S\rangle / \operatorname{Inn} S$.

Proof (1) Let $S = \mathfrak{I}_m$, where $m \in \mathbb{Z}^+$. If $m = 0$ then $\operatorname{Typ} S = \{0\}$. Suppose that $m \neq 0$. Then the idempotents of $\mathfrak{I}_m$ are the identities on the subsets of $[m]$. Furthermore, for all $X, Y \subseteq [m]$, $\mathrm{id}_X \mathscr{D} \mathrm{id}_Y$ iff $\operatorname{card} X = \operatorname{card} Y$. It follows easily that $(\operatorname{Typ} \mathfrak{I}_m, \operatorname{typ}(1)) \cong (\mathbb{Z}^+, m)$ if $m > 0$.

(2) Let $S = \prod_{i=1}^m \mathfrak{I}_{n_i}$, where $m \in \mathbb{Z}^+$ and each $n_i \in \mathbb{N}$. Then, by the proof of (1) together with Proposition 4.1.9(1), we get $(\operatorname{Typ} S, \operatorname{typ}(1)) \cong ((\mathbb{Z}^+)^m, \vec{n})$ where $\vec{n} = (n_1, \dots, n_m)$.

(3) Consider a directed colimit $S = \varinjlim_{i\in I} S_i$ with respect to additive semigroup homomorphisms, where all the S_i are finite fundamental Boolean inverse semigroups. It follows from Proposition 4.1.9(2) that $\operatorname{Typ} S = \varinjlim_{i\in I} \operatorname{Typ} S_i$. By (2) above, each $\operatorname{Typ} S_i$ is a simplicial monoid. Thus, $\operatorname{Typ} S$ is the positive cone of a dimension group.

Now let $a, b \in \operatorname{Idp} S$ such that $a \mathscr{D} b$. There is $u \in S$ such that $a = \mathbf{d}(u)$ and $b = \mathbf{r}(u)$. Let $i \in I$ such that $u \in S_i$ and denote by e_i the unit element of S_i. Since

$$\operatorname{typ}_{S_i}(e_i) = \operatorname{typ}_{S_i}(a) + \operatorname{typ}_{S_i}(e_i \smallsetminus a) = \operatorname{typ}_{S_i}(b) + \operatorname{typ}_{S_i}(e_i \smallsetminus b)\,,$$

it follows from the equation $\operatorname{typ}_{S_i}(a) = \operatorname{typ}_{S_i}(b)$ together with the cancellativity of $\operatorname{Typ} S_i$ (cf. (2) above) that $\operatorname{typ}_{S_i}(e_i \smallsetminus a) = \operatorname{typ}_{S_i}(e_i \smallsetminus b)$, so there is $v \in S_i$ such that $\mathbf{d}(v) = e_i \smallsetminus a$ and $\mathbf{r}(v) = e_i \smallsetminus b$. It follows that $g = u \oplus v$ is an invertible element of S_i and $g\langle a\rangle = b$, so $\operatorname{inn}_g(a) = b$. This proves that the action of $\operatorname{Inn} S$ on $\operatorname{Idp} S$ is exhaustive with respect to the dimension function typ_S. The relation $\operatorname{Typ} S \cong \mathbb{Z}^+ \langle \operatorname{Idp} S\rangle / \operatorname{Inn} S$ follows then immediately from Theorem 4.7.14. □

There is some overlap between some of the conclusions of Proposition 5.1.5(3): since S is locally matricial, the group $\mathrm{Inn}\, S$ is locally finite, thus, since $\mathbb{Z}^+\langle \mathrm{Idp}\, S\rangle$ is an Abelian lattice-ordered group and by Corollary 2.9.8, the monoid $\mathbb{Z}^+\langle \mathrm{Idp}\, S\rangle/\mathrm{Inn}\, S$ is the positive cone of a dimension group. Nevertheless, the proof of Proposition 5.1.5(3) above also yields an explicit representation of $\mathrm{Typ}\, S \cong \mathbb{Z}^+\langle \mathrm{Idp}\, S\rangle/\mathrm{Inn}\, S$ as a directed colimit of simplicial monoids (viz., the monoids $\mathrm{Typ}\, S_i$, for $i \in I$).

We are now aiming at a converse to Proposition 5.1.5. In view of Theorem 4.6.9, there are dimension groups of cardinality $\aleph_2$ whose positive cone is not isomorphic to $\mathrm{Typ}\, S$ for any Boolean inverse semigroup S. For the smaller cardinalities, it is known since the 1979 edition of Goodearl [50] for the countable case, and Goodearl and Handelman [52] for the $\aleph_1$ case, that every positive cone of a dimension group of cardinality at most $\aleph_1$ is isomorphic to $\mathrm{V}(R)$ for some locally matricial von Neumann regular ring R. It turns out that those ring-theoretical proofs can be easily adapted to the context of Boolean inverse semigroups, by observing that all the matrices involved in them have entries either 0 or 1.

The following result apes, both in its statement and its proof, the one in Goodearl [50, Lemma 15.23]. We include a proof for convenience. Recall that the inner endomorphisms ad_x are introduced in Sect. 3.8.

Lemma 5.1.6 *The following statements hold, for any finite fundamental Boolean inverse semigroup S and any Boolean inverse semigroup T:*

(1) *For any $c \in \mathrm{Idp}\, T$ and any monoid homomorphism $\boldsymbol{f}\colon \mathrm{Typ}\, S \to \mathrm{Typ}\, T$ with $\boldsymbol{f}(\mathrm{typ}_S(1)) = \mathrm{typ}_T(c)$, there exists an additive semigroup homomorphism $f\colon S \to T$ such that $f(1) = c$ and $\boldsymbol{f} = \mathrm{Typ} f$.*
(2) *For any additive semigroup homomorphisms $f, g\colon S \to T$, $\mathrm{Typ} f = \mathrm{Typ}\, g$ iff there is $x \in T$ with $\mathbf{d}(x) = f(1)$, $\mathbf{r}(x) = g(1)$, and $g = \mathrm{ad}_x \circ f$.*

Proof By Lawson [75, Theorem 4.18], there are a nonnegative integer m, central idempotents $e_1, \dots, e_m$ of S, and positive integers $n_1, \dots, n_m$ such that $1 = \bigoplus_{i=1}^m e_i$ in S and each $e_i S \cong \mathfrak{I}_{n_i}$. It follows that

$$\mathrm{typ}_T(c) = \boldsymbol{f}(\mathrm{typ}_S(1)) = \sum_{i=1}^{m} \boldsymbol{f}(\mathrm{typ}_S(e_i))\,.$$

By Lemma 4.1.6, there is a decomposition $c = \bigoplus_{i=1}^m c_i$ in $\mathrm{Idp}\, T$ such that each $\boldsymbol{f}(\mathrm{typ}_S(e_i)) = \mathrm{typ}_T(c_i)$. Denote by $\left(e_{j,k}^{(i)} \mid (j,k) \in [n_i] \times [n_i]\right)$ a system of matrix units of $e_i S \cong \mathfrak{I}_{n_i}$ (cf. Example 3.1.8). According to Proposition 5.1.5(2), $\mathrm{Typ}\, S$ is a simplicial monoid, with simplicial basis $\left(\mathrm{typ}_S(e_{1,1}^{(1)}), \dots, \mathrm{typ}_S(e_{1,1}^{(m)})\right)$.

(1) For each $i \in [m]$, $\mathrm{typ}_T(c_i) = \boldsymbol{f}(\mathrm{typ}_S(e_i)) = n_i \cdot \mathrm{typ}_S(e_{1,1}^{(i)})$, thus, again by Lemma 4.1.6, there is a decomposition $c_i = \bigoplus_{j=1}^{n_i} c_{j,j}^{(i)}$ such that each $\mathrm{typ}_T(c_{j,j}^{(i)}) = \boldsymbol{f}\left(\mathrm{typ}_S(e_{j,j}^{(i)})\right)$. Whenever $2 \leq j \leq n_i$, the relation $c_{j,j}^{(i)}\ \mathscr{D}_T\ c_{1,1}^{(i)}$ holds, thus there is $c_{j,1}^{(i)} \in T$ such that $\mathbf{d}(c_{j,1}^{(i)}) = c_{1,1}^{(i)}$ and $\mathbf{r}(c_{j,1}^{(i)}) = c_{j,j}^{(i)}$. Now set $c_{1,j}^{(i)} = \left(c_{j,1}^{(i)}\right)^{-1}$ whenever

$1 \leq j \leq n_i$, and further, set

$$c_{j,k}^{(i)} = c_{j,1}^{(i)} c_{1,k}^{(i)}, \quad \text{whenever } j, k \in [n_i].$$

It is straightforward to verify that the $c_{j,k}^{(i)}$ satisfy the defining relations of the matrix units of $\mathfrak{I}_{n_i}$ (viz., the relations (3.1.4) and (3.1.5), where n is replaced by n_i and $e_{j,k}$ by $c_{j,k}^{(i)}$). The elements of $e_i S \cong \mathfrak{I}_{n_i}$ are exactly the orthogonal joins of elements of the form $e_{j,k}^{(i)}$, that is, all the elements of S of the form $\bigoplus_{j \in \mathrm{dom}(x)} e_{j,x(j)}^{(i)}$, where $x \in \mathfrak{I}_{n_i}$. This makes it possible to define an additive semigroup homomorphism $f_i \colon e_i S \to T$ by the rule

$$f_i\Bigl(\bigoplus_{j \in \mathrm{dom}(x)} e_{j,x(j)}^{(i)}\Bigr) = \bigoplus_{j \in \mathrm{dom}(x)} c_{j,x(j)}^{(i)}, \quad \text{for all } x \in \mathfrak{I}_{n_i}.$$

From the equations $c_{j,k}^{(i)} = c_i c_{j,k}^{(i)} c_i$ it follows, in particular, that $f_i(x) \in c_i T c_i$. Since the central elements c_i are pairwise orthogonal, it is possible to define an additive semigroup homomorphism $f \colon S \to T$ by the rule

$$f(x) = \bigoplus_{i=1}^{m} f_i(e_i x), \quad \text{for all } x \in S.$$

For every $i \in [m]$, $(\mathrm{Typ} f)\bigl(\mathrm{typ}_S(e_{1,1}^{(i)})\bigr) = \mathrm{typ}_T(c_{1,1}^{(i)}) = \boldsymbol{f}\bigl(\mathrm{typ}_S(e_{1,1}^{(i)})\bigr)$. Therefore, $\boldsymbol{f} = \mathrm{Typ} f$.

(2) Suppose first that $g = \mathrm{ad}_x \circ f$ for some $x \in T$ with $\mathbf{d}(x) = f(1)$ and $\mathbf{r}(x) = g(1)$. For every $a \in \mathrm{Idp}\, S$, $g(a) = x f(a) x^{-1} = x f(a) \bigl(x f(a)\bigr)^{-1}$ while, using the inequality $f(a) \leq f(1) = \mathbf{d}(x)$, we get $f(a) = \bigl(x f(a)\bigr)^{-1} x f(a)$, so $f(a) \mathrel{\mathscr{D}} g(a)$, and so $(\mathrm{Typ} f)(\mathrm{typ}_S(a)) = (\mathrm{Typ}\, g)(\mathrm{typ}_S(a))$. This proves that $\mathrm{Typ} f = \mathrm{Typ}\, g$.

Suppose, conversely, that $\mathrm{Typ} f = \mathrm{Typ}\, g$. Set $a = f(1)$ and $b = g(1)$. Set also $f_{j,k}^{(i)} = f\bigl(e_{j,k}^{(i)}\bigr)$ and $g_{j,k}^{(i)} = g\bigl(e_{j,k}^{(i)}\bigr)$, for all $i \in [m]$ and all $j \in [n_i]$. From $\mathrm{Typ} f = \mathrm{Typ}\, g$ it follows that for all $i \in [m]$, the relation $f_{1,1}^{(i)} \mathrel{\mathscr{D}} g_{1,1}^{(i)}$ holds within T, thus there is $x_i \in T$ such that $\mathbf{d}(x_i) = f_{1,1}^{(i)}$ and $\mathbf{r}(x_i) = g_{1,1}^{(i)}$. For each $j \in [n_i]$, the product $g_{j,1}^{(i)} x_i f_{1,j}^{(i)}$ is a trace product, with domain $f_{j,j}^{(i)}$ and range $g_{j,j}^{(i)}$. It follows that the element $x = \bigoplus_{i \in [m],\, j \in [n_i]} g_{j,1}^{(i)} x_i f_{1,j}^{(i)}$ is well defined, and $x^{-1} = \bigoplus_{i \in [m],\, j \in [n_i]} f_{j,1}^{(i)} x_i^{-1} g_{1,j}^{(i)}$. Now for all $i \in [m]$ and all $p, q \in [n_i]$, the only value of $j \in [n_i]$ such that $f_{1,j}^{(i)} f_{p,q}^{(i)} \neq 0$ is $j = p$. Hence

$$x f_{p,q}^{(i)} = g_{p,1}^{(i)} x_i f_{1,p}^{(i)} f_{p,q}^{(i)} = g_{p,1}^{(i)} x_i f_{1,q}^{(i)}.$$

Now the only value of $j \in [n_i]$ such that $f_{1,q}^{(i)} f_{j,1}^{(i)} \neq 0$ is $j = q$. Hence

$$x f_{p,q}^{(i)} x^{-1} = g_{p,1}^{(i)} x_i f_{1,q}^{(i)} f_{q,1}^{(i)} x_i^{-1} g_{1,q}^{(i)} = g_{p,1}^{(i)} x_i f_{1,1}^{(i)} x_i^{-1} g_{1,q}^{(i)} = g_{p,1}^{(i)} g_{1,1}^{(i)} g_{1,q}^{(i)} = g_{p,q}^{(i)}.$$

This holds for all $i \in [m]$ and all $j \in [n_i]$, thus $g = \mathrm{ad}_x \circ f$. □

The inverse semigroup analogue of Goodearl and Handelman [52, Lemma 1.3], itself arising from Kado [64, Lemma 3], is then the following.

Lemma 5.1.7 *Let S be an AF inverse monoid, let T be a Boolean inverse semigroup, let $c \in T$, and let $\boldsymbol{g}\colon \mathrm{Typ}\, S \to \mathrm{Typ}\, T$ be a monoid homomorphism such that $\boldsymbol{g}(\mathrm{typ}_S(1)) = \mathrm{typ}_T(c)$. Then there exists an additive semigroup homomorphism $g\colon S \to T$ such that $g(1) = c$ and $\boldsymbol{g} = \mathrm{Typ}(g)$.*

Proof By virtue of Proposition 5.1.2, we can write $S = \bigcup_{n\in\mathbb{Z}^+} S_n$ (directed union), with each S_n a finite fundamental Boolean inverse semigroup which is additive in S. We may assume that $1 \in S_0$. Denote by $e_n\colon S_n \hookrightarrow S_{n+1}$ and $f_n\colon S_n \hookrightarrow S$ the inclusion maps, for $n \in \mathbb{Z}^+$. Then $\boldsymbol{g} \circ \mathrm{Typ} f_0\colon \mathrm{Typ}\, S_0 \to \mathrm{Typ}\, T$ is a monoid homomorphism sending $\mathrm{typ}_{S_0}(1)$ to $\mathrm{typ}_T(c)$, thus, by Lemma 5.1.6(1), there is an additive semigroup homomorphism $g_0\colon S_0 \to T$ sending 1 to c such that $\boldsymbol{g} \circ \mathrm{Typ} f_0 = \mathrm{Typ}\, g_0$. Let $n \in \mathbb{Z}^+$ and suppose having constructed an additive semigroup homomorphism $g_n\colon S_n \to T$ sending 1 to c such that $\boldsymbol{g} \circ \mathrm{Typ} f_n = \mathrm{Typ}\, g_n$. By Lemma 5.1.6(1), there is an additive semigroup homomorphism $h\colon S_{n+1} \to T$ sending 1 to c such that $\boldsymbol{g} \circ \mathrm{Typ} f_{n+1} = \mathrm{Typ}\, h$. It follows that $h \circ e_n$ and g_n are both additive semigroup homomorphisms from S_n to T, sending 1 to c, with $\mathrm{Typ}(h \circ e_n) = \mathrm{Typ}\, g_n$. By Lemma 5.1.6(2), there exists $x \in T$ such that $\mathbf{d}(x) = \mathbf{r}(x) = c$ and $g_n = \mathrm{ad}_x \circ h \circ e_n$. Set $g_{n+1} = \mathrm{ad}_x \circ h$. Hence $g_n = g_{n+1} \circ e_n = g_{n+1} \restriction_{S_n}$. Moreover, $\mathrm{Typ}\, g_{n+1} = \mathrm{Typ}\, h = \boldsymbol{g} \circ \mathrm{Typ} f_{n+1}$ and $g_{n+1}(1) = c$, thus completing the inductive construction of the g_n. The common extension $g\colon S \to T$ of all g_n is an additive semigroup homomorphism sending 1 to c, and $\boldsymbol{g} = \mathrm{Typ}\, g$. □

It will follow from Example 5.1.12 that the assumption that S be AF cannot be dropped from the statement of Lemma 5.1.7.

We obtain the following inverse semigroup analogue of Goodearl and Handelman [52, Theorem 1.5]. The last part of our statement involves the concept of group-measurability introduced in Definition 4.8.1.

Theorem 5.1.8 *Let $(G, \boldsymbol{u})$ be a dimension group with order-unit. If $\mathrm{card}\, G \leq \aleph_1$, then there exists a unital locally matricial fundamental Boolean inverse semigroup S such that $(G^+, \boldsymbol{u}) \cong (\mathrm{Typ}\, S, \mathrm{typ}_S(1))$. Consequently, $(G^+, \boldsymbol{u})$ is group-measurable, via a locally finite group.*

Proof We first deal with the case where G is simplicial. Let $(G, \boldsymbol{u}) = (\mathbb{Z}^m, \vec{n})$ where $\vec{n} = (n_1, \dots, n_m)$, with all $n_i \in \mathbb{N}$. Setting $S = \prod_{i=1}^m \mathfrak{I}_{n_i}$, then $(\mathrm{Typ}\, S, \mathrm{typ}_S(1)) \cong (G^+, \boldsymbol{u})$ (cf. Proposition 5.1.5(2)).

Next, we deal with the case where G is countable. By the directed colimit representation theorem of Grillet [57] and Effros et al. [38], we can write $(G, \boldsymbol{u}) = \varinjlim \vec{G}$, where $\vec{G}$ is a direct system of simplicial groups with order-unit $(G_n, \boldsymbol{u}_n)$ and positive homomorphisms. Denote by $\boldsymbol{f}_n\colon (G_n^+, \boldsymbol{u}_n) \to (G_{n+1}^+, \boldsymbol{u}_{n+1})$ the restrictions to positive cones of the corresponding transition maps. By the simplicial case, for each $n \in \mathbb{Z}^+$, there are a finite fundamental Boolean inverse semigroup S_n and an isomorphism $\varepsilon_n\colon (\operatorname{Typ} S_n, \operatorname{typ}_{S_n}(1)) \to (G_n^+, \boldsymbol{u}_n)$. By Lemma 5.1.6(1), there is an additive semigroup homomorphism $f_n\colon S_n \to S_{n+1}$, preserving the unit, such that $\varepsilon_{n+1}^{-1}\boldsymbol{f}_n\varepsilon_n = \operatorname{Typ} f_n$. Denote by $\vec{S}$ the direct system of finite fundamental Boolean inverse semigroups and additive semigroup homomorphisms consisting of the S_n and the f_n. Then the sequence $(\varepsilon_n \mid n \in \mathbb{Z}^+)$ defines a natural equivalence from $\operatorname{Typ}\vec{S}$ to $\vec{G}$. Setting $S = \varinjlim \vec{S}$, it follows from Proposition 4.1.9(2) that $(\operatorname{Typ} S, \operatorname{typ}_S(1)) \cong (G^+, \boldsymbol{u})$.

Finally, we deal with the case where $\operatorname{card} G = \aleph_1$. Denoting by ω_1 the first uncountable ordinal, a standard Löwenheim-Skolem type argument enables us to construct an increasing ω_1-sequence $(G_\xi \mid \xi < \omega_1)$ of countable dimension subgroups of G, with $\boldsymbol{u} \in G_0$, such that $G_\lambda = \bigcup_{\xi<\lambda} G_\xi$ for every countable limit ordinal λ. Denote by $\boldsymbol{f}_\xi^\eta\colon G_\xi^+ \to G_\eta^+$ the inclusion map, for any $\xi \le \eta < \omega_1$. By the countable case, for each $\xi < \omega_1$, there is a unital AF inverse semigroup S_ξ such that $(\operatorname{Typ} S_\xi, \operatorname{typ}_{S_\xi}(1)) \cong (G_\xi^+, \boldsymbol{u})$. We construct inductively a direct system of unit-preserving additive transition maps $f_\xi^\eta\colon S_\xi \to S_\eta$, together with a natural isomorphism $(\varepsilon_\xi \mid \xi < \omega_1)$ with $\varepsilon_\xi\colon (\operatorname{Typ} S_\xi, \operatorname{typ}_{S_\xi}(1)) \to (G_\xi^+, \boldsymbol{u})$, as follows. At stage 0, just pick any isomorphism $\varepsilon_0\colon (\operatorname{Typ} S_0, \operatorname{typ}_{S_0}(1)) \to (G_0^+, \boldsymbol{u})$. If everything is defined up to stage ξ, then pick any isomorphism

$$\varepsilon_{\xi+1}\colon (\operatorname{Typ} S_{\xi+1}, \operatorname{typ}_{S_{\xi+1}}(1)) \to (G_{\xi+1}^+, \boldsymbol{u})$$

and further, by virtue of Lemma 5.1.7, pick any additive monoid homomorphism $f_\xi^{\xi+1}\colon S_\xi \to S_{\xi+1}$ such that $\operatorname{Typ} f_\xi^{\xi+1} = \varepsilon_{\xi+1}^{-1} \circ \boldsymbol{f}_\xi^{\xi+1} \circ \varepsilon_\xi$; then set $f_\eta^{\xi+1} = f_\xi^{\xi+1} \circ f_\eta^\xi$ for each $\eta \le \xi$, and, of course, $f_{\xi+1}^{\xi+1} = \operatorname{id}_{S_{\xi+1}}$.

Finally, let $\lambda < \omega_1$ such that everything is defined below λ. Define

$$S_\lambda = \varinjlim_{\xi<\lambda} S_\xi\,. \tag{5.1.1}$$

By virtue of the natural transformation $(\varepsilon_\xi \mid \xi < \lambda)$, we get an isomorphism ε_λ from $\operatorname{Typ} S_\lambda = \varinjlim_{\xi<\lambda} \operatorname{Typ} S_\xi$ onto $\varinjlim_{\xi<\lambda} G_\xi^+ = \bigcup_{\xi<\lambda} G_\xi^+ = G_\lambda^+$ (in order not to overload the notation we drop the units). For $\xi < \lambda$, define f_ξ^λ as the limiting map of the directed colimit (5.1.1). This completes the inductive construction of the desired direct system and natural isomorphism. Defining $S = \varinjlim_{\xi<\omega_1} S_\xi$, it follows that $(\operatorname{Typ} S, \operatorname{typ}_S(1)) \cong (G^+, \boldsymbol{u})$. □

Together with results from earlier sections, this enables us to confirm a guess from the first version of Kudryavtseva et al. [71, § 2]. We quote the following statement from their paper:

> We do not believe that all 0-simplifying AF inverse monoids are necessarily UHF.

Here, a Boolean inverse monoid S is *0-simplifying* if it has no non-trivial additive ideals. By Proposition 4.2.4, this is, for non-trivial S, equivalent to saying that $\operatorname{Typ} S$ has no non-trivial o-ideals, that is, $\operatorname{Typ} S$ is a simple refinement monoid (cf. Sect. 1.5). A Boolean inverse monoid is *UHF* if it is a countable directed colimit, via additive transition maps, of finite symmetric inverse monoids.

Example 5.1.9 A 0-simplifying AF Boolean inverse monoid, which is not UHF.

Proof Pick any simple dimension group G of rank greater than 1 (i.e., with two elements with no nonzero common integer multiple). For example, let $G = \mathbb{Q} \times \mathbb{Q}$, with positive cone $\{(0,0)\} \cup (\mathbb{Q}^{++} \times \mathbb{Q}^{++})$. By Theorem 5.1.8, there is a unital fundamental Boolean inverse semigroup S such that $\operatorname{Typ} S \cong G^+$. Since G is simple and by Proposition 4.2.4, S is 0-simplifying. However, the type monoid of an UHF semigroup is easily seen to have rank at most 1. Hence, S is not UHF. □

Theorem 5.1.8 also has a non-unital version.

Theorem 5.1.10 *Let G be a dimension group. If* $\operatorname{card} G \leq \aleph_1$, *then there exists a locally matricial fundamental Boolean inverse semigroup S such that* $G^+ \cong \operatorname{Typ} S$. *Consequently, G^+ is group-measurable, via a locally finite group.*

Proof We first embed G as an ideal into a unital dimension group H with $\operatorname{card} H \leq \aleph_1$. For example, define H as the lexicographical product $\mathbb{Z} \times_{\mathrm{lex}} G$, which admits $(1,0)$ as an order-unit. By Theorem 5.1.8, there are a locally matricial fundamental Boolean inverse semigroup T and an isomorphism $\varepsilon\colon \operatorname{Typ} T \to H^+$. By Proposition 4.2.4, the set $S = \{x \in T \mid \varepsilon(\operatorname{typ}_T(x)) \in G^+\}$ is an additive ideal of T, thus it is a locally matricial fundamental Boolean inverse semigroup (cf. Corollary 5.1.3), and $\operatorname{Typ} S \cong G^+$. □

By modifying *mutatis mutandis* the proof of Elliott's Theorem stating the uniqueness of an AF C*-algebra with given partially ordered K_0 group, we can also obtain the following uniqueness result. Its proof is a back-and-forth variant of the one of Lemma 5.1.7 and we omit it, referring the reader to Theorems 4.3 and 5.3 in Elliott [39], see also Goodearl [50, Theorem 15.26] for the regular rings analogue.

Theorem 5.1.11 *Let S and T be unital AF inverse semigroups. Then for any isomorphism* $f\colon (\operatorname{Typ} S, \operatorname{typ}_S(1)) \to (\operatorname{Typ} T, \operatorname{typ}_T(1))$, *there exists an isomorphism* $f\colon S \to T$ *of inverse semigroups such that* $f = \operatorname{Typ} f$.

If we drop the assumption that S and T be AF, then counterexamples arise.

Example 5.1.12 Countable, fundamental Boolean inverse monoids S_0 and S_1 such that $(\operatorname{Typ} S_0, \operatorname{typ}_{S_0}(1)) \cong (\operatorname{Typ} S_1, \operatorname{typ}_{S_1}(1)) \cong (\mathbb{Q}^+, 1)$, S_0 is AF, and S_1 is not locally finite (thus $S_0 \not\cong S_1$).

Proof Denote by B the Boolean subalgebra of the powerset algebra of the real interval $I = (0,1]$ generated by all intervals $(x,y]$, where $0 \leq x \leq y \leq 1$ and $x, y \in \mathbb{Q}$. The restriction μ, of the Lebesgue measure on $\mathbb{R}$, to B, is a $\mathbb{Q}^+$-valued V-measure. In particular, $\mu((x,y]) = y - x$, whenever $0 \leq x \leq y \leq 1$ and $x, y \in \mathbb{Q}$. Set $S = \operatorname{Inv}(B, \mu)$ and denote by S_0 the inverse subsemigroup of S consisting of all piecewise translations via rational scalars. By considering all subdivisions

$0 < \frac{1}{n!} < \cdots < \frac{n!-1}{n!} < 1$, it is not hard to see that S_0 is AF, and in fact it is a directed union of finite fundamental Boolean inverse semigroups of the form $\mathfrak{I}_{n!}$. In particular, by Theorem 4.7.13, $(\operatorname{Typ} S_0, \operatorname{typ}_{S_0}(1)) \cong (\mathbb{Q}^+, 1)$.

On the other hand, pick any irrational number α such that $0 < \alpha < 1$ and let $\tau: I \to I$ defined by the rule

$$\tau(x) = \begin{cases} \alpha + x, & \text{if } \alpha + x < 1, \\ \alpha + x - 1, & \text{otherwise.} \end{cases}$$

Then $\tau \in \operatorname{Aut}(B, \mu)$, thus $\tau \in S$, and $\tau^n \neq \mathrm{id}$ for every nonzero integer n. Let S_1 be a countable additive inverse subsemigroup of S such that $S_0 \cup \{\tau\} \subseteq S_1$. By Theorem 4.7.13, $(\operatorname{Typ} S_1, \operatorname{typ}_{S_1}(1)) \cong (\mathbb{Q}^+, 1)$. Nevertheless, S_0 is locally finite while S_1 is not. □

Example 5.1.12 is to be put in contrast with Theorem 4.6.8: the latter implies the uniqueness of the *countable V-measure* representing $(\mathbb{Q}^+, 1)$ (or any given countable conical refinement monoid with order-unit), while the former implies the non-uniqueness of the *countable fundamental Boolean inverse semigroup* representing $(\mathbb{Q}^+, 1)$.

5.2 Representing Positive Cones of Abelian Lattice-Ordered Groups

We know from Theorem 5.1.10 that the positive cone of every dimension group with up to $\aleph_1$ elements is group-measurable. By Theorem 4.6.9, this result does not extend to dimension groups with at least $\aleph_2$ elements. Nevertheless, we shall see with Theorem 5.2.7 that for the special case of lattice-ordered groups, there is no need for any cardinality restriction.

5.2.1 *The Enveloping Boolean Ring of a Distributive Lattice with Zero*

Throughout Sect. 5.2.1 we shall fix a distributive lattice D with zero. The *enveloping Boolean ring* of D is defined as the unique (up to isomorphism) Boolean ring $\mathrm{BR}(D)$ such that D embeds into $\mathrm{BR}(D)$ as a 0-sublattice, and such that D generates $\mathrm{BR}(D)$ as a Boolean ring (cf. Grätzer [53, Theorem 158]).

Details about the construction of $\mathrm{BR}(D)$, along with its basic properties, can be found in Grätzer [53, § II.4]. It is well known (cf. Grätzer [53, Lemma 155]) that the elements of $\mathrm{BR}(D)$ are exactly those of the form

$$x = a_0 + a_1 + \cdots + a_{n-1}, \quad \text{where } a_0 \leq a_1 \leq \cdots \leq a_{n-1} \text{ in } D \tag{5.2.1}$$

(cf. (1.4.2) for the definition of $x + y$ in terms of the Boolean algebra operations). By possibly adding a null summand to the left in (5.2.1), we may assume there that $n = 2m$ for some m, and then

$$x = \sum_{i=0}^{m-1} (a_{2i} + a_{2i+1}) .$$

Now for each $i \in \{0, \dots, m-1\}$, it follows from the inequality $a_{2i} \leq a_{2i+1}$ that $a_{2i} + a_{2i+1} = a_{2i+1} \smallsetminus a_{2i}$. Moreover, the elements $a_{2i+1} \smallsetminus a_{2i}$ are pairwise disjoint, so their sums are in fact orthogonal joins, and so

$$x = \bigoplus_{i=0}^{m-1} (a_{2i+1} \smallsetminus a_{2i}) .$$

Write $D^{[2]} = \{(x, y) \in D \times D \mid x \leq y\}$, and call the elements of the form $b \smallsetminus a$, where $(a, b) \in D^{[2]}$, the *elementary generators* of $\mathrm{BR}(D)$ (relatively to D).

Lemma 5.2.1 *Let $(a, b), (a', b') \in D^{[2]}$. Then $b \smallsetminus a \leq b' \smallsetminus a'$ iff $a' \wedge b \leq a$ and $b \leq a \vee b'$. If this statement holds, then $b \smallsetminus a = \overline{b} \smallsetminus \overline{a}$ for some $\overline{a}, \overline{b} \in D$ with $a' \leq \overline{a} \leq \overline{b} \leq b'$.*

Proof $b \smallsetminus a \leq b' \smallsetminus a'$ iff $b \smallsetminus a \leq b'$ and $(b \smallsetminus a) \wedge a' = 0$, iff $b \leq a \vee b'$ and $b \wedge a' \leq a$.

Suppose, conversely, that $b \leq a \vee b'$ and $b \wedge a' \leq a$. Then $b \smallsetminus a = (b \smallsetminus a) \wedge (b' \smallsetminus a') = (b \wedge b') \smallsetminus \big(b \wedge b' \wedge (a \vee a')\big)$, so we may replace b by $b \wedge b'$ and thus assume that $b \leq b'$. Then a direct application of the paragraph above yields $b \smallsetminus a = (a' \vee b) \smallsetminus (a' \vee a)$. Hence $\overline{a} = a' \vee a$ and $\overline{b} = a' \vee b$ are as required. □

Lemma 5.2.2 *Let $a, b, c \in D$ with $a \leq b$. Then $(b \smallsetminus a) \wedge c = 0$ iff there are $a', b' \in D$ such that $c \leq a' \leq b'$ and $b \smallsetminus a = b' \smallsetminus a'$.*

Proof Suppose first that $(b \smallsetminus a) \wedge c = 0$, that is, $b \wedge c \leq a$. Let $a' = a \vee c$ and $b' = b \vee c$; so $c \leq a' \leq b'$, and a direct application of Lemma 5.2.1 yields $b \smallsetminus a = b' \smallsetminus a'$.

If, conversely, $b \smallsetminus a = b' \smallsetminus a'$ with $c \leq a' \leq b'$, then $(b \smallsetminus a) \wedge c \leq (b' \smallsetminus c) \wedge c = 0$. □

Set $B \downarrow \neg c = \{x \in B \mid x \wedge c = 0_B\}$, for any element c in a Boolean ring B (this is consistent with the notation $B \downarrow b$ in case B has a unit, and it makes sense even in case B has no unit).

Lemma 5.2.3 *For each $c \in D$, there exists a unique isomorphism of Boolean rings from $\mathrm{BR}(D \uparrow c)$ onto $\mathrm{BR}(D) \downarrow \neg c$, sending $y \smallsetminus x$ (within $\mathrm{BR}(D \uparrow c)$) to $y \smallsetminus x$ (within $\mathrm{BR}(D)$) for all $(x, y) \in (D \uparrow c)^{[2]}$.*

Proof The inclusion embedding $D \uparrow c \hookrightarrow D$ extends to a unique embedding $\eta\colon \mathrm{BR}(D\uparrow c) \hookrightarrow \mathrm{BR}(D)$ of Boolean rings (cf. Grätzer [53, Corollary 160]). It follows from Lemma 5.2.2 that the range of η is $\mathrm{BR}(D) \downarrow \neg c$. □

5.2.2 Representing Lattice-Ordered Groups

For an arbitrary Abelian lattice-ordered group G, we shall apply the results of Sect. 5.2.1 to the distributive lattice $D_G = G \sqcup \{\bot\}$, where $\bot$ denotes a new zero element. In particular, D_G embeds, as a sublattice with the same zero element (viz. $\bot$), into $\overline{B}_G = \mathrm{BR}(D_G)$. The subset

$$B_G = \{x \in \overline{B}_G \mid (\exists c \in G)(x \wedge c = \bot)\} = \bigcup_{c \in G} (\overline{B}_G \downarrow \neg c) \tag{5.2.2}$$

is an ideal of the Boolean ring $\overline{B}_G$. It follows from Lemma 5.2.3 that the elements of B_G are exactly the finite joins (equivalently, the finite orthogonal joins), within $\overline{B}_G$, of elements of the form $y \smallsetminus x$, where $(x, y) \in G^{[2]}$.

For each $a \in G$, the translation $D_G \to D_G$ that sends $\bot$ to $\bot$ and any $x \in G$ to $x + a$ is a lattice automorphism. This map extends to a unique automorphism $\overline{\tau}_a$ of the Boolean ring $\overline{B}_G$. From $\overline{\tau}_a(c) = a + c$ whenever $c \in G$, it follows that $\overline{\tau}_a$ restricts to an automorphism τ_a, or τ_a^G in case G needs to be specified, of the Boolean ring B_G. For any $a, b \in G$, the automorphisms τ_{a+b} and $\tau_a \circ \tau_b$ agree on D_G, thus they are equal. Hence, the assignment $c \mapsto \tau_c$ defines an isomorphism from G onto the subgroup $\overline{G} = \{\tau_c \mid c \in G\}$ of $\mathrm{Aut}\, B_G$.

We shall now define a G^+-valued V-measure on B_G. It is convenient to apply here some results of Wehrung [122]. Endowing G with its underlying lattice structure, the *dimension monoid* of G is the commutative monoid $\mathrm{Dim}\, G$ defined by the generators $\Delta(x, y)$, for $(x, y) \in G^{[2]}$, and the relations (D0) $\Delta(x, x) = 0$ for each $x \in G$, (D1) $\Delta(x, z) = \Delta(x, y) + \Delta(y, z)$ whenever $x \leq y \leq z$ in G, and (D2) $\Delta(x \wedge y, x) = \Delta(y, x \vee y)$ for all $x, y \in G$. The lattice denoted by $\mathbf{B}(G)$ on [122, p. 272] is equal, modulo an obvious correction,[2] to the Boolean ring B_G defined in (5.2.2).

Since G is distributive, it follows from [122, Theorem 2.13] that there is a unique isomorphism $\varepsilon_G\colon \mathrm{Dim}\, G \to \mathbb{Z}^+\langle B_G\rangle$ such that $\varepsilon_G \Delta(a, b) = b \smallsetminus a$ whenever $(a, b) \in D_G^{[2]}$.

---

[2]The definition of $\mathbf{B}(G)$ given at the bottom of Wehrung [122, p. 272] is misformulated. Namely, since G has no least element unless it is trivial, $x \smallsetminus \bot$ does not belong to B_G as a rule, so $x \mapsto x \smallsetminus \bot$ does not embed D_G into B_G. What matters here is that the elements of B_G are exactly the finite (orthogonal) joins of elements of the form $b \smallsetminus a$, where $(a, b) \in G^{[2]}$. The correct definition of $B_G = \mathbf{B}(G)$ that ensures this is given by (5.2.2).

Now we take advantage of the additive structure of G. Since G is an Abelian lattice-ordered group, the assignment $G^{[2]} \to G^+$, $(x, y) \mapsto y - x$ satisfies (D0)–(D2) above. By the universal property of the dimension monoid, there is a unique monoid homomorphism $\psi_G \colon \mathrm{Dim}\, D_G \to G^+$ such that $\psi_G \Delta(x, y) = y - x$ whenever $(x, y) \in D_G^{[2]}$. The map $\overline{\mu}_G = \psi_G \circ \varepsilon_G^{-1}$ is a monoid homomorphism from $\mathbb{Z}^+ \langle B_G \rangle$ to G^+, and $\overline{\mu}_G(y \smallsetminus x) = y - x$ whenever $(x, y) \in G^{[2]}$. Since $\overline{\mu}_G$ is a monoid homomorphism and $\overline{\mu}_G^{-1}\{0\} = \{0\}$, the restriction μ_G of $\overline{\mu}_G$ to B_G is a G^+-valued measure on B_G.

Lemma 5.2.4 *For all $z \in B_G$ and all $a, b \in G^+$ such that $\mu_G(z) = a + b$, there is a decomposition $z = x \oplus y$ in B_G such that $\mu_G(x) = a$ and $\mu_G(y) = b$, such that if z is an elementary generator of B_G, then so are x and y. In particular, μ_G is a V-measure.*

Proof Write $z = \bigoplus_{i=1}^n (y_i \smallsetminus x_i)$, where each $(x_i, y_i) \in G^{[2]}$. Our assumption means that $\sum_{i=1}^n (y_i - x_i) = a + b$. Since G^+ is a refinement monoid, there are $a_i, b_i \in G^+$, for $i \in [n]$, such that each $y_i - x_i = a_i + b_i$ while $a = \sum_{i=1}^n a_i$ and $b = \sum_{i=1}^n b_i$. From $x_i \leq x_i + a_i \leq y_i$ it follows that

$$y_i \smallsetminus x_i = \big((x_i + a_i) \smallsetminus x_i\big) \oplus \big(y_i \smallsetminus (x_i + a_i)\big),$$

for each $i \in [n]$. Therefore, setting

$$x = \bigoplus_{i=1}^n \big((x_i + a_i) \smallsetminus x_i\big) \text{ and } y = \bigoplus_{i=1}^n \big(y_i \smallsetminus (x_i + a_i)\big),$$

we get $z = x \oplus y$, $\mu_G(x) = \sum_{i=1}^n a_i = a$, and $\mu_G(y) = \sum_{i=1}^n b_i = b$. □

Lemma 5.2.5 *$\mu_G(\tau_c(z)) = \mu_G(z)$, whenever $c \in G$ and $z \in B_G$.*

Proof Write $z = \bigoplus_{i=1}^n (y_i \smallsetminus x_i)$, where each $(x_i, y_i) \in G^{[2]}$. Then $\tau_c(z) = \bigoplus_{i=1}^n \big((y_i + c) \smallsetminus (x_i + c)\big)$, thus

$$\mu_G(\tau_c(z)) = \sum_{i=1}^n \big((y_i + c) - (x_i + c)\big) = \sum_{i=1}^n (y_i - x_i) = \mu_G(z).$$

□

We set $S_G = \mathrm{Inv}(B_G, \overrightarrow{G})$ (cf. Example 4.4.15). In particular, S_G is a fundamental Boolean inverse semigroup. It follows from Lemma 5.2.5 that all elements of S_G are μ_G-invariant, that is, $S_G \subseteq \mathrm{Inv}(B_G, \mu_G)$ (cf. Notation 4.7.2).

Lemma 5.2.6 *The fundamental Boolean inverse semigroup S_G is μ_G-exhaustive.*

Proof Let $u, v \in B_G$ such that $\mu_G(u) = \mu_G(v)$. We need to find decompositions of u and v matching via transformations of the form τ_a. Write $u = \bigoplus_{i<m} (y_i \smallsetminus x_i)$ and $v = \bigoplus_{j<n} (y'_j \smallsetminus x'_j)$. By assumption, $\sum_{i<m} (y_i - x_i) = \sum_{j<n} (y'_j - x'_j)$. Since G^+

is a refinement monoid, there is a refinement matrix of the form

	$y'_j - x'_j (j < n)$
$y_i - x_i (i < m)$	$c_{i,j}$

, where all $c_{i,j} \in G^+$.

By Lemma 5.2.4, there are decompositions $y_i \smallsetminus x_i = \bigoplus_{j<n} w_{i,j}$ in B_G, where each $w_{i,j}$ is an elementary generator and $\mu_G(w_{i,j}) = c_{i,j}$. Likewise, there are decompositions $y'_j \smallsetminus x'_j = \bigoplus_{i<m} w'_{i,j}$ in B_G, where each $w'_{i,j}$ is an elementary generator and $\mu_G(w'_{i,j}) = c_{i,j}$. Writing $w_{i,j} = q_{i,j} \smallsetminus p_{i,j}$ and $w'_{i,j} = q'_{i,j} \smallsetminus p'_{i,j}$, with $p_{i,j} \leq q_{i,j}$ and $p'_{i,j} \leq q'_{i,j}$, we get $c_{i,j} = q_{i,j} - p_{i,j} = q'_{i,j} - p'_{i,j}$, thus, setting $d_{i,j} = q'_{i,j} - q_{i,j}$, we get $w'_{i,j} = \tau_{d_{i,j}}(w_{i,j})$. Since $u = \bigoplus_{i<m, j<n} w_{i,j}$ and $v = \bigoplus_{i<m, j<n} w'_{i,j}$, we are done. □

A direct application of Theorem 4.7.13 and Proposition 4.8.5 thus yields the following extension of Dobbertin [35, Theorem 13], from V-measures to type monoids of Boolean inverse semigroups.

Theorem 5.2.7 *For every Abelian lattice-ordered group G, there exists a fundamental Boolean inverse semigroup S_G such that* $\operatorname{Typ} S_G \cong G^+$. *That is, G^+ is group-measurable.*

The canonical isomorphism $\eta_G \colon \operatorname{Typ} S_G \to G^+$ is given by

$$\eta_G(\tilde{a}) = \mu_G(a), \quad \text{for every } a \in B_G, \tag{5.2.3}$$

where we set $\tilde{a} = \operatorname{typ}_{S_G}(\mathrm{id}_{B\downarrow a})$.

5.2.3 When is $\mathbf{Inv}(B_G, \overline{G})$ an Inverse Meet-Semigroup?

We shall investigate in which cases the fundamental Boolean inverse semigroup $S_G = \operatorname{Inv}(B_G, \overline{G})$, constructed for the proof of Theorem 5.2.7, is a meet-semilattice under its natural ordering (in which case it is a fundamental Boolean inverse meet-semigroup). Following the standard notation in lattice-ordered groups, we set $|x| = x \vee (-x)$, for any $x \in G$. For every $a \in G$, the set $a^\perp = \{x \in G \mid a \wedge |x| = 0\}$ is an ℓ-ideal of G, that is, an order-convex additive subgroup closed under $x \mapsto |x|$. As usual, G is *projectable* if for all $a, b \in G^+$, the set $a^\perp \downarrow b$ has a largest element. We refer to any textbook on lattice-ordered groups, for example Anderson and Feil [4] or Bigard et al. [22], for more details. An example of a non-projectable, Abelian lattice-ordered group can be found in [4, E 26].

Lemma 5.2.8 *The following statements hold, for any Abelian lattice-ordered group G, with the Boolean ring B_G as defined in* (5.2.2)*:*

(1) *Let $a, b, a', b', h \in G$ such that $h \geq 0$ and $b - a = b' - a' = h$. Then $b \smallsetminus a \leq b \smallsetminus a'$ iff $b \smallsetminus a = b \smallsetminus a'$, iff $|b' - b| \wedge h = 0$.*

(2) *Let* $a, b \in G$ *and let* $(p, q) \in G \times G$. *Then* τ_a *and* τ_b *agree on* $B_G \downarrow (q \smallsetminus p)$ *iff* $\tau_a(q \smallsetminus p) = \tau_b(q \smallsetminus p)$, *iff* $|a - b| \wedge |q - (p \wedge q)| = 0$.
(3) *Let* $a, b \in G$ *and let* $w \in B_G$. *Then* $\tau_a \restriction_{B_G \downarrow w} = \tau_b \restriction_{B_G \downarrow w}$ *iff* $|a - b| \wedge \mu_G(w) = 0$.

Proof (1) Set $c = a' - a = b' - b$. By Lemma 5.2.1, $b \smallsetminus a \leq b' \smallsetminus a'$ iff $a' \wedge b \leq a$ and $b \leq a \vee b'$, that is, since the translations of G are lattice automorphisms, $c \wedge h \leq 0 \leq (-h) \vee c$, or, equivalently, $c \wedge h \leq 0$ and $(-c) \wedge h \leq 0$. By the distributivity of the underlying lattice of G, this is equivalent to $|c| \wedge h \leq 0$, that is, since $0 \leq |c|$ and $0 \leq h$, $|c| \wedge h = 0$.

(2) We may assume that $p = p \wedge q$, that is, $p \leq q$. It follows from (1) that

$$\tau_a(q \smallsetminus p) = \tau_b(q \smallsetminus p) \text{ iff } |a - b| \wedge (q - p) = 0\,. \tag{5.2.4}$$

Suppose that $|a-b| \wedge (q-p) = 0$ and let $w \in B_G \downarrow (q \smallsetminus p)$. There is a decomposition of the form $w = \bigvee_{i<n}(q_i \smallsetminus p_i)$, where all $(p_i, q_i) \in G^{[2]}$. From $q_i \smallsetminus p_i \leq q \smallsetminus p$ if follows (for example by applying μ_G) that $q_i - p_i \leq q - p$. Since $|a-b| \wedge (q-p) = 0$, we get $|a - b| \wedge (q_i - p_i) = 0$. By (5.2.4), it follows that $\tau_a(q_i \smallsetminus p_i) = \tau_b(q_i \smallsetminus p_i)$. Joining those equations over all $i < n$, we get $\tau_a(w) = \tau_b(w)$.

(3) Write $w = \bigoplus_{i<n}(q_i \smallsetminus p_i)$, where each $(p_i, q_i) \in G^{[2]}$. If τ_a and τ_b agree on $B_G \downarrow w$, then they agree on each $q_i \smallsetminus p_i$, thus, by (2), $|a - b| \wedge (q_i - p_i) = 0$. Since $|a-b|^{\perp}$ is an additive subgroup of G, it follows that $|a-b| \wedge \sum_{i<n}(q_i - p_i) = 0$, that is, $|a-b| \wedge \mu_G(w) = 0$. Conversely, if $|a-b| \wedge \mu_G(w) = 0$, then $|a-b| \wedge (q_i - p_i) = 0$ for each $i < n$, thus, by (2), τ_a and τ_b agree on $B_G \downarrow (q_i \smallsetminus p_i)$. Since $w = \bigvee_{i<n}(q_i \smallsetminus p_i)$ and by the distributivity of B_G, it follows that τ_a and τ_b agree on $B_G \downarrow w$. □

Theorem 5.2.9 *Let* G *be an Abelian lattice-ordered group. The fundamental Boolean inverse semigroup* S_G *is an inverse meet-semigroup iff* G *is projectable. Consequently, if* G *is projectable, then* $G^+ \cong \operatorname{Typ} S$ *for some fundamental Boolean inverse meet-semigroup* S.

Proof Suppose first that S_G is an inverse meet-semigroup and let $b, v \in G^+$. We prove that $b^{\perp} \downarrow v$ has a largest element. By assumption, there is a largest $w \in B_G$ such that the maps $\tau_0 = \mathrm{id}$ and τ_b agree on $B_G \downarrow w$. By Lemma 5.2.8, this means that $\mu_G(w) \in b^{\perp}$. We claim that $\mu_G(w)$ is the largest element of $b^{\perp} \downarrow v$. We need to prove that any $h \in b^{\perp} \downarrow v$ is beneath $\mu_G(v)$. We may assume that $h \geq 0$. It follows that $b \wedge h = 0$, thus, by Lemma 5.2.8, τ_0 and τ_b agree on $B_G \downarrow (h \smallsetminus 0)$. By the definition of w, it follows that $h \smallsetminus 0 \leq w$, thus $h = \mu_G(h \smallsetminus 0) \leq \mu_G(w)$, as required.

Suppose, conversely, that G is projectable. We must prove that any two elements $x, y \in S_G$ have a meet in S_G. Call a map $x\colon B_G \downarrow u \to B_G \downarrow v$, where u and v are both elementary generators of B_G, *elementary*, if there exists $c \in G$ such that $\tau_c(u) = v$ and $x(t) = \tau_c(t)$ whenever $t \in B_G \downarrow u$. By the definition of S_G, the elements of S_G are exactly the finite orthogonal joins of elementary functions. By Corollary 3.1.10, it thus suffices to consider the case where x and y are both elementary. This means that x and y are the restrictions to elementary generators of B_G of translations τ_a and τ_b, respectively, where $a, b \in G$. The meet of two elementary generators of B_G is an elementary generator of B_G, thus, by precomposing x and y with a suitable translation, we may assume that the domain of x and the one of y intersect in $B_G \downarrow (v \smallsetminus 0)$, for some $v \in G^+$. Since G is projectable, $|a-b|^{\perp} \downarrow v$ has a largest element, say h.

We claim that $h \smallsetminus 0$ is the largest element of $B_G \downarrow (v \smallsetminus 0)$ such that τ_a and τ_b agree on $B_G \downarrow w$. First, it follows from Lemma 5.2.8 that τ_a and τ_b agree on $B_G \downarrow (h \smallsetminus 0)$. Conversely, let $w \in B_G \downarrow (v \smallsetminus 0)$ such that τ_a and τ_b agree on $B_G \downarrow w$; we must prove that $w \leq h \smallsetminus 0$. We may assume that w is an elementary generator of B_G. By Lemma 5.2.1, $w = q \smallsetminus p$ for some $p, q \in G$ such that $0 \leq p \leq q \leq v$. We must prove that $q \smallsetminus p \leq h \smallsetminus 0$, that is, by Lemma 5.2.1 and since $p \geq 0$, that $q \leq p \vee h$.

By assumption and by Lemma 5.2.8, $q - p \in |a-b|^{\perp}$. Since

$$0 \leq (q \vee h) - (p \vee h) = q - \big(q \wedge (p \vee h)\big) \leq q - p\,,$$

it follows that $(q \vee h) - (p \vee h) \in |a-b|^{\perp}$, and thus the element h', defined as $h' = h + \big((q \vee h) - (p \vee h)\big)$, belongs to $|a-b|^{\perp}$. Since $h \leq h' \leq q \vee h \leq v$ and by the definition of h, it follows that $h = h'$, that is, $q \vee h = p \vee h$, so $q \leq p \vee h$, thus completing the proof of our claim.

It follows that $x \wedge y$ is the restriction of τ_a (equivalently, τ_b) to $B_G \downarrow (h \smallsetminus 0)$. □

Recall that an inverse monoid S is *factorizable* if for every $x \in S$ there is a unit g of S such that $x \leq g$. In case S is Boolean, it is not hard to see that this is equivalent to saying that the partial monoid $\operatorname{Int} S$ is cancellative, and thus, since $\operatorname{Typ} S$ is a refinement monoid and by Corollary 2.7.4, that the monoid $\operatorname{Typ} S$ is cancellative. In Lawson and Scott [77], factorizable Boolean inverse monoids are called *Foulis monoids*. It follows immediately from Lawson [73, Proposition 3.2.8] that in every Foulis monoid, $\mathscr{D} = \mathscr{J}$, so the poset $S/\mathscr{J}$ of all principal ideals of S is isomorphic to the poset $S/\mathscr{D} = \operatorname{Int} S$, and so it is naturally endowed with a structure of a conical partial refinement monoid. Lawson and Scott state on [77, p. 7] the question whether every countable MV-algebra (cf. Example 2.2.8) is isomorphic to $\operatorname{Int} S$ for some Foulis monoid S. Their main result is a positive answer to that question, with the additional information that S is AF.

Our results make it possible to remove the countability assumption from Lawson and Scott's result (of course losing AF), with a twist at cardinalities beyond $\aleph_2$.

Theorem 5.2.10 *Every MV-algebra A is isomorphic to* $\operatorname{Int} S$ *(thus to $S/\mathscr{J}$) for some Foulis monoid S. Moreover, if* $\operatorname{card} A \leq \aleph_1$*, then S may be taken locally matricial.*

Proof It follows from Mundici [86, Theorem 3.8] that there are an Abelian lattice-ordered group G and an order-unit $\boldsymbol{u}$ of G such that $A \cong [0, \boldsymbol{u}]$. By Theorem 5.2.7, there is a Boolean inverse semigroup S such that $\operatorname{Typ} S \cong G^+$. By Theorem 4.2.7, we may assume that S is unital and $(\operatorname{Typ} S, \operatorname{typ}_S(1)) \cong (G^+, \boldsymbol{u})$. In particular, S is a Foulis monoid.

Since $\operatorname{Int} S$ is isomorphic to the interval $[0, \operatorname{typ}_S(1)]$ of $\operatorname{Typ} S$, it follows that $\operatorname{Int} S \cong [0, \boldsymbol{u}] \cong A$.

If, in addition, $\operatorname{card} A \leq \aleph_1$, then $\operatorname{card} G \leq \aleph_1$ as well (for G is the universal group of $\mathrm{U}_{\mathrm{mon}}(A)$, see Example 2.2.8) and we may use Theorem 5.1.8, instead of Theorem 5.2.7, to represent directly $(\operatorname{Typ} S, \operatorname{typ}_S(1)) \cong (G^+, \boldsymbol{u})$, getting the additional information that S is locally matricial. □

We do not know whether S can be made locally matricial in all cardinalities. An equivalent form of that question is stated, in Chap. 7, as Problem 5. On the other hand, the question obtained by changing "MV-algebra" to "effect algebra with refinement" has a negative answer: if G is the dimension group of cardinality $\aleph_2$, constructed in Wehrung [124], mentioned in Theorem 4.6.9, and $\boldsymbol{u}$ is an order-unit of G, then there is no Foulis monoid S such that $S/\mathscr{J} \cong [0, \boldsymbol{u}]$.

The similarity between the relation $\mathscr{J}$ and the relation $\mathscr{D}$ (e.g., $\mathscr{D}$ is contained in $\mathscr{J}$, and often identical to it) suggests that a type theory might be built for the former relation. The following example shows that this could be awkward.

Example 5.2.11 A fundamental Boolean inverse monoid S on which the relation $\mathscr{J}$ is not a V-relation.

Proof Moreira Dos Santos constructs in [85] a countable conical refinement monoid M, with order-unit $\boldsymbol{e}$, such that the quotient $\overline{M}$ of M by the monoid congruence $\equiv$, defined by $x \equiv y$ iff $x \leq^+ y$ and $y \leq^+ x$, is not a refinement monoid.

By possibly enlarging $\boldsymbol{e}$, we may assume that the lower interval $\overline{M} \downarrow (\boldsymbol{e}/{\equiv})$ does not satisfy refinement. Now by Theorem 4.8.7, there is a fundamental Boolean inverse monoid S such that $\big(\operatorname{Typ} S, \operatorname{typ}(1)\big) \cong (M, \boldsymbol{e})$. We may thus assume that $M = \operatorname{Typ} S$ and $\boldsymbol{e} = \operatorname{typ}(1)$.

Towards a contradiction, suppose that the relation $\mathscr{J}$ on S is refining. The map $\varphi\colon \operatorname{Int}(S) \to \overline{M} \downarrow \varphi(\boldsymbol{e})$, $\varphi\colon \boldsymbol{x} \mapsto \boldsymbol{x}/{\equiv}$ is a surjective homomorphism of partial commutative monoids. Now it follows from Lawson [73, Proposition 3.2.8] that $\operatorname{typ}(x) \equiv \operatorname{typ}(y)$ iff $x \mathrel{\mathscr{J}} y$, for all $x, y \in S$. Since $\mathscr{J}$ is assumed to be refining and by Lemma 2.4.5, it follows that φ factors to an isomorphism $\psi\colon S/\mathscr{J} \to \overline{M} \downarrow \varphi(\boldsymbol{e})$. Since S satisfies refinement (cf. Proposition 3.1.9) and $\mathscr{J}$ is a V-relation, $S/\mathscr{J}$ also satisfies refinement (cf. Lemma 2.4.4). Since $\overline{M} \downarrow \varphi(\boldsymbol{e})$ fails refinement, we obtain a contradiction. □

5.2.4 Functoriality of Coordinatization

In this subsection we shall verify that the assignment $G \mapsto S_G$, introduced in Sect. 5.2.2, can be extended in a natural fashion to a functor.

We start by reviewing a few basic, and probably well known, facts about generalized Boolean algebras. We define inductively the *Boolean value* $[\![\mathsf{t}]\!]$, of a term $\mathsf{t}(\mathsf{x}_1, \ldots, \mathsf{x}_n)$ of the language $(\vee, \wedge, \smallsetminus, 0)$ of generalized Boolean algebras, by

$$[\![0]\!] = \varnothing\,;\ [\![\mathsf{x}_i]\!] = \{X \in \operatorname{Pow}[n] \mid i \in X\}\,;$$

$$[\![t_1 \vee t_2]\!] = [\![t_1]\!] \cup [\![t_2]\!]\,;\ [\![t_1 \wedge t_2]\!] = [\![t_1]\!] \cap [\![t_2]\!]\,;\ [\![t_1 \smallsetminus t_2]\!] = [\![t_1]\!] \setminus [\![t_2]\!]\,,$$

whenever $1 \leq i \leq n$ and t_1, t_2 are terms of $(\vee, \smallsetminus, 0)$ (recall that $[n] = \{1, \ldots, n\}$).

For a generalized Boolean algebra B and a finite sequence $\vec{a} = (a_1, \dots, a_n)$ of elements in B, we denote by $B(\vec{a})$ the generalized Boolean subalgebra generated by $\{a_1, \dots, a_n\}$. In particular, the unit of $B(\vec{a})$ is $a_1 \vee \cdots \vee a_n$. It is well known that due to the representation of the elements of $B(\vec{a})$ in disjunctive normal form, the atoms of $B(\vec{a})$ are exactly the nonzero elements of the form $a_{(X)}$, where we set

$$a_{(X)} = \bigwedge_{i \in X} a_i \smallsetminus \bigvee_{i \in [n] \setminus X} a_i \,, \quad \text{for any nonempty } X \subseteq [n] \,,$$

with the usual convention that $\bigvee_{i \in \varnothing} a_i = 0$.

Lemma 5.2.12 *For any term* $\mathsf{t}(\mathsf{x}_1, \dots, \mathsf{x}_n)$ *of the language* $(\vee, \wedge, \smallsetminus, 0)$, *any nonempty subset* X *of* $[n]$, *any generalized Boolean algebra* B, *and any* n*-tuple* $\vec{a} = (a_1, \dots, a_n) \in B^n$, *the inequality* $a_{(X)} \leq \mathsf{t}(\vec{a})$ *holds in* B *iff either* $a_{(X)} = 0$ *or* $X \in [\![t]\!]$.

Proof We argue by induction on the complexity of the term t.

If $\mathsf{t} = 0$ the result is trivial.

Let $\mathsf{t} = \mathsf{x}_i$, for some i. Then $X \in [\![\mathsf{t}]\!]$ iff $i \in X$, which implies in turn that $a_{(X)} \leq \bigwedge_{j \in X} a_j \leq a_i$. On the other hand, if $i \notin X$, then $a_{(X)} \wedge a_i = 0$, thus $a_{(X)} \leq a_i$ iff $a_{(X)} = 0$.

Let $\mathsf{t} = \mathsf{t}_1 \vee \mathsf{t}_2$, for terms t_1 and t_2. Since $a_{(X)}$ is either an atom of $B(\vec{a})$ or zero, $a_{(X)} \leq \mathsf{t}_1(\vec{a}) \vee \mathsf{t}_2(\vec{a})$ iff either $a_{(X)} \leq \mathsf{t}_1(\vec{a})$ or $a_{(X)} \leq \mathsf{t}_2(\vec{a})$, iff either $a_{(X)} = 0$ or $X \in [\![t_1]\!] \cup [\![t_2]\!]$. Now $[\![t]\!] = [\![t_1]\!] \cup [\![t_2]\!]$.

The proofs for the cases $\mathsf{t} = \mathsf{t}_1 \wedge \mathsf{t}_2$ and $\mathsf{t} = \mathsf{t}_1 \smallsetminus \mathsf{t}_2$ are similar. □

Define a *simple atomic formula*, of the language $(\vee, \wedge)$, as a formula of the form $\bigwedge_{i \in X} \mathsf{x}_i \leq \bigvee_{i \in [n] \setminus X} \mathsf{x}_i$, where n is a positive integer and X is a nonempty subset of $[n]$.

Lemma 5.2.13 *Every atomic formula of the language* $(\vee, \wedge, \smallsetminus, 0)$, *is equivalent, within the class of all generalized Boolean algebras, to a conjunction of simple atomic formulas.*

Proof For terms t_1 and t_2 of the language $(\vee, \wedge, \smallsetminus, 0)$, the formula $\mathsf{t}_1 = \mathsf{t}_2$ is equivalent to the formula $\mathsf{t} = 0$, where we set $\mathsf{t} = (\mathsf{t}_1 \smallsetminus \mathsf{t}_2) \vee (\mathsf{t}_2 \smallsetminus \mathsf{t}_1)$. Now for a generalized Boolean algebra B and $\vec{a} = (a_1, \dots, a_n) \in B^n$, it follows from Lemma 5.2.12 that the equation $\mathsf{t}(\vec{a}) = 0$ holds in B iff $a_{(X)} = 0$ whenever $X \in [\![t]\!]$. Now the equation $a_{(X)} = 0$ holds iff $\bigwedge_{i \in X} a_i \leq \bigvee_{i \in [n] \setminus X} a_i$, so it can be expressed by a simple atomic formula. □

By definition, the elements of the Boolean inverse semigroup $S_G = \operatorname{Inv}(B_G, \overline{G})$ constructed for the proof of Theorem 5.2.7 are exactly the finite orthogonal joins of elements of the form $\tau_a \restriction_{b \smallsetminus c} = \tau_a \restriction_{B_G \downarrow (b \smallsetminus c)}$ for $a, b, c \in G$. Accordingly, we say that a finite sequence $\mathfrak{a} = \big((a_i, a'_i, a''_i) \mid 1 \leq i \leq n\big)$ of triples of elements in G is a *G-denotation of length n* if the partial automorphisms $\tau_{a_i} \restriction_{a'_i \smallsetminus a''_i}$, for $1 \leq i \leq n$, are pairwise orthogonal in S_G. In that case, we set $\tau_G(\mathfrak{a}) = \bigoplus_{i=1}^{n} \tau_{a_i} \restriction_{a'_i \smallsetminus a''_i}$. Hence, the elements of S_G are exactly the $\tau_G(\mathfrak{a})$, for G-denotations $\mathfrak{a}$.

Lemma 5.2.14 *The set Σ_G of all G-denotations of length n is the solution set of a finite collection of atomic formulas of the language $(\vee, \wedge, 0, -)$ of lattice-ordered groups.*

Proof A finite sequence $\big((a_i, a_i', a_i'') \mid 1 \leq i \leq n\big)$ is a G-denotation iff

$$(a_i' \smallsetminus a_i'') \wedge (a_j' \smallsetminus a_j'') = ((a_i + a_i') \smallsetminus (a_i + a_i'')) \wedge ((a_j + a_j') \smallsetminus (a_j + a_j'')) = 0\,,$$

whenever $i \neq j$ in $[n]$. By applying Lemma 5.2.13 (or even just the obvious identity $(a \smallsetminus b) \wedge (c \smallsetminus d) = (a \wedge c) \smallsetminus (b \vee d)$), we obtain that the conjunction of those conditions is equivalent to a conjunction of simple atomic formulas of $(\vee, \wedge)$, with parameters from the a_i, a_i', a_i'', $a_i + a_i'$, $a_i + a_i''$. □

Lemma 5.2.15 *The inequality $\tau_G(\mathfrak{a}) \perp \tau_G(\mathfrak{b})$, for G-denotations $\mathfrak{a}$ and $\mathfrak{b}$ of respective lengths m and n, can be expressed by a finite collection of atomic formulas of the language $(\vee, \wedge, 0, -)$ in the entries of $\mathfrak{a}$ and $\mathfrak{b}$.*

Proof Observe that $\tau_G(\mathfrak{a}) \perp \tau_G(\mathfrak{b})$ iff the concatenation $\mathfrak{a} \frown \mathfrak{b}$ is a G-denotation. Then use Lemma 5.2.14. □

Lemma 5.2.16 *The inequality $\tau_G(\mathfrak{a}) \leq \tau_G(\mathfrak{b})$, for G-denotations $\mathfrak{a}$ and $\mathfrak{b}$ of respective lengths m and n, can be expressed by a finite collection of atomic formulas of the language $(\vee, \wedge, 0, -)$ in the entries of $\mathfrak{a}$ and $\mathfrak{b}$.*

Proof Setting $\mathfrak{a} = \big((a_i, a_i', a_i'') \mid 1 \leq i \leq m\big)$ and $\mathfrak{b} = \Big((b_j, b_j', b_j'') \mid 1 \leq j \leq n\Big)$, the inequality $\tau_G(\mathfrak{a}) \leq \tau_G(\mathfrak{b})$ holds iff $\tau_{a_i} \restriction_{a_i' \smallsetminus a_i''} \leq \bigoplus_{j=1}^{n} \tau_{b_j} \restriction_{b_j' \smallsetminus b_j''}$ holds for every $i \in [n]$. For fixed i, the latter inequality is, in turn, equivalent to the conjunction of the inequality

$$a_i' \smallsetminus a_i'' \leq \bigvee_{j=1}^{n} (b_j' \smallsetminus b_j'') \tag{5.2.5}$$

with the relations

$$\tau_{a_i} \restriction_{(a_i' \smallsetminus a_i'') \wedge (b_j' \smallsetminus b_j'')} = \tau_{b_j} \restriction_{(a_i' \smallsetminus a_i'') \wedge (b_j' \smallsetminus b_j'')}\,, \quad \text{for } 1 \leq j \leq n\,.$$

Due to the obvious identity $(x' \smallsetminus x'') \wedge (y' \smallsetminus y'') = (x' \wedge y') \smallsetminus (x'' \vee y'')$, the latter system is equivalent to the system

$$\tau_{a_i} \restriction_{(a_i' \wedge b_j') \smallsetminus (a_i'' \vee b_j'')} = \tau_{b_j} \restriction_{(a_i' \wedge b_j') \smallsetminus (a_i'' \vee b_j'')}\,, \quad \text{for } 1 \leq j \leq n\,.$$

By Lemma 5.2.8, the latter is equivalent to the system

$$|a_i - b_j| \wedge \Big((a_i' \wedge b_j') - \big((a_i' \wedge b_j') \wedge (a_i'' \vee b_j'')\big)\Big) = 0\,, \quad \text{for } 1 \leq j \leq n\,,$$

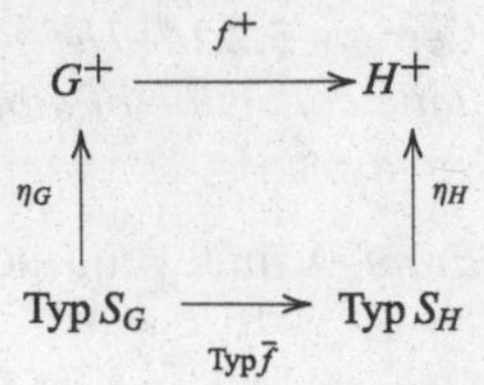

Fig. 5.1 A commutative diagram of commutative monoids

which is a conjunction of atomic formulas of $(\vee, \wedge, 0, -)$. Furthermore, by Lemma 5.2.8, each of the inequalities (5.2.5) is equivalent to a conjunction of simple atomic formulas. □

Lemma 5.2.17 *Let G and H be Abelian lattice-ordered groups and let $f: G \to H$ be an ℓ-homomorphism. Then there exists a unique additive semigroup homomorphism $\overline{f}: S_G \to S_H$ such that*

$$\overline{f}\big(\tau_a^G \restriction_{b \smallsetminus c}\big) = \tau_{f(a)}^H \restriction_{f(b) \smallsetminus f(c)} \quad \text{whenever } a, b, c \in G\,. \tag{5.2.6}$$

Moreover, if we denote by $f^+: G^+ \to H^+$ the domain-range restriction of f to positive cones and by η_G: Typ $G \to G^+$ the canonical isomorphism (given by (5.2.3)*), then the diagram represented in Fig.* 5.1 *commutes.*

Proof Since every element of S_G is a finite orthogonal join of elements of the form $\tau_a^G \restriction_{b \smallsetminus c}$, the uniqueness statement is obvious. For any G-denotation $\mathfrak{a}$, denote by $f(\mathfrak{a})$ the finite sequence obtained by applying f to all entries of $\mathfrak{a}$. Since atomic formulas are preserved under ℓ-homomorphisms, it follows from Lemma 5.2.14 that $f(\mathfrak{a})$ is also a G-denotation. Furthermore, it follows from Lemma 5.2.16 that $\tau_G(\mathfrak{a}) = \tau_G(\mathfrak{b})$ implies that $\tau_H(f(\mathfrak{a})) = \tau_H(f(\mathfrak{b}))$, for all G-denotations $\mathfrak{a}$ and $\mathfrak{b}$. Consequently, there exists a unique map $\overline{f}: S_D \to S_H$ such that

$$\overline{f}(\tau_G(\mathfrak{a})) = \tau_H(f(\mathfrak{a}))\,, \quad \text{for every } \mathfrak{a} \in \Sigma_G\,.$$

Let $a, b \in S_G$ such that $a \perp b$. There are G-denotations $\mathfrak{a}$ and $\mathfrak{b}$ such that $a = \tau_G(\mathfrak{a})$ and $b = \tau_G(\mathfrak{b})$. By Lemma 5.2.15, it follows that $\tau_H(f(\mathfrak{a})) \perp \tau_H(f(\mathfrak{b}))$, that is, $\overline{f}(a) \perp \overline{f}(b)$. Further,

$$\begin{aligned}\overline{f}(a \oplus b) &= \overline{f}\big(\tau_G(\mathfrak{a} \frown \mathfrak{b})\big) = \tau_H\big(f(\mathfrak{a} \frown \mathfrak{b})\big)\\ &= \tau_H\big(f(\mathfrak{a}) \frown f(\mathfrak{b})\big) = \tau_H(f(\mathfrak{a})) \oplus \tau_H(f(\mathfrak{b})) = \overline{f}(a) \oplus \overline{f}(b)\,.\end{aligned}$$

Hence, the map $\overline{f}$ preserves finite orthogonal joins.

Finally, in order to prove that $\overline{f}$ is a semigroup homomorphism, it suffices, by Proposition 3.1.9, to prove that it preserves products of elements of the form $\tau_a^G \restriction_{b \smallsetminus c}$. For $a_i, b_i, c_i \in G$ $(i \in \{0, 1\})$, it is easy to establish the formula

$$\begin{aligned}(\tau_{a_0}^G \restriction_{b_0 \smallsetminus c_0}) \circ (\tau_{a_1}^G \restriction_{b_1 \smallsetminus c_1}) &= \tau_{a_0+a_1}^G \restriction_{(b_1 \smallsetminus c_1) \wedge ((b_0 - a_1) \smallsetminus (c_0 - a_1))} \\ &= \tau_{a_0+a_1}^G \restriction_{(b_1 \wedge (b_0 - a_1)) \smallsetminus (c_1 \vee (c_0 - a_1))}\end{aligned}$$

Hence,

$$\begin{aligned}\overline{f}\Big((\tau_{a_0}^G \restriction_{b_0 \smallsetminus c_0}) \circ (\tau_{a_1}^G \restriction_{b_1 \smallsetminus c_1})\Big) &= \tau_{f(a_0+a_1)}^H \restriction_{f(b_1 \wedge (b_0 - a_1)) \smallsetminus f((c_1 \vee (c_0 - a_1))} \\ &= \tau_{f(a_0)+f(a_1)}^H \restriction_{(f(b_1) \wedge (f(b_0) - f(a_1))) \smallsetminus (f(c_1) \vee (f(c_0) - f(a_1)))} \\ &= \overline{f}(\tau_{a_0}^G \restriction_{b_0 \smallsetminus c_0}) \circ \overline{f}(\tau_{a_1}^G \restriction_{b_1 \smallsetminus c_1}),\end{aligned}$$

thus completing the proof that $\overline{f}$ is an additive semigroup homomorphism.

Finally, the verification of the commutativity of the diagram represented in Fig. 5.1 is straightforward. □

By Lemma 5.2.17, the assignment $G \mapsto S_G, f \mapsto \overline{f}$ defines a functor Υ, from the category **Lgrp** of all Abelian lattice-ordered groups with ℓ-homomorphisms, to the category **Bis** of all Boolean inverse semigroups with additive semigroup homomorphisms: so with the notation above, $\Upsilon(G) = S_G$ and $\Upsilon(f) = \overline{f}$.

Bringing together the results of Sect. 5.2.4, we arrive at the following conclusion, which states that up to the identification between G and G^+, the functor Υ is a functorial left inverse to the type monoid functor.

Theorem 5.2.18 *Denote by* **P** *the* positive cone functor, *which sends every Abelian lattice-ordered group G to its positive cone G^+ and every ℓ-homomorphism f to f^+. Then η is a natural equivalence from* Typ $\circ\,\Upsilon$ *to* **P**.

Hence, *coordinatization of positive cones of Abelian lattice-ordered groups can be done functorially*.

We leave to another place and time the study of the functor Υ: **Lgrp** $\to$ **Bis**.

5.3 Inverse Semigroups with Strongly Separative Type Monoids

In this section we shall isolate a growth type sufficient condition, for a Boolean inverse semigroup, for having its type monoid strongly separative. This growth condition will be called *fork-nilpotence*. We will also relate this condition to the classical one of supramenability for groups.

Definition 5.3.1 Let S be a Boolean inverse semigroup. A *fork* of S is a triple (c, g_1, g_2), where c is an idempotent element of S and $g_1, g_2 \in S$ such that the relations $g_1 \langle c \rangle g_2 \langle c \rangle = 0$ and $c \leq \mathbf{d}(g_i)$ hold whenever $i \in \{1, 2\}$. For every nonnegative integer n, we define $\langle g_1, g_2 \rangle^{-n}(c)$ as the product of all $g^{-1} \langle c \rangle$, where $g = g_{i_1} \cdots g_{i_n}$ with each $i_n \in \{1, 2\}$. (In particular, $\langle g_1, g_2 \rangle^{-0}(c) = c$ and $\langle g_1, g_2 \rangle^{-1}(c) = g_1^{-1} \langle c \rangle g_2^{-1} \langle c \rangle c$.)

A fork (c, g_1, g_2) of S is *nilpotent* if there is a nonnegative integer n such that $\langle g_1, g_2 \rangle^{-n}(c) = 0$. We say that S is *fork-nilpotent* if every fork of S is nilpotent.

The following lemma is a "lifting" of Lemma 2.7.5, where semigroup elements witnessing occasional $\mathscr{D}$-dependencies are stated explicitly.

Lemma 5.3.2 *Let S be a Boolean inverse semigroup, let a, b, c be idempotent elements of S, and let $g_1, g_2 \in S$ such that*

(i) *(c, g_1, g_2) is a fork of S;*
(ii) *$a \oplus c = b \oplus g_1 \langle c \rangle \oplus g_2 \langle c \rangle$.*

Then there are idempotent elements $d, \overline{a}, \overline{b}, \overline{c} \in S$ such that

(0) $\overline{c} = g_1^{-1} \langle c \rangle g_2^{-1} \langle c \rangle c$;
(1) $c = \overline{a} \oplus \overline{c} = \overline{b} \oplus g_1 \langle \overline{c} \rangle \oplus g_2 \langle \overline{c} \rangle$;
(2) $\mathrm{typ}(a) = \mathrm{typ}(d) + \mathrm{typ}(\overline{a})$;
(3) $\mathrm{typ}(b) + \mathrm{typ}(c) = \mathrm{typ}(d) + \mathrm{typ}(\overline{b}) + \mathrm{typ}(\overline{c})$.

Proof We follow the proof of Lemma 2.7.5, keeping track of the elements of S witnessing the relevant $\mathscr{D}$-dependencies.

Any equality of the form $\bigoplus_{i<m} a_i = \bigoplus_{j<n} b_j$ (where $\oplus$ denotes the orthogonal join in S) gives rise to the following refinement matrix (within $(S, \oplus, 0)$):

	$b_j (j < n)$
$a_i (i < m)$	$a_i b_j$

By applying that observation to the equality $a \oplus c = b \oplus g_1 \langle c \rangle \oplus g_2 \langle c \rangle$, we get the following refinement matrix:

	b	$g_1 \langle c \rangle$	$g_2 \langle c \rangle$
a	ab	$ag_1 \langle c \rangle$	$ag_2 \langle c \rangle$
c	bc	$cg_1 \langle c \rangle$	$cg_2 \langle c \rangle$

(5.3.1)

By the same token, applied to $c = g_1^{-1}\langle a\rangle c \oplus g_1^{-1}\langle c\rangle c = g_2^{-1}\langle a\rangle c \oplus g_2^{-1}\langle c\rangle c$, we obtain the following refinement matrix:

	$g_2^{-1}\langle ag_2\langle c\rangle\rangle = g_2^{-1}\langle a\rangle c$	$g_2^{-1}\langle cg_2\langle c\rangle\rangle = g_2^{-1}\langle c\rangle c$
$g_1^{-1}\langle ag_1\langle c\rangle\rangle = g_1^{-1}\langle a\rangle c$	$u = g_1^{-1}\langle a\rangle g_2^{-1}\langle a\rangle c$	$a' = g_1^{-1}\langle a\rangle g_2^{-1}\langle c\rangle c$
$g_1^{-1}\langle cg_1\langle c\rangle\rangle = g_1^{-1}\langle c\rangle c$	$a'' = g_1^{-1}\langle c\rangle g_2^{-1}\langle a\rangle c$	$\overline{c} = g_1^{-1}\langle c\rangle g_2^{-1}\langle c\rangle c$

(5.3.2)

In particular,

$$c = u \oplus a' \oplus a'' \oplus \overline{c}, \tag{5.3.3}$$

hence $c = \overline{a} \oplus \overline{c}$ where we set $\overline{a} = u \oplus a' \oplus a''$. On the other hand, it follows from (5.3.1) and (5.3.2) that

$$c = bc \oplus cg_1\langle c\rangle \oplus cg_2\langle c\rangle = bc \oplus g_1\langle a''\rangle \oplus g_2\langle a'\rangle \oplus g_1\langle \overline{c}\rangle \oplus g_2\langle \overline{c}\rangle,$$

so $c = \overline{b} \oplus g_1\langle \overline{c}\rangle \oplus g_2\langle \overline{c}\rangle$ where we set $\overline{b} = bc \oplus g_1\langle a''\rangle \oplus g_2\langle a'\rangle$.

By combining (5.3.1) and (5.3.2) again, we obtain that

$$a = ab \oplus ag_1\langle c\rangle \oplus ag_2\langle c\rangle = d \oplus \overline{a}^*,$$

where we set $d = ab \oplus g_1\langle u\rangle$ and $\overline{a}^* = g_1\langle a'\rangle \oplus g_2\langle a''\rangle \oplus g_2\langle u\rangle$. From $a' \mathscr{D} g_1\langle a'\rangle$, $a'' \mathscr{D} g_2\langle a''\rangle$, and $u \mathscr{D} g_2\langle u\rangle$, it follows that $\mathrm{typ}(\overline{a}) = \mathrm{typ}(\overline{a}^*)$, whence

$$\mathrm{typ}(a) = \mathrm{typ}(d) + \mathrm{typ}(\overline{a}^*) = \mathrm{typ}(d) + \mathrm{typ}(\overline{a}).$$

Finally, by combining (5.3.1), (5.3.3), and the definitions of d and $\overline{b}$, we get

$$\begin{aligned}
\mathrm{typ}(b) + \mathrm{typ}(c) &= \mathrm{typ}(ab) + \mathrm{typ}(bc) + \mathrm{typ}(u) \\
&\quad + \mathrm{typ}(a') + \mathrm{typ}(a'') + \mathrm{typ}(\overline{c}) \\
&= (\mathrm{typ}(ab) + \mathrm{typ}(u)) \\
&\quad + (\mathrm{typ}(bc) + \mathrm{typ}(a') + \mathrm{typ}(a'')) \\
&\quad + \mathrm{typ}(\overline{c}) \\
&= \mathrm{typ}(d) + \mathrm{typ}(\overline{b}) + \mathrm{typ}(\overline{c}).
\end{aligned}$$

□

Lemma 5.3.3 *Let S be a Boolean inverse semigroup, let a, b, c be idempotent elements of S, and let $g_1, g_2 \in S$ such that*

(i) *(c, g_1, g_2) is a fork of S;*
(ii) *$a \oplus c = b \oplus g_1 \langle c \rangle \oplus g_2 \langle c \rangle$.*

Then there are sequences $(d_n)_{n \in \mathbb{Z}^+}$, $(a_n)_{n \in \mathbb{Z}^+}$, $(b_n)_{n \in \mathbb{Z}^+}$, and $(c_n)_{n \in \mathbb{Z}^+}$ of idempotent elements of S such that:

(0) *$c_0 = a \oplus c$, and $c_{n+1} = \langle g_1, g_2 \rangle^{-n}(c)$ whenever $n \in \mathbb{Z}^+$;*
(1) *$c_n = a_n \oplus c_{n+1} = b_n \oplus g_1 \langle c_{n+1} \rangle \oplus g_2 \langle c_{n+1} \rangle$, whenever $n \in \mathbb{Z}^+$;*
(2) *$\mathrm{typ}(a) = \mathrm{typ}(d_n) + \mathrm{typ}(a_n)$, whenever $n \in \mathbb{Z}^+$;*
(3) *$\mathrm{typ}(b) + \mathrm{typ}(c) = \mathrm{typ}(d_n) + \mathrm{typ}(b_n) + \mathrm{typ}(c_{n+1})$, whenever $n \in \mathbb{Z}^+$.*

Proof We argue by induction on n. We set $a_0 = a$, $b_0 = b$, $c_0 = a \oplus c$, $d_0 = 0$, and $c_1 = c$. Then (0)–(3) trivially hold for $n = 0$. Let $n \in \mathbb{N}$, suppose that (0)–(3) hold at stage $n - 1$, and set $c_n = g_1^{-1} \langle c_{n-1} \rangle g_2^{-1} \langle c_{n-1} \rangle c_{n-1} = \langle g_1, g_2 \rangle^{-n}(c)$. Since $c_{n-1} = a_{n-1} \oplus c_n = b_{n-1} \oplus g_1 \langle c_n \rangle \oplus g_2 \langle c_n \rangle$, it follows from Lemma 5.3.2 that there are $\boldsymbol{d} \in \operatorname{Int} S$ and idempotents a_n, b_n of S such that

$$c_n = a_n \oplus c_{n+1} = b_n \oplus g_1 \langle c_{n+1} \rangle \oplus g_2 \langle c_{n+1} \rangle \,,$$
$$\mathrm{typ}(a_{n-1}) = \boldsymbol{d} + \mathrm{typ}(a_n) \,,$$
$$\mathrm{typ}(b_{n-1}) + \mathrm{typ}(c_n) = \boldsymbol{d} + \mathrm{typ}(b_n) + \mathrm{typ}(c_{n+1}) \,.$$

By using the induction hypothesis, we obtain that

$$\begin{aligned} \mathrm{typ}(a) &= \mathrm{typ}(d_{n-1}) + \mathrm{typ}(a_{n-1}) \\ &= \mathrm{typ}(d_{n-1}) + \boldsymbol{d} + \mathrm{typ}(a_n) \,, \end{aligned}$$

hence (cf. Lemma 4.1.6) $\mathrm{typ}(d_{n-1}) + \boldsymbol{d} = \mathrm{typ}(d_n)$ for some $d_n \leq a$, and

$$\begin{aligned} \mathrm{typ}(b) + \mathrm{typ}(c) &= \mathrm{typ}(d_{n-1}) + \mathrm{typ}(b_{n-1}) + \mathrm{typ}(c_n) \\ &= \mathrm{typ}(d_{n-1}) + \boldsymbol{d} + \mathrm{typ}(b_n) + \mathrm{typ}(c_{n+1}) \\ &= \mathrm{typ}(d_n) + \mathrm{typ}(b_n) + \mathrm{typ}(c_{n+1}) \,. \end{aligned}$$

□

The following result is a Boolean inverse semigroup version of Tarski [109, Theorem 16.10].

Theorem 5.3.4 *Let S be a Boolean inverse semigroup. If S is fork-nilpotent, then the monoid* $\operatorname{Typ} S$ *is strongly separative.*

Proof Since $\operatorname{Typ} S = \mathrm{U}_{\mathrm{mon}}(\operatorname{Int} S)$ and by Corollary 2.7.7, it suffices to prove that $\operatorname{Int} S$ is strongly separative. Let $\boldsymbol{a}, \boldsymbol{b}, \boldsymbol{c} \in \operatorname{Int} S$ such that $\boldsymbol{a} \oplus \boldsymbol{c} = \boldsymbol{b} \oplus 2\boldsymbol{c}$ within $\operatorname{Int} S$. Let $e \in \boldsymbol{a} \oplus \boldsymbol{c}$. By Lemma 2.4.4 (applied to the additive, conical V-equivalence $\mathscr{D}$), there is $(a, c) \in \boldsymbol{a} \times \boldsymbol{c}$ such that $e = a \oplus c$. Since $e \in \boldsymbol{b} \oplus 2\boldsymbol{c}$, it follows from

Lemma 2.4.4 that there is $(b, c_1, c_2) \in \boldsymbol{b} \times \boldsymbol{c} \times \boldsymbol{c}$ such that $e = b \oplus c_1 \oplus c_2$. For each $i \in \{1, 2\}$, it follows from the relation $c \mathrel{\mathscr{D}} c_i$ that there is $g_i \in S$ such that $\mathbf{d}(g_i) = c$ and $\mathbf{r}(g_i) = c_i$. Observe that $g_i \langle c \rangle = g_i g_i^{-1} = c_i$. Hence, (c, g_1, g_2) is a fork of S, so, by assumption, there exists a nonnegative integer n such that $\langle g_1, g_2 \rangle^{-n}(c) = 0$. By Lemma 5.3.3, we obtain, using the notation of that lemma, the equalities

$$c_n = a_n = b_n \qquad \text{(because } c_{n+1} = 0\text{)},$$

$$\boldsymbol{a} = \operatorname{typ}(d_n) + \operatorname{typ}(a_n),$$

$$\boldsymbol{b} + \boldsymbol{c} = \operatorname{typ}(d_n) + \operatorname{typ}(b_n) + \operatorname{typ}(c_{n+1}),$$

whence, using again the equality $c_{n+1} = 0$, we get $\boldsymbol{a} = \boldsymbol{b} + \boldsymbol{c}$. □

We shall now adapt Theorem 5.3.4 to group actions on Boolean rings.

Lemma 5.3.5 *Let (c, g_1, g_2) be a fork in a Boolean inverse semigroup S. We define inductively g_s, for $s \in \{1,2\}^{<\omega}$, by $g_\varnothing = c$, and $g_{(i)^\frown s} = g_i g_s$ for every $(i, s) \in \{1,2\} \times \{1,2\}^{<\omega}$. We set $c_n = \prod_{s \in \{1,2\}^{<n}} g_s^{-1} \langle c \rangle$, for every positive integer n. The following statements hold:*

(1) *$g_i \langle c_{n+1} \rangle \leq c_n$, for all $i \in \{1, 2\}$ and every positive integer n.*
(2) *$g_s \langle c_n \rangle \leq c_{n-\operatorname{len}(s)}$, whenever $s \in \{1,2\}^{<\omega}$ and $\operatorname{len}(s) < n$.*
(3) *Let $n \in \mathbb{Z}^+$ and let $p, q \in \{1,2\}^{<n}$. If none of p and q is a prefix of the other, then $g_p \langle c_n \rangle g_q \langle c_n \rangle = 0$.*

Note The notation $c_n = \prod_{s \in \{1,2\}^{<n}} g_s^{-1} \langle c \rangle$ is consistent with the one used in Lemma 5.3.3.

Proof (1) We compute

$$g_i \langle c_{n+1} \rangle = g_i \big\langle g_1^{-1} \langle c_n \rangle \, g_2^{-1} \langle c_n \rangle \, c_n \big\rangle \leq g_i \big\langle g_i^{-1} \langle c_n \rangle \big\rangle = \mathbf{r}(g_i) c_n \leq c_n \,.$$

(2) follows immediately from (1) via an easy induction argument.

(3) We argue by induction on $\operatorname{len}(p) + \operatorname{len}(q)$. Since none of p and q is a prefix of the other, none of them is the empty sequence, so $p = (i) ^\frown \overline{p}$ and $q = (j) ^\frown \overline{q}$ for some $i, j \in \{1, 2\}$ and $\overline{p}, \overline{q} \in \{1,2\}^{<\omega}$. If $i = j$, then none of $\overline{p}$ and $\overline{q}$ is a prefix of the other, thus, by the induction hypothesis, $g_{\overline{p}} \langle c_n \rangle \, g_{\overline{q}} \langle c_n \rangle = 0$. It follows that

$$g_p \langle c_n \rangle \, g_q \langle c_n \rangle = g_i \big\langle g_{\overline{p}} \langle c_n \rangle \big\rangle \, g_i \big\langle g_{\overline{q}} \langle c_n \rangle \big\rangle = g_i \big\langle g_{\overline{p}} \langle c_n \rangle \, g_{\overline{q}} \langle c_n \rangle \big\rangle = g_i \langle 0 \rangle = 0 \,.$$

Now suppose that $i \neq j$. It follows from (2) above that $g_{\overline{p}} \langle c_n \rangle = c_{n-\operatorname{len}(\overline{p})} \leq c$. Hence, $g_p \langle c_n \rangle = g_i \big\langle g_{\overline{p}} \langle c_n \rangle \big\rangle \leq g_i \langle c \rangle$. Similarly, $g_q \langle c_n \rangle \leq g_j \langle c \rangle$. Therefore,

$$g_p \langle c_n \rangle \, g_q \langle c_n \rangle \leq g_i \langle c \rangle \, g_j \langle c \rangle = 0 \,. \qquad \square$$

The following result is mostly[3] contained in Tarski [109, Theorem 16.10], with a different argument (sketched on pages 224–229 of that reference) and formulation.[4] We remind the reader that the inverse semigroup $\mathrm{Inv}(B, G)$, of partial automorphisms of B piecewise in G, is introduced in Example 4.4.15.

Theorem 5.3.6 *Let G be an exponentially bounded group, acting by automorphisms on a Boolean ring B. Then the monoid $\mathbb{Z}^+\langle B\rangle /\!\!/ G$ is strongly separative.*

Proof By Proposition 4.4.20, $\mathbb{Z}^+\langle B\rangle /\!\!/ G \cong \mathrm{Typ}(\mathrm{Inv}(B, G))$. By Proposition 4.4.16, $\mathrm{Inv}(B, G)$ is isomorphic to the inverse semigroup $S = \mathbf{pHomeo}(\Omega, G)$ of all partial homeomorphisms, of the space Ω of all prime filters of B, which are piecewise in G. Note that the idempotent elements of that semigroup are the identity functions on the compact open subsets of Ω; we shall thus identify those functions with the compact open subsets themselves. With that convention, the equation $g \circ \mathrm{id}_U \circ g^{-1} = \mathrm{id}_{gU}$ enables us to identify $g\langle U\rangle$ with gU, for every $g \in S$ and every compact open subset U of Ω.

Since S is a Boolean inverse semigroup (cf. Example 4.4.14), it suffices, by Theorem 5.3.4, to prove that every fork (c, g_1, g_2) of S is nilpotent. With the identification above, c is now a compact open subset of Ω. Since g_1 and g_2 are piecewise in G, they have a common finite support X (cf. Example 4.4.14). For every positive integer n, we denote by $X^{(n)}$ the set of all products $x_1 \cdots x_n$, where all $x_i \in X$. Since G is exponentially bounded, there exists a positive integer n such that $\mathrm{card}\, X^{(n)} < 2^n$.

Using the notation of Lemma 5.3.5, we shall prove that $c_{n+1} = \varnothing$. Suppose otherwise and let $\mathfrak{p} \in c_{n+1}$. For each $s \in \{1, 2\}^n$, $c_{n+1} \leq g_s^{-1}\langle c\rangle \leq \mathbf{d}(g_s)$, $\mathfrak{p} \in c_{n+1}$, and $X^{(n)}$ is a support of g_s, thus there exists $x_s \in X^{(n)}$ such that $g_s(\mathfrak{p}) = x_s(\mathfrak{p})$. Since $\{1, 2\}^n$ has 2^n elements and $\mathrm{card}\, X^{(n)} < 2^n$, there are distinct $p, q \in \{1, 2\}^n$ such that $x_p = x_q$. Hence, $g_p(\mathfrak{p}) = g_q(\mathfrak{p})$ belongs to $g_p\langle c_{n+1}\rangle \cap g_q\langle c_{n+1}\rangle$, in contradiction with $g_p\langle c_{n+1}\rangle \cap g_q\langle c_{n+1}\rangle = \varnothing$ (cf. Lemma 5.3.5(3)). □

By applying Theorem 5.3.6 to the inverse semigroup $\mathbf{pHomeo}(G, G)$, where G is given the discrete topology and acts on itself by left translations, we obtain immediately the following corollary, first discovered by Lindenbaum and Tarski (cf. Tarski [109, Theorem 16.10]), then again by Rosenblatt [100, Theorem 3.3].

Corollary 5.3.7 *Every exponentially bounded group is supramenable.*

Proof Set $S = \mathbf{pHomeo}(G, G)$. The supramenability of G means that there is no nonzero $\boldsymbol{x} \in \mathrm{Typ}\, S$ such that $2\boldsymbol{x} \leq^+ \boldsymbol{x}$. This is a trivial consequence of the strong separativity of $\mathrm{Typ}\, S$, as given by Theorem 5.3.6. □

---

[3] I believe that Lindenbaum and Tarski's proof, as printed in [109], yields only that the partial commutative monoid $B /\!\!/ G$ satisfies the implication $\boldsymbol{a} + 2\boldsymbol{c} = \boldsymbol{b} + \boldsymbol{c} \Rightarrow \boldsymbol{a} + \boldsymbol{c} \leq^+ \boldsymbol{b}$. However, by Corollary 2.7.7, this still yields the desired conclusion.

[4] Although [109, Theorem 16.10] is stated there for *Abelian G*, it is mentioned on [109, p. 227] that the only consequence of abelianness that is used there is a specific (and unnamed in [109]) growth condition on group words. This condition is, of course, exponential boundedness.

It is a long-standing open problem whether every supramenable group is exponentially bounded. To our knowledge, this question first appeared in print in Wagon [115, Question 12.9(a)]. However, Rosenblatt already asked on [100, p. 51] the question whether the product of two supramenable groups is always supramenable. Since exponential boundedness is trivially preserved under finite product, a negative answer to Rosenblatt's question would imply a negative answer to Wagon's question.

In light of Corollary 5.3.7, it is interesting to ask the following question:

> Can the result of Theorem 5.3.6 (strong separativity of $\mathbb{Z}^+\langle B\rangle /\!\!/ G$) be extended from *exponentially bounded* to *supramenable* groups G?

This question was first raised, in case B is a powerset algebra, by Alexander Pruss (cf. Sect. 1.2.2).

As our next result shows, the answer to this question (for general B) is positive. In view of Corollary 5.3.7, it extends Theorem 5.3.6 (and Tarski [109, Theorem 16.10]) from exponentially bounded to supramenable.

Theorem 5.3.8 *Let G be a supramenable group, acting by automorphisms on a Boolean ring B. Then the monoid $\mathbb{Z}^+\langle B\rangle /\!\!/ G$ is strongly separative. In fact, every fork of* $\operatorname{Inv}(G, B)$ *is nilpotent.*

Proof By Proposition 4.4.20, $\mathbb{Z}^+\langle B\rangle /\!\!/ G \cong \operatorname{Typ}(\operatorname{Inv}(B, G))$. As in the proof of Theorem 5.3.6, we first observe that by Proposition 4.4.16, $\operatorname{Inv}(B, G)$ is isomorphic to the inverse semigroup $S = \mathbf{pHomeo}(\Omega, G)$ of all partial homeomorphisms, of the space Ω of all prime filters of B, which are piecewise in G.

We need to prove that every fork (c, g_1, g_2) of S is nilpotent. We shall use the notation of the proof of Theorem 5.3.6. Toward a contradiction, suppose that $c_{n+1} \neq \varnothing$ for every nonnegative integer n. Since the c_{n+1} form a decreasing sequence of nonempty compact subsets in the Hausdorff space Ω, their intersection c is nonempty. Observe that c is closed, but not open a priori. By Lemma 5.3.5, it follows that

$$g_i\langle c\rangle \leq c \text{ for each } i \in \{1,2\}\,, \text{ while } g_1\langle c\rangle\, g_2\langle c\rangle = 0\,. \tag{5.3.4}$$

Pick $\mathfrak{p} \in c$ and set

$$\varphi(x) = \{g \in G \mid g\mathfrak{p} \in x\}\,,$$

for every $x \subseteq \Omega$. It is straightforward to verify that φ preserves arbitrary unions and intersections, and sends the empty set to itself. Moreover,

(1) $1 \in \varphi(c)$, so $\varphi(c) \neq \varnothing$.
(2) $\varphi(gx) = g\varphi(x)$, for all $g \in G$ and all $x \subseteq \Omega$.

Let $f: a \to b$ in **pHomeo**(Ω, G). There are decompositions $a = \bigsqcup_{i<m} a_i$ and $b = \bigsqcup_{i<m} b_i$, with all a_i and b_i compact open, together with group elements $f_0, \ldots, f_{m-1} \in G$, such that $b_i = f_i a_i$ for each $i < m$, and

$$f(\mathfrak{p}) = f_i(\mathfrak{p}), \quad \text{whenever } i < m \text{ and } \mathfrak{p} \in a_i .$$

For every compact open subset x of Ω contained in the domain of f, $f\langle \mathrm{id}_x \rangle = \mathrm{id}_y$ where $y = \bigoplus_{i<m} f_i(xa_i)$. Assuming the usual identification between x and the identity function id_x, this yields the equality $f\langle x \rangle = \bigoplus_{i<m} f_i(xa_i)$, so x and $f\langle x \rangle$ are G-equidecomposable.

In particular, c and $g_i\langle c \rangle$ are G-equidecomposable, whenever $i \in \{1, 2\}$. Since φ preserves disjoint unions and by (2) above, it follows that the sets $C = \varphi(c)$ and $C_i = \varphi(g_i\langle c \rangle)$ are G-equidecomposable with pieces from the powerset algebra of G. Moreover, from (5.3.4) it follows that $C_1 \cup C_2 \subseteq C$ and $C_1 \cap C_2 = \varnothing$. Since G is supramenable, this implies that $C = \varnothing$, in contradiction with (1) above. □

Putting together several results of this section, we obtain the following result.

Theorem 5.3.9 *The following are equivalent, for every group G:*

(i) *G is supramenable.*
(ii) *Whenever G acts by automorphisms on a Boolean ring B, the monoid* $\mathbb{Z}^+\langle B \rangle /\!/ G$ *has no nonzero idempotents.*
(iii) *Whenever G acts by automorphisms on a Boolean ring B, the monoid* $\mathbb{Z}^+\langle B \rangle /\!/ G$ *is strongly separative.*
(iv) *Whenever G acts by automorphisms on a Boolean ring B, every fork of* $\mathrm{Inv}(B, G)$ *is nilpotent.*

Proof We start recalling that $\mathbb{Z}^+\langle B \rangle /\!/ G \cong \mathrm{Typ}(\mathrm{Inv}(B, G))$ (cf. Proposition 4.4.20).

(i)⇒(iv) is Theorem 5.3.8.

(iv)⇒(iii) follows immediately from Theorem 5.3.4.

(iii)⇒(ii) is trivial.

The particular case of (ii), with G acting on $\mathrm{Pow}\, G$ by left translation, implies that G has no nonempty paradoxical subsets; hence (i) follows. □

In contrast with the equivalence between (ii) and (iii) in Theorem 5.3.9, the following example will show an action of a group G on a Boolean algebra B such that $\mathbb{Z}^+\langle B \rangle /\!/ G$ has no nonzero idempotents, yet it is not order-separative (thus, a fortiori, not strongly separative).

Example 5.3.10 Define $\mathbb{T}$ as the set of all intervals of the rational line $\mathbb{Q}$ of the form either $[0, x]$, with x a nonnegative rational number, or $[0, x)$, with x a positive rational number. Endow $\mathbb{T}$ with the addition given by

$$\boldsymbol{x} + \boldsymbol{y} = \{x + y \mid (x, y) \in \boldsymbol{x} \times \boldsymbol{y}\}, \quad \text{for all } \boldsymbol{x}, \boldsymbol{y} \in \mathbb{T} .$$

Then $\mathbb{T}$ is a countable conical refinement monoid, with order-unit $[0, 1)$. It has no non-trivial idempotents, yet it is not even order-separative (for $[0, 1] + [0, 1) = 2 \cdot [0, 1)$ yet $[0, 1] \not\leq^+ [0, 1))$. By Theorem 4.8.9, there is an action of a group G on a Boolean algebra B such that $\mathbb{T} \cong \mathbb{Z}^+\langle B\rangle /\!\!/ G$.

This monoid $\mathbb{T}$ is identical to the monoid $\mathbb{T}$ introduced in Wehrung [120], minus the top element $[0, \infty)$.

If follows from Theorem 5.3.9 that whenever a supramenable group G acts by automorphisms on a Boolean ring R, the monoid $\mathbb{Z}^+\langle B\rangle /\!\!/ G$ is strongly separative. As witnessed by Example 2.10.8, this result does not extend to $M^+ /\!\!/ G$ for a finite group G acting on a dimension group M.

Another example, with a similar feature, is the following.

Example 5.3.11 Denote by G the group of all self-maps $f_{n,r}\colon \mathbb{Q} \to \mathbb{Q}, x \mapsto 2^n x + r$, where n is an integer and r is a rational number. The assignment $f_{n,r} \mapsto n$ defines a surjective homomorphism $\pi\colon G \twoheadrightarrow \mathbb{Z}$, and the assignment $r \mapsto f_{0,r}$ defines an embedding $\varepsilon\colon \mathbb{Q} \hookrightarrow G$, in such a way that we get an exact sequence of groups

$$1 \longrightarrow \mathbb{Q} \xrightarrow{\ \varepsilon\ } G \xrightarrow{\ \pi\ } \mathbb{Z} \longrightarrow 1\,.$$

In particular, G is metabelian. It is well known that G is not supramenable. In fact, G contains a copy of the free semigroup on two generators: for example, $x \mapsto 2x$ and $x \mapsto 2x + 1$ are such generators. In particular, denoting by $\mathrm{S}(\Omega, \mathbb{Z}^+)$ the commutative monoid of all bounded maps $\Omega \to \mathbb{Z}^+$, the monoid $\mathrm{S}(G, \mathbb{Z}^+) /\!\!/ G$ contains a nonzero idempotent element. Setting $M = \mathrm{S}(G, \mathbb{Z}^+)/\mathbb{Q}$, it follows from Proposition 2.8.6 that $\mathrm{S}(G, \mathbb{Z}^+) /\!\!/ G \cong M /\!\!/ (G/\mathbb{Q}) = M /\!\!/ \mathbb{Z}$ (as $G/\mathbb{Q} \cong \mathbb{Z}$). Since $\mathbb{Q}$ is Abelian, it is supramenable, thus, by Theorem 5.3.9, M is strongly separative.

5.4 Type Monoids with Completeness Conditions

5.4.1 Antisymmetry

The following result is one of the many reformulations, in many different contexts, of the Schröder-Bernstein Theorem. The present formulation and proof outline originate in Banach [17, Théorème 1]. We include a proof for convenience.

Theorem 5.4.1 *Let B be a Boolean ring, let M be a conical refinement monoid, and let $\mu\colon B \to M$ be a groupoid-induced V-measure. If B is conditionally σ-complete, then M is antisymmetric.*

Proof It is easy to verify directly that the Boolean ring $B^{(\mathbb{Z}^+)}$, of all almost zero sequences of elements of B, is also conditionally σ-complete. By Proposition 4.7.11, we may thus assume that μ is both surjective and induced by the action of a group G of automorphisms of B. Let $\boldsymbol{a}, \boldsymbol{b}, \boldsymbol{c} \in M$ such that $\boldsymbol{a}+\boldsymbol{b}+\boldsymbol{c} = \boldsymbol{c}$. We must prove that $\boldsymbol{a}+\boldsymbol{c} = \boldsymbol{c}$. Pick $c_0 \in B$ such that $\mu(c_0) = \boldsymbol{c}$. From $\boldsymbol{a}+\boldsymbol{b}+\boldsymbol{c} = \boldsymbol{c}$ it follows that there is a decomposition $c_0 = a_0 \oplus b_0 \oplus c_1$ in B such that $\mu(a_0) = \boldsymbol{a}$, $\mu(b_0) = \boldsymbol{b}$, and $\mu(c_1) = \boldsymbol{c}$. Since $\mu(c_0) = \mu(c_1) = \boldsymbol{c}$, there is $g \in G$ such that $g(c_0) = c_1$. The latter equation implies that we may define $a_n = g^n(a_0)$, $b_n = g^n(b_0)$, and $c_n = g^n(c_0)$, for each $n \in \mathbb{Z}^+$. Observe that $c_n = a_n \oplus b_n \oplus c_{n+1}$, for each $n \in \mathbb{Z}^+$. In particular, the a_m and the b_n are pairwise orthogonal. Since B is conditionally σ-complete, we may define $\overline{c} = \bigwedge_{n\geq 0} c_n$, $a' = \bigoplus_{n\geq 0} a_n$, $a'' = \bigoplus_{n\geq 1} a_n$, and $b' = \bigoplus_{n\geq 0} b_n$, and then the equation $c_0 = a' \oplus b' \oplus \overline{c}$ holds.

Since $g(a_n) = a_{n+1}$ for each n, we get $g(a') = a''$, thus $\mu(a') = \mu(a'')$, and thus, setting $c' = a'' \oplus b' \oplus \overline{c}$, we get $\mu(c') = \mu(c_0)$. Since $\mu(c_0) = \boldsymbol{c}$ and $c_0 = a_0 \oplus c'$, it follows that $\boldsymbol{a} + \boldsymbol{c} = \boldsymbol{c}$. □

The assumption in Theorem 5.4.1, that B be conditionally σ-complete, cannot be dropped. Indeed, by Theorem 4.8.9, every countable conical refinement monoid is group-measurable. In particular, $G^{\sqcup 0}$ is group-measurable, for every non-trivial countable Abelian group G. The monoid $G^{\sqcup 0}$ is not antisymmetric.

Since the canonical V-measure on a Boolean inverse semigroup is groupoid-induced (cf. Example 4.6.4), we get immediately the following consequence.

Corollary 5.4.2 *Let S be a Boolean inverse semigroup. If* Idp S *is conditionally σ-complete, then* Typ S *is antisymmetric.*

By Theorem 4.8.6, every V-measure on a countable unital Boolean ring is groupoid-induced. This may suggest that the assumption in Theorem 5.4.1, that μ be groupoid-induced, can be dropped. The following example proves that guess wrong.

Example 5.4.3 A conical refinement monoid M, a complete atomic Boolean algebra B, a V-measure $\mu: B \to M$, and $\boldsymbol{a} \in M$ such that the element $\boldsymbol{e} = \mu(1)$ is an order-unit of M, $2\boldsymbol{a} + \boldsymbol{e} = \boldsymbol{e}$, and $\boldsymbol{a} + \boldsymbol{e} \neq \boldsymbol{e}$.

Proof Throughout the proof, define B as the powerset algebra of $\mathbb{Z}^+$. The commutative monoid M will be defined as the enveloping monoid of $B/{\simeq}$, for a suitably defined conical V-equivalence $\simeq$ on B; then μ will be the canonical projection $B \twoheadrightarrow B/{\simeq}$.

Denote by $X \mathbin{\triangle} Y$ the symmetric difference of any sets X and Y. We say that a set is *even* (resp., *odd*) if its cardinality is finite and even (resp., finite and odd).

Claim 1 $\operatorname{card}(X \mathbin{\triangle} Y) \equiv \operatorname{card} X + \operatorname{card} Y \pmod 2$, for any finite sets X and Y.

Proof Just observe that the following relation holds:

$$\operatorname{card}(X \mathbin{\triangle} Y) = \operatorname{card} X + \operatorname{card} Y - 2\operatorname{card}(X \cap Y). \qquad \square \text{ Claim 1.}$$

We define a binary relation $\simeq$ on B, by letting $X \simeq Y$ hold if either X and Y are both finite and $\operatorname{card} X = \operatorname{card} Y$, or X and Y are both infinite and $X \mathbin{\triangle} Y$ is even. Observe that, by Claim 1, $X \simeq Y$ always implies that $X \mathbin{\triangle} Y$ is even. Furthermore, if $X \simeq Y$ and X and Y are both infinite, then $X \cap Y$ is infinite as well.

Claim 2 The binary relation $\simeq$ is an additive and conical V-equivalence on B.

Proof of Claim. It is trivial that $\simeq$ is both reflexive and symmetric. Let $X, Y, Z \subseteq \mathbb{Z}^+$ such that $X \simeq Y$ and $Y \simeq Z$; we must prove that $X \simeq Z$. The conclusion is trivial if one of the sets X, Y, Z is finite (in which case they are all finite). If X, Y, Z are all infinite, then the set $X \mathbin{\triangle} Z = (X \mathbin{\triangle} Y) \mathbin{\triangle} (Y \mathbin{\triangle} Z)$ is, by Claim 1, even, thus $X \simeq Z$. Hence, $\simeq$ is an equivalence relation on B. It is trivially conical.

Let $X_0, X_1, Y_0, Y_1 \in B$ such that $X_0 \cap X_1 = Y_0 \cap Y_1 = \varnothing$ while $X_0 \simeq Y_0$ and $X_1 \simeq Y_1$. Then $(X_0 \sqcup X_1) \mathbin{\triangle} (Y_0 \sqcup Y_1) = (X_0 \mathbin{\triangle} X_1) \mathbin{\triangle} (Y_0 \mathbin{\triangle} Y_1) = (X_0 \mathbin{\triangle} Y_0) \mathbin{\triangle} (X_1 \mathbin{\triangle} Y_1)$ is, by Claim 1, even, so $X_0 \sqcup X_1 \simeq Y_0 \sqcup Y_1$. This proves that $\simeq$ is additive.

Finally we prove that $\simeq$ is refining. Let $X = X_0 \sqcup X_1$ and Y such that $X \simeq Y$. We must find a decomposition $Y = Y_0 \sqcup Y_1$ such that each $X_i \simeq Y_i$.

Suppose first that either X_0 or X_1 (say X_0) is finite. Since either $\operatorname{card} Y = \operatorname{card} X$ or Y is infinite, Y has a subset Y_0 of the same cardinality as X_0; whence $X_0 \simeq Y_0$. Set $Y_1 = Y \setminus Y_0$. If X is finite, then all X_i and Y_i are finite, and $X_i \simeq Y_i$, so we are done. Suppose that X (thus X_1) is infinite. From $X \mathbin{\triangle} Y = X_0 \mathbin{\triangle} Y_0 \mathbin{\triangle} X_1 \mathbin{\triangle} Y_1$ it follows that $X_1 \mathbin{\triangle} Y_1 = (X \mathbin{\triangle} Y) \mathbin{\triangle} (X_0 \mathbin{\triangle} Y_0)$, thus, since $X \simeq Y$ and $X_0 \simeq Y_0$, we get, using Claim 1, the relation $X_1 \simeq Y_1$.

Suppose from now on that X_0 and X_1 are both infinite. Pick an element $z \in X_0 \cap Y$. The following set

$$Y_0 = \begin{cases} X_0 \cap Y\,, & \text{if } X_0 \setminus Y \text{ is even}\,,\\ X_0 \cap Y \setminus \{z\}\,, & \text{if } X_0 \setminus Y \text{ is odd}\,. \end{cases}$$

is contained in Y. Moreover, $Y_0 \subseteq X_0$, thus $X_0 \mathbin{\triangle} Y_0 = X_0 \setminus Y_0$, and thus

$$X_0 \mathbin{\triangle} Y_0 = \begin{cases} X_0 \setminus Y\,, & \text{if } X_0 \setminus Y \text{ is even}\,,\\ (X_0 \setminus Y) \sqcup \{z\}\,, & \text{if } X_0 \setminus Y \text{ is odd}\,. \end{cases}$$

In particular,

$$X_0 \simeq Y_0\,. \tag{5.4.1}$$

Now set $Y_1 = Y \setminus Y_0$. By definition, $Y = Y_0 \sqcup Y_1$. From $X \mathbin{\triangle} Y = X_0 \mathbin{\triangle} Y_0 \mathbin{\triangle} X_1 \mathbin{\triangle} Y_1$ it follows that $X_1 \mathbin{\triangle} Y_1 = (X \mathbin{\triangle} Y) \mathbin{\triangle} (X_0 \mathbin{\triangle} Y_0)$, thus, since $X \simeq Y$ and by (5.4.1), we get, using Claim 1, the relation $X_1 \simeq Y_1$. Therefore, $\simeq$ is refining. □ Claim 2.

By Claim 2 together with Lemma 2.4.4, we get a natural structure of a conical partial refinement monoid on the quotient set $P = B/{\simeq}$. By Proposition 2.1.8, P embeds, as a lower interval, into its enveloping monoid $M = \mathrm{U_{mon}}(P)$. By Theorem 2.2.3, M is a conical refinement monoid.

Denote by $\mu(X)$ the equivalence class of a set X relatively to $\simeq$. Setting $\boldsymbol{a} = \mu(\{0\})$ and $\boldsymbol{e} = \mu(\mathbb{Z}^+)$, it follows from the relation $\mathbb{Z}^+ \simeq 2 + \mathbb{Z}^+$ that $\boldsymbol{e} = \mu(2 + \mathbb{Z}^+)$. Since $\{0\} \simeq \{1\}$, we get $\boldsymbol{a} = \mu(\{1\})$, thus $2\boldsymbol{a} + \boldsymbol{e} = \mu(\{0\}) + \mu(\{1\}) + \mu(2 + \mathbb{Z}^+) = \mu(\mathbb{Z}^+) = \boldsymbol{e}$. On the other hand, $\boldsymbol{a} + \boldsymbol{e} = \mu(\{1\}) + \mu(2 + \mathbb{Z}^+) = \mu(1 + \mathbb{Z}^+)$, thus, since $1 + \mathbb{Z}^+ \not\simeq \mathbb{Z}^+$, it follows that $\boldsymbol{a} + \boldsymbol{e} \neq \boldsymbol{e}$. □

Although Example 5.4.3 might suggest that the completeness of B puts no strain on the range of any V-measure with domain B, the following example shows that this is not the case either.

Example 5.4.4 A countable conical refinement monoid M with order-unit such that there are no conditionally σ-complete Boolean ring B and no V-measure $\mu\colon B \to M$ with generating range.

Proof We consider again the conical refinement monoid P, introduced in Example 4.7.9, defined by the generators ε, 1 and the relation $\varepsilon + 1 = 1$.

Suppose that there are a conditionally σ-complete Boolean ring B and a V-measure $\mu\colon B \to P$ with generating range. Since 1 is a finite sum of elements of the range of μ, it necessarily belongs to the range of μ, that is, $1 = \mu(e_0)$ for some $e_0 \in B$. Since $\mu(e_0) = 1 = \varepsilon + 1$ and μ is a V-measure, there are $a_0, e_1 \in B$ such that $e_0 = a_0 \oplus e_1$, $\mu(a_0) = \varepsilon$, and $\mu(e_1) = 1$. Continuing in this manner, we get elements a_n and e_n, for $n \in \mathbb{Z}^+$, such that each $\mu(a_n) = \varepsilon$, each $\mu(e_n) = 1$, and each $e_n = a_n \oplus e_{n+1}$. Since B is conditionally σ-complete, the elements $\overline{e} = \bigwedge_{n\in\mathbb{Z}^+} e_n$ and $a_{(X)} = \bigvee_{n\in X} a_n$, for $X \subseteq \mathbb{Z}^+$, are well defined, and $e_0 = a_{(\mathbb{Z}^+)} \oplus \overline{e}$. The latter equation can be written $e_0 = a_{(2\mathbb{Z}^+)} \oplus a_{(1+2\mathbb{Z}^+)} \oplus \overline{e}$. It follows that

$$1 = \mu(a_{(2\mathbb{Z}^+)}) \oplus \mu(a_{(1+2\mathbb{Z}^+)}) \oplus \mu(\overline{e})\,. \tag{5.4.2}$$

Since $a_{(2\mathbb{Z}^+)}$ is an infinite orthogonal join of elements all sent by μ to ε, we get $n\varepsilon \leq^+ \mu(a_{(2\mathbb{Z}^+)})$ for each $n \in \mathbb{Z}^+$, thus $1 \leq^+ \mu(a_{(2\mathbb{Z}^+)})$. Likewise, $1 \leq^+ \mu(a_{(1+2\mathbb{Z}^+)})$. By (5.4.2), it follows that $2 \cdot 1 \leq^+ 2 \cdot 1 + \mu(\overline{e}) \leq^+ 1$ within P, a contradiction. □

5.4.2 Power Cancellation and Unperforation

Let us put even more conditions on the Boolean ring B, by requiring B be complete atomic (i.e., B is the powerset algebra of a given set). The following result was established, using Hall's Matching Theorem, by Miklós Laczkovich (private communication). For a proof, see Wehrung [120, Corollary 1.4].

Theorem 5.4.5 (Laczkovich) *Let B be a Boolean ring, let M be a conical refinement monoid, and let $\mu\colon B \to M$ be a groupoid-induced V-measure with generating range. If B is complete atomic, then M is unperforated.*

The same way as Theorem 5.4.1, Theorem 5.4.5 can be reformulated in terms of Boolean inverse semigroups.

Corollary 5.4.6 *Let S be a Boolean inverse semigroup. If* Idp *S is complete atomic, then* Typ *S is unperforated.*

In particular, whenever a group G acts by automorphisms on a complete atomic Boolean ring B, the monoid $\mathbb{Z}^+\langle B\rangle /\!\!/ G$ is power cancellative. The question, whether the assumption on B could be relaxed to mere completeness, was asked in Wagon [115, Problem 14, p. 231]. Truss solved that problem in the negative, in [111, Theorem 1.1].

Theorem 5.4.7 (Truss) *Let B be the Boolean ring of all Borel subsets of the Cantor space, modulo the sets of first Baire category. Then there are a group G of automorphisms of B and distinct elements $\boldsymbol{a}, \boldsymbol{b} \in \mathbb{Z}^+\langle B\rangle /\!\!/ G$ such that $2\boldsymbol{a} = 2\boldsymbol{b}$.*

Truss also proves in [111, Theorem 4.1] the following negative cancellation result.

Theorem 5.4.8 (Truss) *Let B be the powerset algebra of $\mathbb{Z}^+$. There is a group G of permutations of $\mathbb{Z}^+$ such that $\mathbb{Z}^+\langle B\rangle /\!\!/ G$ has elements $\boldsymbol{a}$, $\boldsymbol{b}$, $\boldsymbol{c}$ such that $\boldsymbol{a} + \boldsymbol{c} = \boldsymbol{b} + \boldsymbol{c}$ but there are no $\boldsymbol{x}, \boldsymbol{y}, \boldsymbol{z} \in \mathbb{Z}^+\langle B\rangle /\!\!/ G$ such that $\boldsymbol{a} = \boldsymbol{x} + \boldsymbol{z}$, $\boldsymbol{b} = \boldsymbol{y} + \boldsymbol{z}$, and $\boldsymbol{x} + \boldsymbol{c} = \boldsymbol{y} + \boldsymbol{c} = \boldsymbol{c}$.*

5.4.3 Refinement Algebras

The following definition is equivalent, for full conical monoids, to the one given in Tarski [109, Definition 11.26].

Definition 5.4.9 A conical refinement monoid M is a *refinement algebra* if for all $\boldsymbol{a}_0, \boldsymbol{a}_1, \boldsymbol{b}, \boldsymbol{c} \in M$, if $\boldsymbol{a}_0 + \boldsymbol{a}_1 + \boldsymbol{c} = \boldsymbol{b} + \boldsymbol{c}$, then there are $\boldsymbol{b}_0, \boldsymbol{b}_1, \boldsymbol{c}_0, \boldsymbol{c}_1 \in M$ such that $\boldsymbol{b} = \boldsymbol{b}_0 + \boldsymbol{b}_1$, $\boldsymbol{c} = \boldsymbol{c}_0 + \boldsymbol{c}_1$, and $\boldsymbol{a}_i + \boldsymbol{c}_i = \boldsymbol{b}_i + \boldsymbol{c}_i$ whenever $i \in \{0, 1\}$.

If $\boldsymbol{a}_0 + \boldsymbol{a}_1 + \boldsymbol{c} = \boldsymbol{c}$, then, setting $\boldsymbol{b} = 0$ in the definition above, it follows from the conicality of M that $\boldsymbol{b}_i = 0$, thus $\boldsymbol{a}_i + \boldsymbol{c}_i = \boldsymbol{c}_i$, and thus $\boldsymbol{a}_i + \boldsymbol{c} = \boldsymbol{c}$. Hence, *every refinement algebra is antisymmetric*. In particular, the conical refinement monoid M of Example 5.4.3, although it is generated by the range of a V-measure on a complete atomic Boolean ring, is not a refinement algebra. Another example of a conical refinement monoid which is not a refinement algebra is the primitive monoid of Example 5.4.4 (consider the equation $\varepsilon + \varepsilon + 1 = 0 + 1$).

On the positive side, the following result is a particular case of Tarski [109, Theorem 11.12]. The reduction of our context to the one of [109] works the same way as the beginning of the proof of Theorem 5.4.1.

Theorem 5.4.10 (Tarski) *Let B be a Boolean ring, let M be a conical refinement monoid, and let $\mu\colon B \twoheadrightarrow M$ be a groupoid-induced V-measure. If B is conditionally σ-complete, then M is a refinement algebra.*

The assumption in Theorem 5.4.10, that B be conditionally σ-complete, cannot be dropped. Indeed, by Theorem 4.8.9, every countable conical refinement monoid is group-measurable, and the (countable) conical refinement monoid of Example 5.4.4 is not a refinement algebra.

Corollary 5.4.11 *Let S be a Boolean inverse semigroup. If* $\operatorname{Idp} S$ *is conditionally σ-complete, then* $\operatorname{Typ} S$ *is a refinement algebra.*

5.4.4 Conditionally Countably Closed Boolean Inverse Semigroups

The results of the present section are more conveniently formulated in the language of (finitely closed, generalized) cardinal algebras, as defined in Tarski [109], as opposed to mere commutative monoids. By definition, a generalized cardinal algebra, or GCA, is a partial commutative monoid endowed with an infinitary partial addition, defined on countable sequences, satisfying certain attributes that ought to be satisfied by any well behaved infinitary addition. It turns out that those infinitary axioms reflect in a strong way on the first-order structure of the partial monoid, making it possible to establish, in a non-trivial way, such results as antisymmetry or unperforation.

Since the reference [109] is not easy to find, we also refer the reader to Wehrung [122, § 3.4] for a brief outline of those concepts. Nevertheless, a full length treatment of the matter would take up too much space and we shall only provide outlines of the results.

Definition 5.4.12 A Boolean inverse semigroup S is *countably closed*, or *σ-closed*, if every countable orthogonal subset of S has a join.

Example 5.4.13 Let B be a σ-complete Boolean ring. Then the Boolean inverse semigroup $\operatorname{Inv}(B)$ (cf. Example 4.4.9) is σ-closed. The join f of a countable orthogonal sequence $(f_n \mid n \in \mathbb{Z}^+)$ of elements of $\operatorname{Inv}(B)$ is given by the relations $\mathbf{d}(f) = \bigoplus_{n\in\mathbb{Z}^+} \mathbf{d}(f_n)$ and $\mathbf{r}(f) = \bigoplus_{n\in\mathbb{Z}^+} \mathbf{r}(f_n)$, together with

$$f(x) = \bigoplus_{n\in\mathbb{Z}^+} f_n\bigl(x\,\mathbf{d}(f_n)\bigr)\,, \quad \text{for all } x \in B \downarrow \mathbf{d}(f)\,.$$

More generally, for every group G of automorphisms of B, the set $\operatorname{Inv}^{\sigma}(B, G)$, of all countable orthogonal joins of elements of $\operatorname{Inv}(B, G)$ (cf. Example 4.4.15), is a σ-closed Boolean inverse semigroup. This extends naturally to any additive inverse subsemigroup of $\operatorname{Inv}(B)$ (by Theorem 4.4.17, every fundamental Boolean inverse semigroup has this form). The latter result can be obtained by a straightforward application of Tarski [109, Theorem 11.23] to B viewed as a GCA.

The proof of the following result is similar to the one of Wehrung [122, Proposition 13.10], and its origin can be traced back to Tarski [109, Theorem 11.32].

Theorem 5.4.14 *Let S be a σ-closed Boolean inverse semigroup. Then* $\operatorname{Typ} S$ *is a GCA.*

A counterpart of Theorem 5.4.14, for every ring R which is either $\aleph_0$-left continuous or $\aleph_0$-right continuous (or even every quotient of such a ring), is stated in Wehrung [122, Corollary 13.14]: namely, the underlying monoid $\mathrm{V}(R)$ of the nonstable K-theory of R is a GCA.

Chapter 6
Constructions Involving Involutary Semirings and Rings

The axioms of ring theory, when deprived of the existence of additive inverses, yield the axioms of *semirings*. When endowed with an additional involutary anti-automorphism (we will talk about *involutary semirings*), semirings will enjoy quite a fruitful interaction with Boolean inverse semigroups, the basic idea being to have the multiplications agree and the inversion map correspond to the involution.

In Sect. 6.1, we will set the basic framework for involutary semirings, enabling us to detect Boolean inverse semigroups in both involutary semirings and involutary rings.

Section 6.2 will introduce our prototype of involutary semiring, obtained via the natural expansion of the enveloping monoid $\mathrm{U}_{\mathrm{mon}}(S)$ of a Boolean inverse semigroup S, endowed with its operation of orthogonal addition.

One of the side products of that construction, discussed in Sect. 6.3, will be a convenient introduction of the *additive enveloping K-algebra* $K\langle S\rangle$ of a Boolean inverse semigroup S, for any unital ring K. This object is the universal K-algebra, containing S as a subsemigroup, with range centralizing K, in such a way that finite orthogonal sums in the Boolean inverse semigroup are turned to finite sums in the ring.

The presentation of the additive enveloping K-algebra given in Sect. 6.3 is not sufficient, a priori, to establish even basic structural results of $K\langle S\rangle$. Section 6.4 partly fills this gap, in particular listing a few sufficient conditions, for an additive semigroup embedding $S \hookrightarrow T$ of Boolean inverse semigroups, to extend to an embedding $K\langle S\rangle \hookrightarrow K\langle T\rangle$ of K-algebras.

In Sect. 6.5, we present a general argument showing, in particular, that Leavitt path algebras of quivers are particular cases of the $K\langle S\rangle$ construction. Analogues of those results for C*-algebras are also presented.

In Sect. 6.6, we show how to canonically adjoin a unit to any Boolean inverse semigroup, by using the unitization constructions for rings and for generalized Boolean algebras.

F. Wehrung, *Refinement Monoids, Equidecomposability Types, and Boolean Inverse Semigroups*, Lecture Notes in Mathematics 2188,
DOI 10.1007/978-3-319-61599-8_6

For a Boolean inverse semigroup S and a unital ring K, there is a canonical monoid homomorphism f from the type monoid of S to the nonstable K-theory of the ring $K\langle S\rangle$. Although we will prove that f can be, in some exceptional cases, an isomorphism, we will show in Sect. 6.7 a few counterexamples showing that this statement does not hold in full generality.

While Boolean inverse semigroups mimick, in many of their aspects, von Neumann regular rings or, more generally, exchange rings, we will show in Sect. 6.8 a feature of Boolean inverse semigroups absent from those ring-theoretical contexts: namely, they afford a natural definition of a tensor product. This study will be pursued in Sect. 6.9, where we will prove that the type monoid functor and the tensor product bifunctor commute.

6.1 Inverse Semigroups in Involutary Semirings

We begin with a classical definition.

Definition 6.1.1

- A *semiring* is a structure $(M, +, 0, \cdot)$, where $(M, +, 0)$ is a commutative monoid, $(M, \cdot)$ is a semigroup, $x \cdot 0 = 0 \cdot x = 0$, $x \cdot (y + z) = (x \cdot y) + (x \cdot z)$, and $(x + y) \cdot z = (x \cdot z) + (y \cdot z)$, for all $x, y, z \in M$.
- An *involutary semiring* is a structure $(M, +, 0, \cdot, ^*)$, where $(M, +, 0, \cdot)$ is a semiring and * is a unary operation on M (the *involution* of our structure) such that $(x + y)^* = x^* + y^*$, $(x \cdot y)^* = y^* \cdot x^*$, and $(x^*)^* = x$, for all $x, y \in M$. An element $x \in M$ is *self-adjoint* if $x = x^*$.
- Following the terminology in use for involutary rings, we say that an involutary semiring M is *proper* if $x^*x = 0$ implies that $x = 0$, for each $x \in M$.

Following a widespread convention, we will usually write xy instead of $x \cdot y$. We will often extend, to involutary semirings, attributes of commutative monoids, by simply applying those attributes to the underlying commutative monoid. For example, an involutary semiring is cancellative (resp., conical) if its underlying commutative monoid is cancellative (resp., conical).

Definition 6.1.2 Two elements x and y in an involutary semiring M are *orthogonal*, in notation $x \perp y$, if $x^*y = xy^* = 0$. Further, let $z = x \oplus y$ hold if $z = x + y$ and $x \perp y$, for all $x, y, z \in M$. We will call $\oplus$ the *orthogonal addition in* M.

It is obvious that the orthogonality relation is symmetric (i.e., $x \perp y$ iff $y \perp x$). There are trivial examples where none of the relations of orthogonality and meet-orthogonality in M (cf. Definition 2.3.1) contains the other. However, in most contexts that we shall encounter, orthogonality implies meet-orthogonality.

Lemma 6.1.3 *The following statements hold, for any involutary semiring M:*

(1) $x \perp z$ *and* $y \perp z$ *implies that* $x + y \perp z$, *for all* $x, y, z \in M$.
(2) *Suppose that M is conical. Then* $x \perp y$, $u \leq^+ x$, *and* $v \leq^+ y$ *implies that* $u \perp v$, *for all* $x, y, u, v \in M$. *Furthermore, the orthogonal addition* $\oplus$ *endows M with a structure of a partial commutative monoid.*
(3) *Suppose that M is both conical and proper, and let* $x, y \in M$. *If* $x \perp y$, *then* $x \wedge y = 0$ *within* $(M, \leq^+)$.

Proof (1) follows trivially from the distributivity of the multiplication of M with respect to its addition.

(2) There are elements $\overline{u}, \overline{v} \in M$ such that $x = u + \overline{u}$ and $y = v + \overline{v}$. It follows that $0 = x^*y = u^*v + u^*\overline{v} + \overline{u}^*v + \overline{u}^*\overline{v}$, thus, since M is conical, $u^*v = 0$. The proof that $uv^* = 0$ is similar.

The final statement of (2) follows trivially.

(3) Let $z \in M$ such that $z \leq^+ \begin{matrix} x \\ y \end{matrix}$. It follows from (2) above that $z \perp z$, whence $z^*z = 0$. Since M is proper, $z = 0$. □

Definition 6.1.4 A nonempty subset S in an involutary semiring M is an *inverse semigroup in M* if S is a multiplicative subsemigroup of M and $x^* = x^{-1}$ for all $x \in S$.

In particular, if S is an inverse semigroup in an involutary semiring M, with the same zero, then the orthogonality relation of S is the restriction to S of the orthogonality relation of M: that is, $x \perp y$ iff $x^{-1}y = xy^{-1} = 0$, iff $x^*y = xy^* = 0$, for all $x, y \in S$.

Lemma 6.1.5 *Let S be an inverse semigroup in a conical involutary semiring M. Then the set $S^{\oplus}$, of all finite orthogonal sums of elements of M, is an inverse semigroup in M.*

Proof It is trivial that $S^{\oplus}$ is closed under the involution of M.

Claim 1 The set $S^{\oplus}$ is a multiplicative subsemigroup of M.

Proof of Claim. Let $x, y \in S^{\oplus}$. There are decompositions $x = \bigoplus_{i=1}^m x_i$ and $y = \bigoplus_{j=1}^n y_j$, with $m, n \in \mathbb{Z}^+$ and elements $x_i, y_j \in S$. Since $xy = \sum_{(i,j)\in[m]\times[n]} x_iy_j$, all is left to prove is that $x_iy_j \perp x_{i'}y_{j'}$ whenever $(i,j) \neq (i',j')$. In that case, either $m \geq 2$ or $n \geq 2$, thus $0 \in S$. First observe that

$$(x_iy_j)^*x_{i'}y_{j'} = y_j^*x_i^*x_{i'}y_{j'}\,. \tag{6.1.1}$$

If $i \neq i'$, then $x_i^*x_{i'} = 0$, thus, by (6.1.1), $(x_iy_j)^*x_{i'}y_{j'} = 0$. Suppose now that $i = i'$. Then $x_i^*x_{i'} = \mathbf{d}(x_i)$ is idempotent, thus, by (6.1.1), $(x_iy_j)^*x_{i'}y_{j'} \leq y_j^*y_{j'}$ (with respect to the natural ordering of S), thus, as $y_j^*y_{j'} = 0$, we get again $(x_iy_j)^*x_{i'}y_{j'} = 0$. Likewise, $x_iy_j(x_{i'}y_{j'})^* = 0$, so $x_iy_j \perp x_{i'}y_{j'}$. □ Claim 1.

Claim 2 The subset $\overline{S} = \{(x, y) \in M \times M \mid x = xx^*x\}$ is closed under finite orthogonal sums. In particular, it contains $S^{\oplus}$.

Proof of Claim. Let $x, y \in \overline{S}$ be orthogonal and let $z = x + y$. Since $x = xx^*x$ and $y = yy^*y$, we get $zz^*z = z + xx^*y + xy^*x + xy^*y + yx^*x + yx^*y + yy^*x$. From $x \perp y$ it follows that $xx^*y = xy^*x = xy^*y = yx^*x = yx^*y = yy^*x = 0$. Therefore, $zz^*z = z$ belongs to $\overline{S}$. □ Claim 2.

Claim 3 The elements x^*x and y^*y commute, for all $x, y \in S^{\oplus}$.

Proof of Claim. Write again $x = \bigoplus_{i=1}^{m} x_i$ and $y = \bigoplus_{j=1}^{n} y_j$, with $m, n \in \mathbb{Z}^+$ and elements $x_i, y_j \in S$. Since the x_i (resp., y_j) are pairwise orthogonal, we obtain

$$x^*x = \sum_{i=1}^{m} x_i^*x_i \text{ and } y^*y = \sum_{j=1}^{n} y_j^*y_j\,. \tag{6.1.2}$$

Since all elements $x_i^*x_i = \mathbf{d}(x_i)$ and $y_j^*y_j = \mathbf{d}(y_j)$ are idempotent elements of S, they pairwise commute. By (6.1.2), it follows that x^*x and y^*y commute. □ Claim 3.

The claims above enable us to apply Lemma 3.1.1 to the structure $(S^{\oplus}, \cdot, ^*)$, thus completing the proof. □

Theorem 6.1.6 *Let M be a conical refinement involutary semiring and let S be an inverse semigroup in M, satisfying the following conditions:*

(1) *Every element of S has index at most* 1 *in M.*
(2) *S is a lower subset of $(M, \leq^+)$.*
(3) Idp *S is an upward directed subset of $(M, \leq^+)$.*
(4) *S is closed under finite orthogonal sums within M.*

Then S is a Boolean inverse semigroup, with the same zero as M. Furthermore, S and Idp *S are both lower subsets of $(M, \leq^+)$, the algebraic preordering $\leq^+$ restricts, on S, to the natural ordering of S, and the orthogonal addition of S is the restriction to S of the orthogonal addition of M.*

Proof It follows from (4) (applied to the empty sum) that $0 \in S$, so S is an inverse semigroup with zero.

Claim 1 Let $x \in M$ and $y \in S$. If $x + y = y$, then $x = 0$.

Proof of Claim. From $y = x + y$ it follows that $y = 2x + y$, thus, by (1), $x = 0$. □ Claim 1.

Claim 2 Let $x, y \in S$. If $x \perp y$, then $x \wedge y = 0$ within $(M, \leq^+)$.

Proof of Claim. Let $z \in M$ such that $z \leq^+ \begin{matrix} x \\ y \end{matrix}$. It follows from (2) that $z \in S$. Further, it follows from Lemma 6.1.3 that $z \perp z$, whence $z^*z = 0$. Hence, $z = zz^*z = 0$. □ Claim 2.

Claim 3 Let $x, y \in S$. Then $x + y \in S$ iff $x \perp y$ (within the involutary semiring M) iff $x \perp y$ (within the inverse semigroup S).

Proof of Claim. The statements $x \perp y$ (within M) and $x \perp y$ (within S) are clearly equivalent. If this statement holds, then, by (4), $x + y \in S$.

Suppose, conversely, that the element $z = x + y$ belongs to S. Since $x = xx^*x$ and $y = yy^*y$, we get $z = zz^*z = z + xx^*y + xy^*x + xy^*y + yx^*x + yx^*y + yy^*x$, thus, by Claim 1 together with the conicality of M, $xx^*y = xy^*y = 0$. It follows that $x^*y = x^*xx^*y = 0$ and $xy^* = xy^*yy^* = 0$, so $x \perp y$. □ Claim 3.

Claim 4 Denote by $\leq$ the natural ordering of S and let $x, y \in S$. Then $x \leq^+ y$ iff $x \leq y$, for all $x, y \in S$.

Proof of Claim. Suppose first that $x \leq^+ y$. There is $z \in M$ such that $x + z = y$. It follows from (2) that $z \in S$. By Claim 3, $x \perp z$. Therefore, $y\,\mathbf{d}(x) = yx^*x = xx^*x + zx^*x = x + 0 = x$, that is, $x \leq y$.

Suppose, conversely, that $x \leq y$. By (3), there is $e \in \operatorname{Idp} S$ such that $\begin{matrix}\mathbf{d}(x)\\ \mathbf{d}(y)\end{matrix} \leq^+ e$. By the paragraph above, $\mathbf{d}(y) \leq e$, that is, $\mathbf{d}(y) = \mathbf{d}(y)e$, thus, multiplying on the left by y, we get $y = ye$. Let $x' \in M$ such that $e = \mathbf{d}(x) + x'$. We get

$$y = ye = y(\mathbf{d}(x) + x') = y\,\mathbf{d}(x) + yx' = x + yx',$$

whence $x \leq^+ y$. □ Claim 4.

It follows from (2) that S and $\operatorname{Idp} S$, endowed with the restrictions of the addition of M, are lower intervals of $(M, +, 0)$. In particular, S and $\operatorname{Idp} S$, endowed with this addition, are conical partial refinement monoids.

Moreover, by Claim 3, the addition operations in S and $\operatorname{Idp} S$ are the restrictions of the orthogonal addition in M. By Claim 2, those additions are multiple-free. By applying (3), together with Proposition 2.3.10, to $\operatorname{Idp} S$, we obtain that $\operatorname{Idp} S$ is a generalized Boolean algebra.

Finally let $x, y \in S$ be orthogonal. By Claim 3, the element $z = x + y$ belongs to S. By Claim 4, both inequalities $x \leq z$ and $y \leq z$ hold within S. Now let $t \in S$ such that $\begin{matrix}x\\ y\end{matrix} \leq t$ within S. By (2) together with Claim 4, it follows that $\begin{matrix}x\\ y\end{matrix} \leq^{\oplus} t$ within $(S, \oplus, 0)$. By applying Lemma 2.3.9 within $(S, \oplus, 0)$, it follows that $z \leq^{\oplus} t$ within $(S, \oplus, 0)$. By Claim 4, it follows that $z \leq t$ within S. Therefore, z is the orthogonal join of x and y within S, thus completing the verification that S is a Boolean inverse semigroup. □

We conclude this section with the following analogue of Theorem 6.1.6 for involutary rings. This result says that the embedding problem of an inverse semigroup into an involutary ring R (via an involutary semigroup homomorphism) is, essentially, the same as the embedding problem of a *Boolean* inverse semigroup into R, with preservation of orthogonal sums.

Theorem 6.1.7 *Let R be an involutary ring, let S be an inverse semigroup in R, and let A be a commutative subring of self-adjoint elements in R, containing* $\operatorname{Idp} S$, *such that $xAx^{-1} \subseteq A$ for every $x \in S$. Then $S \cup A$ is contained in a Boolean inverse* *semigroup $\overline{S}$ in R, with the same idempotents as A. Furthermore, orthogonal join in $\overline{S}$ is induced by orthogonal addition in R.*

Proof It is well known since Foster [46] that the set B of all idempotent elements of A is a generalized Boolean algebra under the operations given by

$$x \wedge y = xy\,,\; x \vee y = x + y - xy\,,\; \text{and } x \smallsetminus y = x - xy\,, \quad \text{for all } x, y \in B\,.$$

We set $B^{\sqcup 1} = B \sqcup \{1\}$ for a new element 1 such that $x1 = 1x = x$ for all $x \in R$, and we define

$$\overline{S} = \left\{x \in R \mid xx^*x = x \text{ and } \{xbx^*, x^*bx\} \subseteq B \text{ for all } b \in B^{\sqcup 1}\right\}.$$

It is obvious that $\overline{S}$ is closed under the involution $x \mapsto x^*$. Now let $x, y \in \overline{S}$. From $y \in \overline{S}$ it follows that $yby^* \in B$. Since $x \in \overline{S}$, it follows that $xyby^*x^* \in B$, that is, $(xy)b(xy)^* \in B$. Likewise, we can prove that $(xy)^*b(xy) \in B$. For the quasi-inverse property, we observe that since yy^* and x^*x both belong to B, they commute, thus

$$xy(xy)^*xy = x(yy^*)(x^*x)y = xx^*xyy^*y = xy\,.$$

We have thus proved that $\overline{S}$ is a subsemigroup of R, closed under the involution $x \mapsto x^*$. The elements of the form x^*x, for $x \in \overline{S}$, all belong to B, thus they commute pairwise. By Lemma 3.1.1, it follows that $\overline{S}$ *is an inverse semigroup in the involutary ring R*. Furthermore, for any $x \in S$ and any $b \in B^{\sqcup 1}$, it follows from our assumption that the element xbx^{-1} belongs to A. Since this element is idempotent (for b and $x^{-1}x$ commute, thus $xbx^{-1}xbx^{-1} = xx^{-1}xb^2x^{-1} = xbx^{-1}$), we get $xbx^{-1} \in B$. Likewise, $x^{-1}bx \in B$, so we have proved that $\overline{S}$ *contains S*.

The subset $\overline{S}$ obviously contains B. Conversely, every idempotent $x \in \overline{S}$ satisfies $x = x^*x \in B$, whence the subset $\operatorname{Idp} \overline{S} = B$ is a generalized Boolean algebra.

Now let x and y be orthogonal elements in $\overline{S}$. For every $b \in B^{\sqcup 1}$, $x^*by \leq x^*y = 0$ (where $\leq$ denotes the natural ordering of $\overline{S}$), thus $x^*by = 0$. Likewise, $y^*bx = 0$. It follows that

$$(x+y)^*b(x+y) = x^*bx + x^*by + y^*bx + y^*by = x^*bx + y^*by\,.$$

From $x, y \in \overline{S}$ it follows that x^*bx and y^*by both belong to B, thus, in particular, they are idempotent. Furthermore, from $x^*y = xy^* = 0$ it follows that $(x^*bx)(y^*by) = (y^*by)(x^*bx) = 0$, whence $x^*bx + y^*by$ is idempotent. Since $x^*bx + y^*by \in A$, it follows that $x^*bx + y^*by \in B$, that is, $(x+y)^*b(x+y) \in B$. A similar proof yields that $(x+y)b(x+y)^* \in B$. Further, it follows again from the orthogonality of x and y that $(x+y)(x^* + y^*) = xx^* + yy^*$, and thus, by the same token,

$$(x+y)(x^* + y^*)(x+y) = xx^*x + yy^*y = x + y\,,$$

thus completing the verification that $x + y \in \overline{S}$. The elements $a = x^*x$ and $b = y^*y$ both belong to B. From $x = (x + y)a$ and $y = (x + y)b$ it follows that $x \leq x + y$ and $y \leq x + y$. Let $z \in \overline{S}$ such that $x \leq z$ and $y \leq z$, that is, $x = za$ and $y = zb$. Since a and b are orthogonal elements of B, their sum $a + b$ belongs to B, whence $x + y = z(a + b) \leq z$ (the inequality holding within the inverse semigroup $\overline{S}$), therefore completing the verification that $x+y$ is the orthogonal join of a and b. □

Throughout this work, we will use Theorem 6.1.7 only in the case where A is the subring of R generated by Idp S (clearly, $xAx^{-1} \subseteq A$ for any $x \in S$).

The study of C*-algebras generated, in various ways, by inverse semigroups, is quite an active topic of research, see, for example, Duncan and Paterson [36], Paterson [93, § 2.1], Exel [40, 42]. As an immediate consequence of Theorem 6.1.7, observe the following.

Corollary 6.1.8 *Let S be an inverse semigroup. If an involutary ring (resp., a C*-algebra) R is generated by S, then it is also generated by a Boolean inverse semigroup containing S, in which orthogonal join is induced by orthogonal addition in R.*

6.2 The Additive Enveloping Involutary Semiring of a Boolean Inverse Semigroup

Recall from Sect. 3.1.2 that any Boolean inverse semigroup S, endowed with its partial operation of orthogonal join, is a conical partial refinement monoid (cf. Proposition 3.1.9). This entitles us to define the enveloping monoid $\mathrm{U_{mon}}(S)$ (cf. Sect. 2.1). It follows from Theorem 2.2.3 that $\mathrm{U_{mon}}(S)$ is a conical refinement monoid. As the following result shows, more can be said about the monoid structure of $\mathrm{U_{mon}}(S)$.

A commutative monoid M is *Archimedean* if its algebraic preordering $\leq^+$ is antisymmetric (i.e., it is an ordering) and for all $x, y \in M$, if $nx \leq^+ (n + 1)y$ for all $n \in \mathbb{N}$, then $x \leq^+ y$. Equivalently (cf. Wehrung [122, Lemma 3.4]), M embeds, with its algebraic preordering, into a power of $[0, \infty]$ endowed with its natural ordering. In particular, M is unperforated.

Proposition 6.2.1 *Let S be a Boolean inverse semigroup. Then every element of S (resp.,* $\mathrm{U_{mon}}(S)$*) has index at most* 1 *(resp., finite index) in* $\mathrm{U_{mon}}(S)$*, and* $\mathrm{U_{mon}}(S)$ *is the positive cone of an Archimedean dimension group.*

Proof We start by proving that every element $a \in S$ has index at most 1 within $\mathrm{U_{mon}}(S)$. Let $a = 2x + y$ within $\mathrm{U_{mon}}(S)$, where $x, y \in \mathrm{U_{mon}}(S)$. Since S is a lower subset of $\mathrm{U_{mon}}(S)$ (cf. Proposition 2.1.8), both elements x and $2x$ belong to S, and $2x = x \oplus x$ (orthogonal join within S). Hence $\mathbf{d}(x) = x^*x = 0$, so $x = 0$.

Since $\mathrm{U_{mon}}(S)$ is a conical refinement monoid, additively generated by S, it follows from Wehrung [122, Lemma 3.11] that every element of $\mathrm{U_{mon}}(S)$ has finite index. It follows then from [122, Proposition 3.13] that $\mathrm{U_{mon}}(S)$ is the positive cone of an Archimedean dimension group. □

Proposition 6.2.2 *Let S be a Boolean inverse semigroup. Then there is a unique involutary semiring structure on $\mathrm{U_{mon}}(S)$ whose multiplication and involution both extend those of S. The pair $(S, \mathrm{U_{mon}}(S))$ satisfies the properties* (1)–(4) *stated in Theorem* 6.1.6. *Furthermore, $\mathrm{U_{mon}}(S)$ is proper.*

We shall call $\mathrm{U_{mon}}(S)$, endowed with the above-mentioned structure of involutary semiring, the *additive enveloping semiring of S*.

Proof The uniqueness statement follow trivially from the fact that $\mathrm{U_{mon}}(S)$ is additively generated by S.

For any $x \in S$, it follows from Proposition 3.1.9 that the left multiplication $S \to \mathrm{U_{mon}}(S)$, $y \mapsto xy$ is a homomorphism of partial commutative monoids from S to $\mathrm{U_{mon}}(S)$. By the universal property of $\mathrm{U_{mon}}(S)$ (cf. Proposition 2.1.7), this map extends to a unique monoid endomorphism λ_x of $\mathrm{U_{mon}}(S)$. Again by Proposition 3.1.9, the assignment $x \mapsto \lambda_x$ is a homomorphism of partial commutative monoids from $(S, \oplus, 0)$ to the endomorphism monoid of $\mathrm{U_{mon}}(S)$, thus, by the universal property of $\mathrm{U_{mon}}(S)$, it extends to a unique homomorphism λ from $\mathrm{U_{mon}}(S)$ to its endomorphism monoid. Set $x \cdot y = \lambda(x)(y)$, for all $x, y \in \mathrm{U_{mon}}(S)$. By construction, the binary operation $\cdot$ is a monoid bimorphism (cf. Sect. 2.5). Since the operation $\cdot$ obviously extends the multiplication of S, we obtain the formula

$$\Big(\sum_{i\in I} x_i\Big) \cdot \Big(\sum_{j\in J} y_j\Big) = \sum_{(i,j)\in I\times J} x_i y_j \,, \tag{6.2.1}$$

for all finite sets I, J and all elements $x_i, y_j \in S$ (for $i \in I$ and $j \in J$). Now it follows easily from (6.2.1), together with Proposition 3.1.9, that the operation $\cdot$ is both associative and distributive over the addition of $\mathrm{U_{mon}}(S)$.

The map $S \to \mathrm{U_{mon}}(S)$, $x \mapsto x^*$ is a homomorphism of partial commutative monoids, which extends to a unique monoid endomorphism of $\mathrm{U_{mon}}(S)$, which we shall denote by $x \mapsto x^*$. Observe that $\big(\sum_{i\in I} x_i\big)^* = \sum_{i\in I} x_i^*$, for every finite set I and all elements x_i, for $i \in I$, of S. By using (6.2.1), the verification that $\mathrm{U_{mon}}(S)$ is an involutary semiring is routine. Now it is obvious that S is an inverse semigroup in $\mathrm{U_{mon}}(S)$. That the property (1), stated in Theorem 6.1.6, is satisfied follows from Proposition 6.2.1. The verifications of the properties (2)–(4) are trivial.

Let $x \in \mathrm{U_{mon}}(S)$ such that $x^*x = 0$. We can write $x = \sum_{i\in[n]} x_i$, where $n \in \mathbb{Z}^+$ and each $x_i \in S$. It follows that $\sum_{(i,j)\in[n]\times[n]} x_i^* x_j = 0$, thus, since $\mathrm{U_{mon}}(S)$ is conical, each $x_i^* x_i = 0$, and thus, since $x_i \in S$, we get $x_i = x_i x_i^* x_i = 0$. This holds for each $i \in [n]$, so $x = 0$. □

Recall that the tensor product $M \otimes N$, of two commutative monoids M and N, is defined in Sect. 2.5. The following result extends that construction to either semirings or involutary semirings.

Lemma 6.2.3 *The following statements hold, for any semirings M and N:*

(1) *There is a unique semiring structure on the commutative monoid $M \otimes N$ such that*

$$(x \otimes y) \cdot (x' \otimes y') = (xx') \otimes (yy'), \quad \text{for all } x, x' \in M \text{ and all } y, y' \in N. \tag{6.2.2}$$

(2) *Suppose that, in addition, M and N are both involutary semirings. Then there is a unique involution on $M \otimes N$ such that $(x \otimes y)^* = x^* \otimes y^*$ whenever $(x, y) \in M \times N$. Furthermore, if M and N are both conical and proper, then so is $M \otimes N$.*

Proof The proof is similar to the classical proof for rings. Since $M \otimes N$ is additively generated by $\{x \otimes y \mid (x, y) \in M \times N\}$, the uniqueness statements of (1) and (2) are both trivial.

Now we deal with the existence statements. For each $(x, y) \in M \times N$, the assignment $(x', y') \mapsto (xx') \otimes (yy')$ defines a bimorphism from $M \times N$ to $M \otimes N$, thus there is a unique monoid endomorphism $\tau_{x,y}$ of $M \otimes N$ such that $\tau_{x,y}(x' \otimes y') = (xx') \otimes (yy')$ for each $(x', y') \in M \times N$. The assignment $(x, y) \mapsto \tau_{x,y}$ is a bimorphism from $M \times N$ to the endomorphism monoid of $M \otimes N$, thus there is a unique monoid homomorphism τ from $M \otimes N$ to its endomorphism monoid such that $\tau(x \otimes y) = \tau_{x,y}$ for each $(x, y) \in M \times N$. Set $z \cdot z' = \tau(z)(z')$, for all $z, z' \in M \otimes N$.

The construction of the involution is similar, and even easier.

Now suppose that M and N are both conical and proper. By Lemma 2.5.1, $M \otimes N$ is also conical. Let $z \in M \otimes N$ such that $z^*z = 0$. Write $z = \sum_{i<n}(x_i \otimes y_i)$, where $n \in \mathbb{Z}^+$ and each $(x_i, y_i) \in (M \setminus \{0\}) \times (N \setminus \{0\})$. Then $0 = z^*z = \sum_{i,j<n}\big((x_i^*x_j) \otimes (y_i^*y_j)\big)$, thus, since $M \otimes N$ is conical, $(x_i^*x_i) \otimes (y_i^*y_i) = 0$ for each $i < n$. Since M and N are both proper, $x_i^*x_i \neq 0$ and $y_i^*y_i \neq 0$ for each $i < n$. By Lemma 2.5.1, it follows that $n = 0$, that is, $z = 0$. □

6.3 The Additive Enveloping K-Algebra of a Boolean Inverse Semigroup

This section presents a preliminary study, for a unital ring K and a Boolean inverse semigroup S, of the K-algebra $K\langle S\rangle$ defined by generators S forming a multiplicative subsemigroup centralizing K, subjected to the relations $z = x + y$, within $K\langle S\rangle$, whenever $z = x \oplus y$, within S.

Definition 6.3.1 Let K be a unital ring. A *K-algebra*[1] is a ring R, endowed with a structure of a bimodule over K, such that the equations

$$(\lambda x)y = \lambda(xy), \quad (x\lambda)y = x(\lambda y), \quad (xy)\lambda = x(y\lambda) \tag{6.3.1}$$

[1] Such a structure is called a *K-ring* in Cohn [31, Sect. 1].

are satisfied for all $x, y \in R$ and all $\lambda \in K$. If, in addition, K and R are both involutary rings, we say that R is an *involutary K-algebra* if $(\lambda x)^* = x^* \lambda^*$ whenever $(\lambda, x) \in K \times R$.

For a K-algebra R, a subset X in R *centralizes* K if $x\lambda = \lambda x$ whenever $(\lambda, x) \in K \times X$. A map f with values in R centralizes K, if the range of f centralizes K.

Observe that the definition above does not require K be a subring of R (i.e., R may not be unital).

Definition 6.3.2 Let S be a Boolean inverse semigroup and let R be a ring. A map $f: S \to R$ is an *additive measure* if f is a semigroup embedding from S to the multiplicative semigroup of R and $f(x \oplus y) = f(x) + f(y)$ whenever x and y are orthogonal elements of S. (In particular, $f(0) = 0$.) If, in addition, R is an involutary ring, we say that f is a *$*$-additive measure* if it is an additive measure and $f(x^{-1}) = f(x)^*$ for all $x \in S$.

In particular, in the context of Theorem 6.1.7, the inclusion map from $\overline{S}$ into R is a $*$-additive measure. The following result shows that additive measures relate type theory of Boolean inverse semigroups and nonstable K-theory of rings.

Proposition 6.3.3 *Let S be a Boolean inverse semigroup, let R be a ring, and let $f: S \to R$ be an additive measure. Then there is a unique monoid homomorphism $f: \mathrm{Typ}\, S \to \mathrm{V}(R)$ such that*

$$f\big(\mathrm{typ}_S(a)\big) = [f(a)]_R \quad \textit{whenever } a \in \mathrm{Idp}\, S\,.$$

Proof Since $\mathrm{Typ}\, S$ is additively generated by all $\mathrm{typ}_S(a)$ for $a \in \mathrm{Idp}\, S$, the uniqueness statement is trivial.

As to the existence, it follows from the additivity of f that the assignment $\varphi: \mathrm{Idp}\, S \to \mathrm{V}(R)$, $a \mapsto [f(a)]_R$ defines a homomorphism of partial monoids. For any idempotent elements a and b of S, it $a \mathrel{\mathscr{D}_S} b$, that is, $a = x^{-1}x$ and $b = xx^{-1}$ for some $x \in S$, then $f(a) = f(x^{-1})f(x)$ and $f(b) = f(x)f(x^{-1})$, thus $f(a)$ and $f(b)$ are Murray–von Neumann equivalent in R, that is, $\varphi(a) = \varphi(b)$. It follows that φ factors, through the map $a \mapsto \mathrm{typ}_S(a)$, to a homomorphism $f: \mathrm{Typ}\, S \to \mathrm{V}(R)$. □

Notation 6.3.4 Let K be a unital ring and let S be a Boolean inverse semigroup. We define the semiring tensor product $K\langle S\rangle = K \otimes \mathrm{U_{mon}}(S)$ (cf. Lemma 6.2.3).

In a number of cases, K will be endowed with an involution, so $K\langle S\rangle$ will be an involutary K-algebra (via Lemma 6.2.3).

Proposition 6.3.5 *Let K be a unital ring and let S be a Boolean inverse semigroup. Then there is a unique structure of K-algebra on $K\langle S\rangle$ such that*

$$\lambda(1 \otimes x) = (1 \otimes x)\lambda = \lambda \otimes x \quad \textit{for all } (\lambda, x) \in K \times \mathrm{U_{mon}}(S)\,. \tag{6.3.2}$$

Furthermore, the canonical map $j_S: S \to K\langle S\rangle$, $x \mapsto 1 \otimes x$ *is an additive measure, and it centralizes K. If, in addition, K is an involutary ring and $K\langle S\rangle$ is endowed with the corresponding involutary ring structure, then j_S is a $*$-additive measure.*

Proof By definition, $K\langle S\rangle$ is a semiring. Since $0 = (\lambda \otimes x) + ((-\lambda) \otimes x)$ for each $(\lambda, x) \in K \times S$, the additive monoid of $K\langle S\rangle$ is a group; whence $K\langle S\rangle$ is a ring. A standard argument about universal objects, similar to the one used in the proof of Lemma 6.2.3, yields the existence of a unique structure of K-bimodule on $K\langle S\rangle$ such that

$$\alpha(\beta\otimes x) = (\alpha\otimes x)\beta = (\alpha\beta)\otimes x \quad \text{whenever } \alpha, \beta \in K \text{ and } x \in \mathrm{U_{mon}}(S)\,. \tag{6.3.3}$$

This implies (6.3.2) trivially; in particular, j_S centralizes K. The verification of Eq. (6.3.1) in $K\langle S\rangle$ is straightforward; whence $K\langle S\rangle$ is a K-algebra. Conversely, any structure of K-bimodule on $K\langle S\rangle$ satisfying (6.3.2) also satisfies (6.3.3): for example, $\alpha(\beta \otimes x) = \alpha(\beta(1 \otimes x)) = (\alpha\beta)(1 \otimes x) = (\alpha\beta) \otimes x$. This yields the uniqueness statement about the bimodule structure on $K\langle S\rangle$.

By the definition of the multiplication in $K\langle S\rangle$ (cf. Lemma 6.2.3), j_S defines a semigroup homomorphism from S to the multiplicative semigroup of $K\langle S\rangle$. For all $x, y, z \in S$, if $z = x \oplus y$ within S, then $z = x + y$ within $\mathrm{U_{mon}}(S)$, thus $1 \otimes z = (1 \otimes x) + (1 \otimes y)$ within $K\langle S\rangle$. This means that j_S is an additive measure. □

By virtue of Proposition 6.3.5, we will call $K\langle S\rangle$ the *additive enveloping K-algebra of S*. The universal property of $K\langle S\rangle$ is contained in the following result.

Proposition 6.3.6 *Let S be a Boolean inverse semigroup and let K be a unital ring. Then the canonical map $j_S\colon S \to K\langle S\rangle$ is universal among all the additive measures (resp., $*$-additive measures), centralizing K, from S to a K-algebra (resp., to an involutary K-algebra).*

Proof We verify the result for K-algebras; the result for involutive K-algebras follows immediately.

Let R be a K-algebra and let $f\colon S \to R$ be an additive measure centralizing K. We must prove that there is a unique homomorphism $g\colon K\langle S\rangle \to R$ of K-algebras such that $f = g \circ j_S$. Necessarily, $g(\lambda \otimes x) = g(\lambda j_S(x)) = \lambda f(x)$ whenever $(\lambda, x) \in K \times S$; the uniqueness statement for g follows immediately.

Since $f\colon S \to R$ turns finite orthogonal sums (within S) to sums (within R), it extends to a unique monoid homomorphism $\overline{f}\colon \mathrm{U_{mon}}(S) \to R$. Since f is a multiplicative semigroup homomorphism, so is $\overline{f}$. The assignment $K \times \mathrm{U_{mon}}(S) \to R$, $(\lambda, x) \mapsto \lambda\overline{f}(x)$ defines a monoid bimorphism, thus there is a unique monoid homomorphism $g\colon K\langle S\rangle \to R$ such that $g(\lambda \otimes x) = \lambda\overline{f}(x)$ whenever $(\lambda, x) \in K \times \mathrm{U_{mon}}(S)$. From the assumption that f centralizes K it follows in a routine manner that g is a homomorphism of K-algebras. □

Notation 6.3.7 For a Boolean inverse semigroup S and a unital ring K, we shall write λx, or, equivalently, $x\lambda$, instead of $\lambda \otimes x$, whenever $(\lambda, x) \in K \times S$. With this notational convention, every element of $K\langle S\rangle$ can be written in the form $\sum_{i<n} \lambda_i x_i$, where $n \in \mathbb{Z}^+$ and each $(\lambda_i, x_i) \in K \times S$. Such an expression is not unique, as whenever $z = x \oplus y$ in S, the elements z and $x + y$ of $K\langle S\rangle$ are identical.

Keeping in mind the conventions introduced in Notation 6.3.7, Proposition 6.3.6 says that $K\langle S\rangle$ is exactly the K-algebra defined by the set of generators S, subjected to the relations stating that S is a multiplicative subsemigroup centralizing K together with the relations $z = x + y$ whenever $x, y, z \in S$ such that $z = x \oplus y$ within S. We will see in Theorem 6.4.7 that if $K \neq \{0\}$, then j_S is one-to-one.

Remark 6.3.8 This material bears close connections with work by Steinberg [104, 105], as follows. For an ample topological groupoid $\mathcal{G}$ and a commutative, unital ring $\Bbbk$, Steinberg defines in [104] a certain $\Bbbk$-algebra, called the *étale groupoid algebra of* $\mathcal{G}$ and denoted there by $\Bbbk\mathcal{G}$. Further, for an inverse semigroup S with zero, Steinberg defines a topological groupoid called the *universal tight groupoid* $\mathcal{U}_{\mathrm{T}}(S)$ of S. Putting those concepts together, he then defines the étale groupoid algebra $\Bbbk\,\mathcal{U}_{\mathrm{T}}\,(S)$, and states, in [105, Corollary 5.3], a characterization of $\Bbbk\,\mathcal{U}_{\mathrm{T}}\,(S)$ via generators and relations. This characterization is stated under the assumption that S is a so-called *Hausdorff inverse semigroup* , which means that $(S \downarrow x) \cap (S \downarrow y)$ is a finitely generated lower subset, for any $x, y \in S$. If S is a Boolean inverse semigroup, then this is equivalent to S being an inverse meet-semigroup (cf. Definition 3.7.7). Hence, for a Boolean inverse semigroup S, Steinberg's characterization, given in [105, Corollary 5.3], yields the isomorphism $\Bbbk\,\mathcal{U}_{\mathrm{T}}\,(S) \cong \Bbbk\langle S\rangle$ (still under the additional assumption that S be an inverse meet-semigroup).

As the following result shows, bias generation implies ring generation.

Lemma 6.3.9 *Let X be a subset in a Boolean inverse semigroup S and let K be a unital involutary ring. If X generates S as a bias, then $j_S[X]$ generates $K\langle S\rangle$ as an involutary subring.*

Proof It suffices to prove that $j_S(s)$ belongs to the involutary subring of $K\langle S\rangle$ generated by X, for any $s \in S$. Since s can be expressed as the evaluation, at a finite sequence of elements of X, of a term, in the similarity type $\mathcal{L}_{\mathrm{BIS}}$ (cf. Sect. 3.2.2), it suffices to prove that all the operations of $\mathcal{L}_{\mathrm{BIS}}$ can be expressed as terms of the language of involutary rings. For 0, the product, and the involution, nothing needs to be done. The skew difference $\oslash$ and the skew join $\triangledown$ can be expressed by the following terms:

$$\begin{aligned} \mathsf{x} \oslash \mathsf{y} &= (\mathsf{x}\mathsf{x}^* - \mathsf{x}\mathsf{x}^*\mathsf{y}\mathsf{y}^*)\mathsf{x}(\mathsf{x}^*\mathsf{x} - \mathsf{x}^*\mathsf{x}\mathsf{y}^*\mathsf{y})\,, \\ \mathsf{x} \triangledown \mathsf{y} &= (\mathsf{x} \oslash \mathsf{y}) + \mathsf{y}\,. \end{aligned} \tag{6.3.4}$$

□

Example 6.3.10 If $K = \mathbb{Z}$, then the construction of $K\langle S\rangle$ can be simplified. Indeed, by Proposition 6.2.1, $\mathrm{U}_{\mathrm{mon}}(S)$ is the positive cone of a dimension group G. Hence, $\mathbb{Z}\langle S\rangle$ is exactly that dimension group G (i.e., the Grothendieck group of $\mathrm{U}_{\mathrm{mon}}(S)$), endowed with the unique ring structure extending the semiring structure of $\mathrm{U}_{\mathrm{mon}}(S)$. Since $\mathrm{U}_{\mathrm{mon}}(S)$ is a conical commutative monoid, it is also the positive cone of a structure of an ordered ring on $\mathbb{Z}\langle S\rangle$. Accordingly, we shall often emphasize the additional involutary semiring structure on $\mathrm{U}_{\mathrm{mon}}(S)$, by denoting it $\mathbb{Z}^+\langle S\rangle$ instead. This notation is consistent with the notation $\mathbb{Z}^+\langle B\rangle$ introduced in Example 2.2.7

(view the Boolean ring B as a Boolean inverse semigroup). The elements of $\mathbb{Z}^+\langle S\rangle$ can be recognized, within the ring $\mathbb{Z}\langle S\rangle$, as the finite sums of elements of S.

Example 6.3.11 The monoid $G^{\sqcup 0}$ (cf. Definition 1.5.1) is a Boolean inverse meet-semigroup, for any group G. The orthogonal joins in $G^{\sqcup 0}$ are all trivial. Hence, the additive enveloping K-algebra $K\langle G^{\sqcup 0}\rangle$ of $G^{\sqcup 0}$ is nothing else as the group algebra $K[G]$, for any unital ring K: in notation, $K\langle G^{\sqcup 0}\rangle = K[G]$.

Example 6.3.12 Let K be a unital ring and let n be a positive integer. Then $K\langle \mathfrak{I}_n\rangle$ is isomorphic to the K-algebra $\mathrm{M}_n(K)$ of all $n \times n$ matrices over K. For an infinite analogue of this result, see Theorem 6.4.5.

6.4 Embedding Properties of the Additive Enveloping Algebra

We begin with an example, which shows that the assignment $S \mapsto K\langle S\rangle$ does not preserve embeddings as a rule.

Example 6.4.1 A finite symmetric inverse monoid T and an additive Boolean inverse submonoid S of T such that the canonical map $\mathbb{Z}\langle S\rangle \to \mathbb{Z}\langle T\rangle$ is not one-to-one.

Proof We let $T = \mathfrak{I}_4$, and we denote by G the Klein subgroup of T with generators the transpositions $g_0 = \begin{pmatrix}1 & 2\end{pmatrix}$ and $g_1 = \begin{pmatrix}3 & 4\end{pmatrix}$. Then $S = G^{\sqcup 0}$ is an additive Boolean inverse submonoid of T (send 0 to the empty function). Define $f_0 = 1$ (i.e., the identity function on [4]) and set $f_1 = g_0 g_1$. The identity functions u and v on $\{3,4\}$ and $\{1,2\}$, respectively, satisfy $f_0 u = g_0 u$, $f_1 u = g_1 u$, $f_0 v = g_1 v$, $f_1 v = g_0 v$, and $1 = u \oplus v$. It follows that $f_0 + f_1 = g_0 + g_1$ within $\mathbb{Z}\langle \mathfrak{I}_4\rangle$. Nevertheless, f_0, f_1, g_0, g_1 are distinct elements of G, whence $f_0 + f_1 \neq g_0 + g_1$ within $\mathbb{Z}\langle G^{\sqcup 0}\rangle = \mathbb{Z}[G]$. □

We will see that embeddings are nonetheless preserved in many cases. Moreover, the techniques involved to prove this will yield a more precise description of the algebras $K\langle S\rangle$. The following notation will be used throughout this section.

Notation 6.4.2 For a Boolean inverse semigroup S, a family $\vec{x} = (x_i \mid i \in I)$ of elements of S with I finite, and $e \in \operatorname{Idp} S$, we write

$$\mathrm{Z}_{\vec{x}}(e) = \{i \in I \mid x_i e = 0\}\,,$$

$$\Theta_{\vec{x}}(e) = \left\{(i,j) \in (I \setminus \mathrm{Z}_{\vec{x}}(e))^2 \mid x_i e = x_j e\right\}\,,$$

$$\Upsilon_{\vec{x}}(e) = (I \setminus \mathrm{Z}_{\vec{x}}(e))^2 / \Theta_{\vec{x}}(e)\,.$$

Hence, $\Theta_{\vec{x}}$ is an equivalence relation on $I \setminus \mathrm{Z}_{\vec{x}}$ and $\Upsilon_{\vec{x}}$ is the associated partition of $I \setminus \mathrm{Z}_{\vec{x}}$.

For the remainder of this work, we shall denote by $K[S]_0$ the *contracted semigroup algebra, of a semigroup S with zero, over K*, that is, the quotient of the semigroup algebra $K[S]$ by the two-sided ideal generated by the zero element of S.

Lemma 6.4.3 *Let S be a Boolean inverse semigroup, let K be a unital ring, let $(x_i \mid i \in I)$ be a family of elements of S, with I finite, let $(\alpha_i \mid i \in I) \in K^I$, and let $e \in \operatorname{Idp} S$. Then $\sum_{i\in I} \alpha_i x_i e = 0$ within $K[S]_0$ iff $\sum_{i\in X} \alpha_i = 0$ whenever $X \in \Upsilon_{\vec{x}}(e)$.*

Proof Denote by s_X the constant value of $x_i e$ for $i \in X$, whenever $X \in \Upsilon_{\vec{x}}(e)$. Then $\sum_{i\in I} \alpha_i x_i e = \sum_{X\in\Upsilon_{\vec{x}}(e)} \bigl(\sum_{i\in X} \alpha_i\bigr) s_X$. Since the s_X are pairwise distinct elements of $S \setminus \{0\}$, the desired conclusion follows. □

We shall now give a "concrete" description of the additive K-algebra $K\langle \mathfrak{I}_\Omega \rangle$, for any unital ring K and any set Ω. We shall denote by $K^{(\Omega)}$ the set of all maps $\Omega \to K$ with finite support. On some occasions this set will be viewed as a K-bimodule, on some others as a right K-module. We will often identify every $p \in \Omega$ with the corresponding element of $K^{(\Omega)}$. We denote by $\operatorname{End} K_K^{(\Omega)}$ the K-algebra of all endomorphisms of $K^{(\Omega)}$ viewed as a right K-module.

For any $x \in \mathfrak{I}_\Omega$, we denote by $\rho(x)$ the unique endomorphism of $K_K^{(\Omega)}$ defined by

$$\rho(x)(p) = \begin{cases} x(p)\,, & \text{if } p \in \operatorname{dom}(x)\,, \\ 0\,, & \text{otherwise.} \end{cases} \tag{6.4.1}$$

The proof of the following lemma is trivial and we omit it.

Lemma 6.4.4 *The map ρ an additive measure on $\mathfrak{I}_\Omega$, with values in* $\operatorname{End} K_K^{(\Omega)}$.

Using Proposition 6.3.6, it follows that ρ induces a homomorphism $\rho_K \colon K\langle \mathfrak{I}_\Omega \rangle \to \operatorname{End} K_K^{(\Omega)}$ of K-algebras. The following result states that this particular representation of $K\langle \mathfrak{I}_\Omega \rangle$ is faithful.

Theorem 6.4.5 *The map $\rho_K \colon K\langle \mathfrak{I}_\Omega \rangle \to \operatorname{End} K_K^{(\Omega)}$ is an embedding, for any unital ring K and any set Ω.*

Proof Let $(x_i \mid i \in I)$ be a family of elements of $\mathfrak{I}_\Omega$, with I finite, and let $(\alpha_i \mid i \in I)$ be a family of elements of K. We suppose that $\sum_{i\in I} \alpha_i \rho(x_i) = 0$, and we must prove that $\sum_{i\in I} \alpha_i x_i = 0$ within $K\langle \mathfrak{I}_\Omega \rangle$. We set

$$\begin{aligned} Z_{\vec{x}}(p) &= \{i \in I \mid \rho(x_i)(p) = 0\} = \{i \in I \mid p \notin \operatorname{dom}(x_i)\}\,, \\ \Theta_{\vec{x}}(p) &= \bigl\{(i,j) \in (I \setminus Z_{\vec{x}}(p))^2 \mid x_i(p) = x_j(p)\bigr\}\,, \\ \Upsilon_{\vec{x}}(p) &= (I \setminus Z_{\vec{x}}(p))^2/\Theta_{\vec{x}}(p)\,, \end{aligned}$$

for any $p \in \Omega$. Our assumption that $\sum_{i\in I} \alpha_i \rho(x_i) = 0$ means that $\sum_{i\in I} \alpha_i \rho(x_i)(p) = 0$ for any $p \in \Omega$, that is, denoting by $p(X)$ the constant value of $x_i(p)$ for $i \in X$, whenever $X \in \Upsilon_{\vec{x}}(p)$,

$$\sum_{X \in \Upsilon_{\vec{x}}(p)} \Big(\sum_{i \in X} \alpha_i\Big) p(X) = 0\,, \quad \text{for each } p \in \Omega\,.$$

Since the $p(X)$, for $X \in \Upsilon_{\vec{x}}(p)$, are distinct elements of Ω, they are linearly independent over K, so we get

$$\sum_{i \in X} \alpha_i = 0\,, \quad \text{for every } p \in \Omega \text{ and every } X \in \Upsilon_{\vec{x}}(p)\,. \tag{6.4.2}$$

Now we set $D = \bigcup_{i \in I} \operatorname{dom}(x_i)$, and

$$\|x = y\|_D = \{p \in D \cap \operatorname{dom}(x) \cap \operatorname{dom}(y) \mid x(p) = y(p)\}\,, \quad \text{for all } x, y \in \mathfrak{I}_\Omega\,.$$

The Boolean subalgebra $\mathfrak{U}$ of $\operatorname{Pow} D$ generated by $\big\{\|x_i = x_j\| \mid (i,j) \in I \times I\big\}$ is finite.

Claim 1 Let $U \in \operatorname{At}\mathfrak{U}$ and let $p \in U$. Then $\mathrm{Z}_{\vec{x}}(\mathrm{id}_U) = \mathrm{Z}_{\vec{x}}(p)$ and $\Theta_{\vec{x}}(\mathrm{id}_U) = \Theta_{\vec{x}}(p)$ (thus also $\Upsilon_{\vec{x}}(\mathrm{id}_U) = \Upsilon_{\vec{x}}(p)$).

Proof of Claim. It is trivial that $\mathrm{Z}_{\vec{x}}(\mathrm{id}_U) \subseteq \mathrm{Z}_{\vec{x}}(p)$. Now let $i \in \mathrm{Z}_{\vec{x}}(p)$, that is, $p \notin \operatorname{dom}(x_i)$. Since $\operatorname{dom}(x_i) = \|x_i = x_i\|_D \in \mathfrak{U}$ and $p \in U \in \operatorname{At}\mathfrak{U}$, it follows that $U \cap \operatorname{dom}(x_i) = \varnothing$, whence $x_i \circ \mathrm{id}_U = 0$, that is, $i \in \mathrm{Z}_{\vec{x}}(\mathrm{id}_U)$. Hence, $\mathrm{Z}_{\vec{x}}(\mathrm{id}_U) = \mathrm{Z}_{\vec{x}}(p)$. Consequently, $\Theta_{\vec{x}}(\mathrm{id}_U) \subseteq \Theta_{\vec{x}}(p)$. Now let $(i,j) \in \Theta_{\vec{x}}(p)$, that is, $i, j \notin \mathrm{Z}_{\vec{x}}(p)$ and $p \in \|x_i = x_j\|_D$. Since $p \in U \in \operatorname{At}\mathfrak{U}$ and $\|x_i = x_j\|_D \in \mathfrak{U}$, it follows that $U \subseteq \|x_i = x_j\|_D$, so $(i,j) \in \Theta_{\vec{x}}(\mathrm{id}_U)$. □ Claim 1.

Claim 2 $\sum_{i \in I} \alpha_i x_i \mathrm{id}_U = 0$ within $K[\mathfrak{I}_\Omega]_0$, for every $U \in \operatorname{At}\mathfrak{U}$.

Proof of Claim. Pick $p \in U$. It follows from (6.4.2) together with Claim 1 that

$$\sum_{i \in X} \alpha_i = 0\,, \quad \text{for every } X \in \Upsilon_{\vec{x}}(\mathrm{id}_U)\,.$$

The desired conclusion follows then from Lemma 6.4.3. □ Claim 2.

We can now conclude the proof of Theorem 6.4.5. Indeed, since $K\langle\mathfrak{I}_\Omega\rangle$ is, canonically, a quotient algebra of $K[\mathfrak{I}_\Omega]_0$, it follows from Claim 2 that

$$\sum_{i \in I} \alpha_i x_i \mathrm{id}_U = 0 \quad \text{within } K\langle\mathfrak{I}_\Omega\rangle\,, \quad \text{for any } U \in \operatorname{At}\mathfrak{U}\,.$$

By summing up all those equations, over $U \in \operatorname{At}\mathfrak{U}$, and observing that the union of all atoms of $\mathfrak{U}$ is $\bigcup_{i \in I} \operatorname{dom}(x_i)$, the desired conclusion follows. □

Now let S be a Boolean inverse semigroup and denote by Ω the set of all prime ideals of S. Exel's regular representation $\lambda\colon S \hookrightarrow \mathfrak{I}_\Omega$ of S is an additive semigroup embedding of Boolean inverse semigroups (cf. Sect. 3.3). This additive semigroup embedding lifts to a homomorphism $\lambda_K\colon K\langle S\rangle \to K\langle \mathfrak{I}_\Omega\rangle$ of K-algebras. The following result is the main lemma of this section.

Lemma 6.4.6 *The homomorphism λ_K is one-to-one. Furthermore, the following statements are equivalent, for any family $\vec{x} = (x_i \mid i \in I)$ of elements of S, with I finite, and any family $(\alpha_i \mid i \in I)$ of elements of K:*

(i) $\sum_{i\in I} \alpha_i x_i = 0$ *within* $K\langle S\rangle$.
(ii) $\sum_{i\in I} \alpha_i \lambda(x_i) = 0$ *within* $K\langle \mathfrak{I}_\Omega\rangle$.
(iii) *there is a finite Boolean subring U of* $\operatorname{Idp} S$ *such that* $\bigvee_{i\in I} \mathbf{d}(x_i) \le 1_U$ *and* $\sum_{i\in I} \alpha_i x_i u = 0$ *within* $K[S]_0$ *whenever* $u \in \operatorname{At} U$.

Furthermore, if S is a Boolean inverse meet-semigroup, then any finite Boolean subring U of $\operatorname{Idp} S$, *containing* $\{\mathbf{d}(x_i \wedge x_j) \mid i,j \in I\}$, *satisfies* (iii) *above.*

Proof The directions (i)⇒(ii) and (iii)⇒(i) are all trivial. Note that by Theorem 6.4.5, Assumption (ii) is equivalent to saying that $\sum_{i\in I} \alpha_i(\rho\lambda)(x_i) = 0$ within $\operatorname{End} K_K^{(\Omega)}$, where $\rho\colon \mathfrak{I}_\Omega \to \operatorname{End} K_K^{(\Omega)}$ is defined in (6.4.1).

Now we prove (ii)⇒(iii). Set $e = \bigvee_{i\in I} \mathbf{d}(x_i)$ and $B = (\operatorname{Idp} S) \downarrow e$, and denote by $\overline{\Omega}$ the ultrafilter space of B. For each $\mathfrak{p} \in \overline{\Omega}$, the subset $S \uparrow \mathfrak{p}$ is a prime filter of S, and it follows from our assumption that $\sum_{i\in I} \alpha_i(\rho\lambda)(x_i)(S\uparrow\mathfrak{p}) = 0$ within $K^{(\Omega)}$. By definition of λ and ρ, this means, setting $Z_{\vec{x}}(\mathfrak{p}) = \{i \in I \mid 0 \in x_i\mathfrak{p}\}$, that

$$\sum_{i\in I\setminus Z_{\vec{x}}(\mathfrak{p})} \alpha_i(S\uparrow x_i\mathfrak{p}) = 0 \text{ within } K^{(\Omega)}\,, \quad \text{for any } \mathfrak{p} \in \overline{\Omega}\,. \tag{6.4.3}$$

Set $\Theta_{\vec{x}}(\mathfrak{p}) = \{(i,j) \in (I \setminus Z_{\vec{x}}(\mathfrak{p}))^2 \mid S\uparrow x_i\mathfrak{p} = S\uparrow x_j\mathfrak{p}\}$, $\Upsilon_{\vec{x}}(\mathfrak{p}) = (I\setminus Z_{\vec{x}}(\mathfrak{p}))^2/\Theta_{\vec{x}}(\mathfrak{p})$, and denote by $\mathfrak{p}(X)$ the constant value of $S\uparrow x_i\mathfrak{p}$ for $i \in X$, whenever $X \in \Upsilon_{\vec{x}}(\mathfrak{p})$. Then (6.4.3) is equivalent to

$$\sum_{X\in\Upsilon_{\vec{x}}(\mathfrak{p})} \Bigl(\sum_{i\in X} \alpha_i\Bigr)\mathfrak{p}(X) = 0\,, \quad \text{whenever } \mathfrak{p} \in \overline{\Omega}\,. \tag{6.4.4}$$

Since the $\mathfrak{p}(X)$, where $X \in \Upsilon_{\vec{x}}(\mathfrak{p})$, are distinct elements of Ω, they are linearly independent over K, thus (6.4.4) is equivalent to

$$\sum_{i\in X} \alpha_i = 0\,, \quad \text{whenever } \mathfrak{p} \in \overline{\Omega} \text{ and } X \in \Upsilon_{\vec{x}}(\mathfrak{p})\,. \tag{6.4.5}$$

For all $i,j \in I$ and all $\mathfrak{p} \in \overline{\Omega}$, $S\uparrow(x_i\mathfrak{p}) = S\uparrow(x_j\mathfrak{p})$ iff $x_ip = x_jp$ for some $p \in \mathfrak{p}$. Hence, $\Theta_{\vec{x}}(\mathfrak{p})$ is the directed union of all $\Theta_{\vec{x}}(p)$ where $p \in \mathfrak{p}$. Since $Z_{\vec{x}}(\mathfrak{p})$ is, trivially, the directed union of all $Z_{\vec{x}}(p)$ where $p \in \mathfrak{p}$, it follows that

$$\text{For each } \mathfrak{p} \in \overline{\Omega}\,, \text{ there is } p \in \mathfrak{p} \text{ such that } Z_{\vec{x}}(\mathfrak{p}) = Z_{\vec{x}}(p) \text{ and } \Theta_{\vec{x}}(\mathfrak{p}) = \Theta_{\vec{x}}(p)\,. \tag{6.4.6}$$

Set $\Delta = \{p \in \overline{\Omega} \mid \sum_{i\in I} \alpha_i x_i p = 0 \text{ within } K[S]_0\}$. It follows from Lemma 6.4.3 that $\Delta = \{p \in \overline{\Omega} \mid \sum_{i\in X} \alpha_i = 0 \text{ whenever } X \in \Upsilon_{\vec{x}}(p)\}$. Moreover, it follows from (6.4.5) and (6.4.6) that $\overline{\Omega} = \bigcup_{p\in\Delta} \overline{\Omega}(p)$ (where we set $\overline{\Omega}(p) = \{\mathfrak{p} \in \overline{\Omega} \mid p \in \mathfrak{p}\}$). Since $\overline{\Omega}$ is compact (cf. Theorem 1.4.2), there is a finite subset F of Δ such that $e = \bigvee F$. From the definition of Δ it follows that $\sum_{i\in I} \alpha_i x_i p = 0$ within $K[S]_0$, whenever $p \in F$. The Boolean subalgebra U of B generated by F is finite, and every atom of U is beneath some element of F. Hence,

$$\sum_{i\in I} \alpha_i x_i u = 0 \text{ within } K[S]_0 \,, \text{ whenever } u \in \operatorname{At} U \,.$$

This completes the proof of the equivalence of (i)–(iii). In particular, from the equivalence (i)⇔(ii) it follows that λ_K is one-to-one.

Now suppose, in addition, that S is a Boolean inverse meet-semigroup and that (iii) holds at some finite Boolean subring V of $\operatorname{Idp} S$. Let U be a finite Boolean subring of $\operatorname{Idp} S$ such that each $\mathbf{d}(x_i \wedge x_j) \in U$. We must prove that $\sum_{i\in I} \alpha_i x_i u = 0$ within $K[S]_0$ whenever $u \in \operatorname{At} U$. We may replace V by the Boolean subring of $\operatorname{Idp} S$ generated by $U \cup V$, and thus assume that $U \subseteq V$. Hence, for each $u \in \operatorname{At} U$, there exists $v \in \operatorname{At} V$ such that $v \leq u$. By assumption,

$$\sum_{i\in I} \alpha_i x_i v = 0 \text{ within } K[S]_0 \,,$$

that is, by Lemma 6.4.3, $\sum_{i\in X} \alpha_i = 0$ whenever $X \in \Upsilon_{\vec{x}}(v)$. Hence, in order to prove that $\sum_{i\in I} \alpha_i x_i u = 0$ within $K[S]_0$, it suffices to prove that $\Upsilon_{\vec{x}}(u) = \Upsilon_{\vec{x}}(v)$, that is, $\mathrm{Z}_{\vec{x}}(u) = \mathrm{Z}_{\vec{x}}(v)$ and $\Theta_{\vec{x}}(u) = \Theta_{\vec{x}}(v)$.

The containment $\mathrm{Z}_{\vec{x}}(u) \subseteq \mathrm{Z}_{\vec{x}}(v)$ is trivial. Let $i \in \mathrm{Z}_{\vec{x}}(v)$, that is, $\mathbf{d}(x_i)v = 0$. Since $\mathbf{d}(x_i) \in U$ and $0 < v \leq u \in \operatorname{At} U$, it follows that $\mathbf{d}(x_i)u = 0$, that is, $i \in \mathrm{Z}_{\vec{x}}(u)$, thus proving that $\mathrm{Z}_{\vec{x}}(u) = \mathrm{Z}_{\vec{x}}(v)$. It follows immediately that $\Theta_{\vec{x}}(u) \subseteq \Theta_{\vec{x}}(v)$. Now let $(i,j) \in \Theta_{\vec{x}}(v)$, that is, $v \leq \mathbf{d}(x_i)\,\mathbf{d}(x_j)$ and $x_i v = x_j v$. From the latter equation it follows that $x_i v \leq x_i \wedge x_j$, whence $v = \mathbf{d}(x_i v) \leq \mathbf{d}(x_i \wedge x_j)$. Since $\mathbf{d}(x_i \wedge x_j) \in U$ and $0 < v \leq u \in \operatorname{At} U$, it follows that $u \leq \mathbf{d}(x_i \wedge x_j)$, whence $(i,j) \in \Theta_{\vec{x}}(u)$. This completes the proof that $\Theta_{\vec{x}}(u) = \Theta_{\vec{x}}(v)$. □

We can now reap the consequences of Lemma 6.4.6.

Theorem 6.4.7 *Let S be a Boolean inverse semigroup and let K be a unital subring of a unital ring K'. Then the canonical map $\varphi\colon K\langle S\rangle \to K'\langle S\rangle$ is one-to-one. Furthermore, if K is nontrivial, then the canonical additive measure $j_S\colon S \to K\langle S\rangle$ is one-to-one.*

Proof Since the contracted semigroup ring $K[S]_0$ is a subring of $K'[S]_0$ (cf. Lemma 6.4.3), the equivalence (i)⇔(iii) in Lemma 6.4.6 yields immediately that φ is one-to-one. Now suppose that K is nontrivial and let $x, y \in S$ such that $j_S(x) = j_S(y)$, that is, $j_S(x - y) = 0$. By the equivalence (i)⇔(iii) in Lemma 6.4.6, there is a finite Boolean subring U of $\operatorname{Idp} S$ such that $\mathbf{d}(x) \vee \mathbf{d}(y) \leq 1_U$ and $xu = yu$, within $K[S]_0$ (thus, since K is nontrivial, within S), whenever $u \in \mathrm{At}\, U$. It follows that $x = \bigoplus_{u \in \mathrm{At}\, U} xu = \bigoplus_{u \in \mathrm{At}\, U} yu = y$. □

Theorem 6.4.8 *Let S be a lower inverse subsemigroup of a Boolean inverse semigroup T, and let K be a unital ring. Then the canonical map $K\langle S \rangle \to K\langle T \rangle$ is one-to-one.*

Proof Let $(x_i \mid i \in I)$ be a family of elements of S, with I finite, and let $(\alpha_i \mid i \in I) \in K^I$, such that $\sum_{i \in I} \alpha_i x_i = 0$ within $K\langle T \rangle$. Set $e = \bigvee_{i \in I} \mathbf{d}(x_i)$. By Lemma 6.4.6, there is a finite Boolean subring U of $\mathrm{Idp}\, T$ such that $e \leq 1_U$ and $\sum_{i \in I} \alpha_i x_i u = 0$ within $K[T]_0$ whenever $u \in \mathrm{At}\, U$. We may replace U by eU and thus assume that $1_U = e$. Since S is a lower subset of T, it follows that $U \subseteq \mathrm{Idp}\, S$, so $\sum_{i \in I} \alpha_i x_i u = 0$ within $K[S]_0$ whenever $u \in \mathrm{At}\, U$. Therefore, $\sum_{i \in I} \alpha_i x_i u = 0$ within $K\langle S \rangle$. □

Theorem 6.4.9 *Let K be a unital ring and let S be an additive Boolean inverse subsemigroup of a Boolean inverse meet-semigroup T. If S is closed under the meet operation, then the canonical map $K\langle S \rangle \to K\langle T \rangle$ is one-to-one.*

Proof Let $\vec{x} = (x_i \mid i \in I)$ be a family of elements of S, with I finite, and let $(\alpha_i \mid i \in I) \in K^I$. The Boolean subring U of $\mathrm{Idp}\, T$, generated by the subset $\{\mathbf{d}(x_i \wedge x_j) \mid i, j \in I\}$, is finite. It follows from our assumption that U is also a Boolean subring of $\mathrm{Idp}\, S$. Now it follows from the equivalence (i)$\Leftrightarrow$(iii) in Lemma 6.4.6 that $\sum_{i \in I} \alpha_i x_i = 0$ within $K\langle S \rangle$ iff $\sum_{i \in I} \alpha_i x_i u = 0$ within $K[S]_0$ whenever $u \in \mathrm{At}\, U$, iff $\sum_{i \in I} \alpha_i x_i = 0$ within $K\langle T \rangle$. □

We say that an involutary ring K is *positive definite* if $\sum_{i=1}^{n} x_i^* x_i = 0$ implies that all $x_i = 0$, whenever $n \in \mathbb{N}$ and each $x_i \in K$. Every positive definite involutary ring is proper (cf. Definition 6.1.1), but the converse fails for easy examples.

Theorem 6.4.10 *Let S be a Boolean inverse semigroup and let K be a unital involutary ring. If K is positive definite, then so is $K\langle S \rangle$.*

Proof By Lemma 6.4.6, it suffices to prove that $K\langle \mathfrak{I}_\Omega \rangle$ is positive definite, for any set Ω. The range of the map ρ_K of Theorem 6.4.5 is contained in the involutary K-algebra B consisting of all endomorphisms of $K_K^{(\Omega)}$ whose matrix is row- and column-finite. Here the involution on B is defined by applying the involution to the transpose matrix. Now ρ_K becomes an embedding of involutary K-algebras, so it suffices to prove that B is positive definite. Let $x_i = \left(x_{p,q}^i \mid (p,q) \in \Omega \times \Omega\right)$ be elements of B, for $1 \leq i \leq n$, such that $\sum_{i=1}^{n} x_i^* x_i = 0$. Since the (p,p)th entry of $x_i^* x_i$ is $\sum_{q \in \Omega} (x_{q,p}^i)^* x_{q,p}^i$, it follows that $\sum_{(i,q) \in [n] \times \Omega} (x_{q,p}^i)^* x_{q,p}^i = 0$, for every $p \in \Omega$. Since K is positive definite, each $x_{q,p}^i = 0$. □

We can also apply our methods to describe the analogue of the $K\langle S \rangle$ construction in the world of C*-algebras. We denote by $\mathrm{C}^*\langle S \rangle$ the universal C*-algebra containing a copy of S in such a way that S is a multiplicative subsemigroup with the same zero, $x^{-1} = x^*$, and $z = x \oplus y$ within S implies that $z = x + y$ within $\mathrm{C}^*\langle S \rangle$, and we call it the *additive enveloping C*-algebra of S*. Although the existence of $\mathrm{C}^*\langle S \rangle$ follows from general methods, it will also be a consequence of the construction described in our next result.

Theorem 6.4.11 *Let S be a Boolean inverse semigroup. Then $\mathrm{C}^*\langle S\rangle$ has a dense involutive subalgebra isomorphic to $\mathbb{C}\langle S\rangle$.*

Proof It follows from Lemma 6.4.6 that Exel's regular representation $\lambda: S \to \mathfrak{I}_\Omega$ of S extends to an embedding $\lambda_\mathbb{C}: \mathbb{C}\langle S\rangle \hookrightarrow \mathbb{C}\langle \mathfrak{I}_\Omega\rangle$ of involutive $\mathbb{C}$-algebras. Now we use the method of the proof of Theorem 6.4.5 to embed $\mathbb{C}\langle \mathfrak{I}_\Omega\rangle$ into a C*-algebra, with a twist. Instead of defining $\rho(x)$ as an endomorphism of $\mathbb{C}_\mathbb{C}^{(\Omega)}$, we define it, via the same formula (6.4.1), as a bounded endomorphism of the complex Hilbert space $\ell^2(\Omega)$ (of norm 1 if $x \neq 0$). The proof that ρ extends to an embedding of involutive $\mathbb{C}$-algebras $\rho_\mathbb{C}$, from $\mathbb{C}\langle \mathfrak{I}_\Omega\rangle$ into the C*-algebra $\mathbf{B}(\Omega)$ of all bounded endomorphisms of $\ell^2(\Omega)$, is, *mutatis mutandis*, identical to the one of Theorem 6.4.5. Therefore, $\rho_\mathbb{C} \circ \lambda_\mathbb{C}$ is an embedding, of involutive $\mathbb{C}$-algebras, from $\mathbb{C}\langle S\rangle$ into $\mathbf{B}(\Omega)$.

Define a *$*$-representation of* S as a $*$-additive measure $\psi: S \to A$, for a C*-algebra A. For every $x \in \mathbb{C}\langle S\rangle$, we define $\|x\|$ as the supremum of $\|\overline{\psi}(x)\|_A$, where $\psi: S \to A$ ranges over all $*$-representations of S and $\overline{\psi}$ denotes the canonical extension of ψ to $\mathbb{C}\langle S\rangle$ given by Proposition 6.3.6. For any $s \in S$ and every $*$-representation $\psi: S \to A$ of S, $\psi(s)$ is a partial isometry of A, thus $\|\psi(s)\|_A \leq 1$. Since every $x \in \mathbb{C}\langle S\rangle$ is a finite linear combination of elements of S, it follows that $\|x\|$ is a nonnegative real number. Hence, $\|{-}\|$ is a seminorm on $\mathbb{C}\langle S\rangle$. Since $\rho_\mathbb{C} \circ \lambda_\mathbb{C}$ is one-to-one, it follows that $\|{-}\|$ is a norm on $\mathbb{C}\langle S\rangle$. Furthermore, it follows immediately from the definition of that norm that $\|x^*x\| = \|x^2\|$ for every $x \in \mathbb{C}\langle S\rangle$. Hence, the completion C_S of $\mathbb{C}\langle S\rangle$, with respect to that norm, is a C*-algebra.

In order to conclude the proof, it thus suffices to verify that C_S, together with the canonical $*$-additive semigroup embedding $S \hookrightarrow C_S$, satisfies the universal property defining $\mathrm{C}^*\langle S\rangle$ (thus proving that $C_S \cong \mathrm{C}^*\langle S\rangle$). The canonical extension $\overline{\psi}: \mathbb{C}\langle S\rangle \to A$, of a $*$-representation $\psi: S \to A$ of S, is 1-Lipschitz with respect to the norm introduced above on $\mathbb{C}\langle S\rangle$, thus it extends to a unique $*$-homomorphism $C_S \to A$. □

Remark 6.4.12 For an arbitrary (not necessarily Boolean) inverse semigroup S, the universal C*-algebra $\mathrm{C}^*(S)$ of S (cf. Duncan and Paterson [36], Paterson [93, § 2.1]) can be defined as the C*-algebra defined by generators S, subject to the relations stating that S, with its inversion map, is an involutary subsemigroup. The construction $\mathrm{C}^*\langle S\rangle$, for S Boolean, adds all the relations corresponding to the orthogonal joins in S, thus proving that $\mathrm{C}^*\langle S\rangle$ is a quotient of $\mathrm{C}^*(S)$. The analogue of the *reduced C*-algebra* $\mathrm{C}^*_{\mathrm{red}}(S)$ would then be the closure $\mathrm{C}^*_{\mathrm{red}}\langle S\rangle$, of the image of $\mathbb{C}\langle S\rangle$ under $\rho_\mathbb{C} \circ \lambda_\mathbb{C}$, within the C*-algebra $\mathbf{B}(\Omega)$ introduced in the proof of Theorem 6.4.11.

6.5 Path Algebras as Additive Enveloping Algebras

In this section we shall demonstrate that the additive enveloping algebra construct $K\langle S\rangle$ extends classical constructions of path algebras over quivers, such as Leavitt path algebras. We did not attempt to give an exhaustive enumeration of all the known algebras that would fall in the class of all $K\langle S\rangle$, hoping that our chosen example, namely path algebras, would illustrate the general principle appropriately.

Definition 6.5.1 Let S be an inverse semigroup. An *additive equation system* over S is a collection of (formal) equations, each one having the form

$$\bigoplus_{i=1}^{m} x_i = \bigoplus_{j=1}^{n} y_j \,, \tag{6.5.1}$$

where m, n are nonnegative integers and all x_i, y_j belong to S.

While Eq. (6.5.1) can be interpreted, as it stands, in any Boolean inverse semigroup S' with a homomorphism $S \to S'$, it can also be interpreted in any involutary ring R with a $*$-additive measure $S \to R$, by simply interpreting $\oplus$ as the orthogonal addition of R.

Allowing $m = 1$ and $n = 0$ in (6.5.1), we see that equations of the type $x = 0$ can be incorporated to additive equation systems.

The above-mentioned general principle is expressed by the following result.

Theorem 6.5.2 *Let Σ be an additive equation system over a (not necessarily Boolean) inverse semigroup S. Then there exists a Boolean inverse semigroup S^{Σ} such that for every unital involutary ring K, the involutary K-algebra $K(S, \Sigma)$ defined by generators S, centralizing K, subjected to all equations in Σ, is isomorphic to $K\langle S^{\Sigma}\rangle$.*

Proof We define S^{Σ} as the bias (cf. Sect. 3.2) defined by generators S, together with all equations stating that S is an inverse subsemigroup, and the following relations:

$$(\cdots((x_1 \triangledown x_2) \triangledown x_3) \triangledown \cdots) \triangledown x_m = (\cdots((y_1 \triangledown y_2) \triangledown y_3) \triangledown \cdots) \triangledown y_n \,, \tag{6.5.2}$$

$$x_{i_1}^{-1} x_{i_2} = x_{i_1} x_{i_2}^{-1} = 0 \quad \text{whenever } i_1 \neq i_2 \text{ in } [m] \,, \tag{6.5.3}$$

$$y_{j_1}^{-1} y_{j_2} = y_{j_1} y_{j_2}^{-1} = 0 \quad \text{whenever } j_1 \neq j_2 \text{ in } [n] \,, \tag{6.5.4}$$

for every equation $\bigoplus_{i=1}^{m} x_i = \bigoplus_{j=1}^{n} y_j$ in Σ. Denote by $\varepsilon\colon S \to S^{\Sigma}$ the canonical homomorphism of inverse semigroups.

Now let R be an involutary K-algebra and let $f\colon S \to R$ be a homomorphism of involutary K-algebras, with range centralizing K, such that the images under f of the elements of S satisfy Σ. The centralizer R' of K in R is an involutary subring of R (not

Fig. 6.1 The maps $f, \varepsilon, \pi, \varphi$

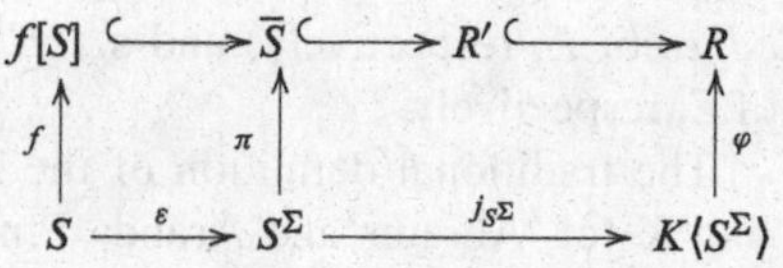

necessarily a K-subalgebra!), containing $f[S]$. By applying Theorem 6.1.7 to $f[S]$ within the ambient involutary ring R', we obtain a Boolean inverse subsemigroup $\overline{S}$ in R' containing S, in which finite orthogonal joins specialize finite orthogonal sums from R'. Since ∇ specializes to orthogonal join on orthogonal elements, it follows from the assumption on f that all equations (6.5.2)–(6.5.4) are satisfied within $\overline{S}$. By the definition of S^{Σ}, there is a unique bias homomorphism (i.e., additive semigroup homomorphism) $\pi: S^{\Sigma} \to \overline{S}$ such that $\pi \circ \varepsilon = f$ (here we are, of course, identifying f with its range restriction to $\overline{S}$). Since $\pi[S^{\Sigma}] \subseteq \overline{S} \subseteq R'$, it follows from Proposition 6.3.6 that the $*$-additive measure $\pi: S^{\Sigma} \to R$ factors, through $j_{S^{\Sigma}}$, to a unique homomorphism $\varphi: K\langle S^{\Sigma}\rangle \to R$ of involutary K-algebras. By definition, $\varphi \circ (j_{S^{\Sigma}} \circ \varepsilon) = f$. The situation can be visualized on Fig. 6.1.

Since S^{Σ} is generated, as a bias, by $\varepsilon[S]$, it follows from Lemma 6.3.9 that $j_{S^{\Sigma}}[S^{\Sigma}]$ is contained in the involutary subring of $K\langle S^{\Sigma}\rangle$ generated by $(j_{S^{\Sigma}} \circ \varepsilon)[S]$. It follows that φ is the unique homomorphism of involutary K-algebras satisfying $\varphi \circ (j_{S^{\Sigma}} \circ \varepsilon) = f$.

We have thus proved that $K\langle S^{\Sigma}\rangle$, together with the homomorphism $j_{S^{\Sigma}} \circ \varepsilon: S \to K\langle S^{\Sigma}\rangle$, satisfies the universal property defining $K(S, \Sigma)$. □

The C*-algebra version of Theorem 6.5.2 runs as follows (see Theorem 6.4.11 for the notation $\mathrm{C}^*\langle S^{\Sigma}\rangle$).

Theorem 6.5.3 *Let Σ be an additive equation system over a (not necessarily Boolean) inverse semigroup S, and denote by S^{Σ} the Boolean inverse semigroup constructed in the proof of Theorem* 6.5.2. *Then the C*-algebra* $\mathrm{C}^*(S, \Sigma)$, *defined by generators S subjected to all equations in Σ, is isomorphic to* $\mathrm{C}^*\langle S^{\Sigma}\rangle$.

Proof It suffices to prove that the canonical $*$-additive measure $S \hookrightarrow \mathrm{C}^*\langle S^{\Sigma}\rangle$ satisfies the universal property defining $\mathrm{C}^*(S, \Sigma)$. The proof runs, *mutatis mutandis*, like the one of Theorem 6.5.2, actually a bit easier since $R' = R$. □

A direct application of Theorems 6.4.11, 6.5.2, and 6.5.3 yields the following.

Corollary 6.5.4 *Let Σ be an additive equation system over a (not necessarily Boolean) inverse semigroup S. Then $\mathbb{C}(S, \Sigma)$ is isomorphic to a dense involutive subalgebra of* $\mathrm{C}^*(S, \Sigma)$.

Let us illustrate the results above on Leavitt path algebras. A *quiver* (or *graph*) is a quadruple $E = (E^0, E^1, \mathbf{s}, \mathbf{t})$, where E^0 and E^1 are disjoint sets (the *vertices* and *edges* of E, respectively) and $\mathbf{s}, \mathbf{t}\colon E^1 \to E^0$, the *source map* and the *target map*[2] of E, respectively.

The traditional definition of the *Leavitt path algebra* $\mathsf{L}_K(E)$, formulated for a ring K (cf. Abrams and Aranda Pino [1, 2] or Ara et al. [14]), can be obviously extended to the context of involutary rings. Namely, for a quiver E as above and an involutary, unital ring K, we fix a set E^{-1}, disjoint from $E^0 \cup E^1$, together with a bijection $(E^1 \to E^{-1}, x \mapsto x^*)$, and then we define $\mathsf{L}_K(E)$ as the involutary K-algebra defined by the generators $E^0 \cup E^1$, subjected to the relations

$$ab = \delta_{a,b} a\,, \tag{6.5.5}$$

$$a^* = a\,, \tag{6.5.6}$$

$$x = \mathbf{s}(x) x \, \mathbf{t}(x)\,, \tag{6.5.7}$$

$$x^* y = \delta_{x,y} \, \mathbf{t}(x)\,, \tag{6.5.8}$$

$$a = \sum \left(zz^* \mid z \in \mathbf{s}^{-1}\{a\} \right) \quad \text{whenever } \mathbf{s}^{-1}\{a\} \text{ is finite nonempty}\,, \tag{6.5.9}$$

for $a, b \in E^0$ and $x, y \in E^1$. As usual, $\delta_{x,y}$ denotes the Kronecker symbol.

We shall prove in Theorem 6.5.6 that $\mathsf{L}_K(E) \cong K\langle \mathsf{G}^{\mathrm{B}}(E) \rangle$, for a suitable Boolean inverse semigroup $\mathsf{G}^{\mathrm{B}}(E)$.

To that end, we first consider the *graph inverse semigroup* $\mathsf{G}(E)$ of E. By definition, $\mathsf{G}(E)$ is the involutary semigroup with zero, defined by the set of generators $E^0 \cup E^1$, subjected to the relations (6.5.5)–(6.5.8). The following result is established in Jones and Lawson [62, § 2], extending work from Ash and Hall [16]. See also Mesyan et al. [83, § 2.2].

Lemma 6.5.5 *The involutary semigroup $\mathsf{G}(E)$ is an inverse semigroup, where $x^{-1} = x^*$ for every $x \in \mathsf{G}(E)$.*

Further properties of $\mathsf{G}(E)$ are established in Jones and Lawson [62], for example that it is an inverse meet-semigroup (cf. [62, Corollary 2.16]). We will not need those additional properties here.

Theorem 6.5.6 *For every quiver E, there is a Boolean inverse semigroup $\mathsf{G}^{\mathrm{B}}(E)$ such that $\mathsf{L}_K(E) \cong K\langle \mathsf{G}^{\mathrm{B}}(E) \rangle$ as involutary K-algebras, for every unital involutary ring K.*

Proof We let $S = \mathsf{G}(E)$, and we let the equation system Σ_E consist of the equation $0_{\mathsf{G}(E)} = 0$, together with all equations of the form

$$a = z_1 z_1^* \oplus \cdots \oplus z_n z_n^*\,, \ \text{whenever } a \in E^0\,,\ n > 0\,, \text{ and } \mathbf{s}^{-1}\{a\} = \{z_1, \dots, z_n\} \text{ with the } z_i \text{ distinct}\,. \tag{6.5.10}$$

[2] Due to a conflict of notation and intuitive meaning with the notations **d** and **r**, for the domain and the range, in inverse semigroups, we kept "s" for the source but we changed the usual "r" to "t" for the target.

Using the notation of the proof of Theorem 6.5.2, we define $\mathsf{G}^{\mathrm{B}}(E) = \mathsf{G}(E)^{\Sigma_E}$. It is easy to verify that $K(\mathsf{G}(E), \Sigma_E) = \mathsf{L}_K(E)$. Hence, the desired conclusion follows from Theorem 6.5.2. □

While Theorem 6.5.6 is stated for involutary K-algebras (for a given unital involutary ring K), a similar result holds for the Leavitt path algebra $\mathsf{L}_K(E)$, now without involution, for any unital ring K (without involution): first state Theorem 6.5.6 for $K = \mathbb{Z}$ (with the identity involution), then observe that $\mathsf{L}_K(E) \cong K \otimes_{\mathbb{Z}} \mathsf{L}_{\mathbb{Z}}(E)$ and $K\langle S\rangle \cong K \otimes_{\mathbb{Z}} \mathbb{Z}\langle S\rangle$.

By using Theorem 6.5.3 instead of Theorem 6.5.2, we can also see that the analogue of Theorem 6.5.6 for *graph C*-algebras* (i.e., the C*-analogues of Leavitt path algebras) applies.

Further generalizations of Theorem 6.5.6 apply to variants of the Leavitt path algebra, for example the *Cohn path algebra* of E, defined by the sets of Eqs. (6.5.5)–(6.5.8). For that construction, we just need to replace $\mathsf{G}^{\mathrm{B}}(E)$ by the bias defined by generators $\mathsf{G}(E)$, subjected to all relations stating that $\mathsf{G}(E)$ is an involutary subsemigroup with the same zero (i.e., the *enveloping bias* of $\mathsf{G}(E)$).

On the other hand, Theorem 6.5.6 does not extend to the *Leavitt path algebras of separated graphs* introduced by Ara and Goodearl [9]. For example, the defining relations of the Leavitt path algebra associated to the separated graph introduced at the beginning of [9, § 5] are those stating that x_1, x_2, y_1, y_2, v are orthogonal idempotents, together with $e_i^* e_i = f_i^* f_i = v$ for $i \in \{1, 2\}$, $e_i^* e_j = f_i^* f_j = 0$ for $i \neq j$, and $v = e_1 e_1^* + e_2 e_2^* = f_1 f_1^* + f_2 f_2^*$. Those relations do not imply that the idempotents $e_1 e_1^*$ and $f_1 f_1^*$ commute, so attempting to extend Lemma 6.5.5, to that example, breaks down.

As pointed to the author by Pere Ara, the latter irregularity is corrected by the construction, introduced in Ara and Exel [7, Notation 5.8], of the involutary K-algebra $\mathsf{L}_K^{\mathrm{ab}}(E, C)$ associated with a separated graph (E, C). Indeed, $\mathsf{L}_K^{\mathrm{ab}}(E, C) \cong \mathsf{L}_K(E, C)/J$, where J is the two-sided ideal generated by the commutators between the sources (or, equivalently, the targets) of finite products of elements of $E^0 \cup E^1$. This definition ensures that Lemma 6.5.5 extends to the involutary subsemigroup of $\mathsf{L}_K^{\mathrm{ab}}(E, C)$ generated by $E^0 \cup E^1$. Hence, $\mathsf{L}_K^{\mathrm{ab}}(E, C) \cong K\langle S\rangle$, *for a suitable Boolean inverse semigroup* S. Unlike the Boolean inverse semigroup $\mathsf{G}^{\mathrm{B}}(E)$ constructed in the proof of Theorem 6.5.6, this semigroup S is not independent of K a priori.

6.6 The Boolean Unitization of a Boolean Inverse Semigroup

The *standard unitization* of a non-unital ring R is the unital ring $\widetilde{R}$, uniquely determined up to isomorphism, such that R is a two-sided ideal of $\widetilde{R}$ and $\widetilde{R} = \mathbb{Z} \oplus R$ as Abelian groups. It can be realized as $\widetilde{R} = \mathbb{Z} \times R$, endowed with componentwise addition together with multiplication defined by

$$(m, x) \cdot (n, y) = (mn, mx + ny + xy)\,, \quad \text{for all } (m, x), (n, y) \in \mathbb{Z} \times R\,. \tag{6.6.1}$$

The ring R is then identified with its isomorphic copy $\{0\} \times R$ within $\widetilde{R}$. We set $\widetilde{R} = R$ in case R is unital.

If R is non-unital, then $\widetilde{R}$ is not Boolean. Hence, for a Boolean ring B, we define the *Boolean unitization* of B as B itself if B is unital, and the product $(\mathbb{Z}/2\mathbb{Z}) \times B$ endowed with componentwise addition together with multiplication as in (6.6.1), otherwise. If B is a non-unital Boolean ring, then the Boolean unitization of B is a unital Boolean ring, in which B is a prime ideal.

The main purpose of the present section is to extend to arbitrary Boolean inverse semigroups the Boolean unitization construction for Boolean rings.

Definition 6.6.1 A *Boolean unitization* of a Boolean inverse semigroup S consists of an additive semigroup embedding from S into a Boolean inverse monoid $\widetilde{S}$ such that the following conditions hold:

(1) If S is unital, then $\widetilde{S} = S$;
(2) Every element of $\widetilde{S}$ has the form $(1 \smallsetminus e) \oplus x$, where $e \in \operatorname{Idp} S$ and $x \in eSe$.

The additive semigroup embedding being often understood, we will then identify the Boolean unitization with its underlying Boolean inverse monoid $\widetilde{S}$. Observe that since $\widetilde{S}$ is a Boolean inverse semigroup, we can write

$$\widetilde{S} = S \cup \{(1 \smallsetminus e) \oplus x \mid e \in \operatorname{Idp} S \text{ and } x \in eSe\} .$$

Theorem 6.6.2 *Every Boolean inverse semigroup has a Boolean unitization.*

Proof Let S be a non-unital Boolean inverse semigroup. In particular, $B = \operatorname{Idp} S$ is a Boolean ring without largest element.

Recall from Example 6.3.10 that the ring $R = \mathbb{Z}\langle S\rangle$ is also the enveloping ring of the enveloping semiring $R^+ = \mathbb{Z}^+\langle S\rangle = \mathrm{U}_{\mathrm{mon}}(S)$ on S. We apply the standard unitization procedure to R, getting the unital ring $\widetilde{R}$. The elements of $\widetilde{R}$ can be uniquely written in the form $x = n + y$, where $(n, y) \in \mathbb{Z} \times R$. The involution of R is extended to an involution of $\widetilde{R}$, by setting $(n + y)^* = n + y^*$.

Furthermore, the Boolean unitization of B can be identified with the subset $\widetilde{B} = B \cup \{1 - x \mid x \in B\}$ of $\widetilde{R}$.

We set $\widetilde{S} = S \cup \{(1 - e) + \overline{x} \mid e \in B \text{ and } x \in eSe\}$. Hence $\widetilde{S}$ is a subset of $\widetilde{R}$.

Observe that for any $x \in \widetilde{S}$, any $e \in B$ and $\overline{x} \in eSe$ such that $x = (1 - e) + \overline{x}$ determine each other. Furthermore, whenever $e' \geq e$ in B, we can also write $x = (1 - e') + (e' - e) + \overline{x}$, that is, $x = (1 - e') + \overline{x}'$ where $\overline{x}' = (e' \smallsetminus e) \oplus \overline{x}$ (within S), an element of $e'Se'$.

Now let $x, y \in \widetilde{S}$. We must prove that $xy \in \widetilde{S}$. This is trivial if $x \in S$ and $y \in S$. If $x \notin S$ and $y \in S$, then $x = (1 - e) + \overline{x}$, where $e \in B$ and $\overline{x} \in eSe$. By the paragraph above, we may replace e by $e \vee \mathbf{d}(y) \vee \mathbf{r}(y)$, so $y = eye$. Then $xy = \overline{x}y \in S$. The proof that $yx \in S$ is symmetric. Finally, if $x \notin S$ and $y \notin S$, then, by the paragraph above, there are $e \in B$ and $\overline{x}, \overline{y} \in eSe$ such that $x = (1 - e) + \overline{x}$ and $y = (1 - e) + \overline{y}$. It follows that $xy = (1 - e) + \overline{x}\,\overline{y}$, with $\overline{x}\,\overline{y} \in eSe$, so $xy \in \widetilde{S}$. Moreover, it is obvious that $\widetilde{S}$ is closed under the involution of $\widetilde{R}$. For $e \in B$ and $\overline{x} \in eSe$, the element $x = (1 - e) + \overline{x} \in \widetilde{S}$ satisfies $x^*x = (1 - e) + \overline{x}^{-1}\overline{x} = (1 - e) + \mathbf{d}(x) = 1 - (e \smallsetminus \mathbf{d}(\overline{x})) \in \widetilde{B}$, thus $xx^*x = (1-e) + \overline{x}\,\overline{x}^{-1}\overline{x} = (1-e) + \overline{x} = x$. Since $\widetilde{B}$ is a commuting subset of $\widetilde{R}$, it thus follows from Lemma 3.1.1 that $\widetilde{S}$ is an inverse semigroup in $\widetilde{R}$, with $\operatorname{Idp}\widetilde{S} = \widetilde{B}$. In particular, $\operatorname{Idp}\widetilde{S}$ is Boolean.

Let $x, y \in \widetilde{S}$ be orthogonal; we prove that $x + y \in \widetilde{S}$. Up to symmetry, there are only two cases to consider. The first case is $\{x, y\} \subseteq S$; then $x + y$ (within $\widetilde{R}$) is identical to $x \oplus y$ (within S), so it belongs to S. The second case is $x = (1-e) + \overline{x}$, where $e \in B$ and $\overline{x} \in eSe$, and $y \in S$. We may replace e by $e \vee \mathbf{d}(y) \vee \mathbf{r}(y)$, and thus assume that $y = eye$. Then $x + y = (1-e) + (\overline{x} \oplus y)$ belongs to $\widetilde{S}$. Hence, in any case, $x + y \in \widetilde{S}$. The same argument as the one at the end of the proof of Theorem 6.1.7 yields then that $x + y$ is the join of $\{x, y\}$ within $\widetilde{S}$. Therefore, $\widetilde{S}$ is a Boolean inverse semigroup, and S is an additive ideal of $\widetilde{S}$. □

Although Theorem 6.6.2 does not state any uniqueness result about the Boolean unitization $\widetilde{S}$, that result is a consequence of the following universal characterization of $\widetilde{S}$. For a semigroup S and a monoid T, we say that a map $f\colon S \to T$ is *unit-preserving* if the image under f of the unit element of S, if it exists, is the unit element of T.

Proposition 6.6.3 *Let $\widetilde{S}$ be a Boolean unitization of a Boolean inverse semigroup S. Then for every Boolean inverse monoid T, every unit-preserving additive semigroup homomorphism $f\colon S \to T$ extends to a unique additive monoid homomorphism $g\colon \widetilde{S} \to T$.*

Proof We may assume that S is not unital. Set $B = \operatorname{Idp} S$. Necessarily,

$$g\big((1 \smallsetminus a) \oplus x\big) = (1_T \smallsetminus f(a)) \oplus f(x) \qquad \text{whenever } a \in \operatorname{Idp} S \text{ and } x \in aSa\,. \tag{6.6.2}$$

In particular, the uniqueness statement about g is obvious. In order to prove that (6.6.2) defines an extension $g\colon \widetilde{S} \to T$ of f, we first need to prove that there is no ambiguity in the expression of an element of $\widetilde{S}$, that is, we need to prove that $(1 \smallsetminus e) \oplus x \neq y$, for all $e \in B$, $x \in eSe$, and $y \in S$. If $(1 \smallsetminus e) \oplus x = y$, then $1 \smallsetminus e = (1 \smallsetminus e)y = y \smallsetminus ey \in S$, thus, as $e \in B$, we get $1 \in S$, a contradiction. Next, we need to prove the implication

$$(1 \smallsetminus a) \oplus x = (1 \smallsetminus b) \oplus y \Longrightarrow (1_T \smallsetminus f(a)) \oplus f(x) = (1_T \smallsetminus f(b)) \oplus f(y)\,,$$

whenever $a, b \in B$, $x \in aSa$, and $y \in bSb$. Replacing b by $a \vee b$ and decreasing y accordingly, we may assume that $a \leq b$. In that case, our assumption $(1 \smallsetminus a) \oplus x = (1 \smallsetminus b) \oplus y$ means that $y = (b \smallsetminus a) \oplus x$, thus, since f is additive, we obtain the equation $f(y) = \big(f(b) \smallsetminus f(a)\big) \oplus f(x)$, from which the desired conclusion $(1_T \smallsetminus f(a)) \oplus f(x) = (1_T \smallsetminus f(b)) \oplus f(y)$ follows easily. This proves that (6.6.2) indeed defines an extension g of f from $\widetilde{S}$ to T. The verification that g is both additive and unit-preserving is routine. □

Corollary 6.6.4 *The Boolean unitization of a Boolean inverse semigroup S is unique up to isomorphism.*

Proposition 6.6.5 *Let $\widetilde{S}$ be the Boolean unitization of a Boolean inverse semigroup S. Then S is an additive ideal in $\widetilde{S}$, and if S is not unital, then $\widetilde{S}/S$ is the two-element inverse semigroup. Furthermore, $\mathbb{Z}\langle\widetilde{S}\rangle$ is the standard unitization of $\mathbb{Z}\langle S\rangle$ and $\operatorname{Idp}\widetilde{S}$ is the Boolean unitization of $\operatorname{Idp} S$.*

Proof By Corollary 6.6.4, it is sufficient to verify Proposition 6.6.5 on the construct $\widetilde{S}$ given by the proof of Theorem 6.6.2. In particular, we keep the notation of that proof (e.g., $R = \mathbb{Z}\langle S\rangle$ and $\widetilde{R}$ is the standard unitization of R). We verified, in the course of the proof of Theorem 6.6.2, that S is an ideal of $\widetilde{S}$ and that $\operatorname{Idp}\widetilde{S}$ is the Boolean unitization of $\operatorname{Idp} S$. Define $\pi\colon \widetilde{S} \to \{0,1\}$ by $\pi\restriction_S = 0$ and $\pi\big((1 \smallsetminus e) \oplus x\big) = 1$, whenever $e \in B$ and $x \in eSe$. Any $x, y \in \widetilde{S}$ such that $\pi(x) = \pi(y)$ satisfy the relation $x \equiv_S y$ introduced in the statement of Proposition 3.4.6 (take $z = 0$ if $x, y \in S$, and $z = 1 \smallsetminus e$ for large enough e if $x, y \notin S$). Hence, π induces an isomorphism from $\widetilde{S}/S$ onto $\{0,1\}$.

Finally, we must prove that for any ring Z, any additive measure $f\colon \widetilde{S} \to Z$ extends to a unique ring homomorphism $\overline{f}\colon \widetilde{R} \to Z$. By Proposition 6.3.6, the restriction of f to S extends to a unique ring homomorphism $f_1\colon R \to Z$. Further, f_1 extends to a unique ring homomorphism $\overline{f}\colon \widetilde{R} \to Z$ such that $\overline{f}(1) = f(1)$. Since $\widetilde{R}$ is generated by $R \cup \{1\}$ (as a subring) and R is generated by S, $\widetilde{R}$ is generated by $\widetilde{S}$; this proves the uniqueness of $\overline{f}$. Therefore, $\widetilde{R} \cong \mathbb{Z}\langle\widetilde{S}\rangle$. □

6.7 From Type Monoid to Nonstable K-Theory

Recall from Proposition 6.3.5 that for every Boolean inverse semigroup S and every unital ring K, the canonical map $j_S\colon S \to K\langle S\rangle$ is an additive measure. By Proposition 6.3.3, there is a unique monoid homomorphism $f\colon \operatorname{Typ} S \to \mathrm{V}(K\langle S\rangle)$ such that

$$f\big(\operatorname{typ}_S(x)\big) = [a]_R \quad \text{whenever } a \in \operatorname{Idp} S\,.$$

We will call f the *canonical map from* $\operatorname{Typ} S$ *to* $\mathrm{V}(K\langle S\rangle)$.

By virtue of Example 6.3.11, the following result yields a class of Boolean inverse meet-semigroups S for which the canonical map $\operatorname{Typ} S \to \mathrm{V}(\Bbbk\langle S\rangle)$, with $\Bbbk$ a division ring, is not surjective.

Proposition 6.7.1 *Let $\Bbbk$ be a division ring, and let G be a group containing a torsion element of order not a power of the characteristic of $\Bbbk$. Then the canonical map $f\colon \operatorname{Typ} G^{\sqcup 0} \to \mathrm{V}(\Bbbk[G])$ is one-to-one, but not surjective.*

Proof Since $G^{\sqcup 0}$ is a Boolean inverse semigroup with exactly two idempotents, its type monoid is isomorphic to $\mathbb{Z}^+$, with $\operatorname{typ}_{G^{\sqcup 0}}(1) = 1$. The augmentation map $\pi\colon \Bbbk[G] \twoheadrightarrow \Bbbk$ (i.e., the unique homomorphism of $\Bbbk$-algebras sending every element of G to 1) is a surjective homomorphism of unital rings, thus, for all $m, n \in \mathbb{Z}^+$, if $f(m) = f(n)$, that is, $m \cdot [1]_{\Bbbk[G]} = n \cdot [1]_{\Bbbk[G]}$, then, applying $\mathrm{V}(\pi)$, we get $m \cdot [1]_{\Bbbk} = n \cdot [1]_{\Bbbk}$, so $\Bbbk^m \cong \Bbbk^n$, whence, as $\Bbbk$ is a division ring, $m = n$. Hence f is one-to-one.

Now suppose that f is surjective.

Claim The group algebra $\Bbbk[G]$ has no non-trivial idempotents.

Proof of Claim. Let e be a non-trivial idempotent of $\Bbbk[G]$. Since f is surjective, $[e]_{\Bbbk[G]}$ belongs to the range of f, that is, since $G^{\sqcup 0}$ has no non-trivial idempotents, there exists $n \in \mathbb{Z}^+$ such that $[e]_{\Bbbk[G]} = n \cdot [1]_{\Bbbk[G]}$. From $e \neq 0$ it follows that $n \geq 1$. If $n \geq 2$, then $2 \cdot [1]_{\Bbbk[G]} \leq^+ [e]_{\Bbbk[G]} \leq^+ [1]_{\Bbbk[G]}$ within $\mathrm{V}(\Bbbk[G])$, thus, applying again $\mathrm{V}(\pi)$, we get $2 \cdot [1]_{\Bbbk} \leq^+ [1]_{\Bbbk}$ within $\mathrm{V}(\Bbbk)$, a contradiction since $\mathrm{V}(\Bbbk) \cong \mathbb{Z}^+$. Therefore, $n = 1$, so $e \sim_{\Bbbk[G]} 1$.

Now suppose that $e \neq 1$. By applying the result of the paragraph above to $1 - e$, we obtain that $1 - e \sim_{\Bbbk[G]} 1$. Since $1 = e \oplus (1 - e)$, it follows that $[1]_{\Bbbk[G]} = 2 \cdot [1]_{\Bbbk[G]}$, a contradiction. □ Claim.

So far we have not used our assumption about torsion elements in G. Now we do so. Our assumption means that there are a prime number q, distinct from the characteristic of $\Bbbk$, and an element g of G of order q. Since $q \cdot 1_{\Bbbk}$ has a multiplicative inverse in $\Bbbk$, the element $e = (1/q)\sum_{k=0}^{q-1} g^k$ is well defined and distinct from both 0 and 1. Obviously, e is idempotent. By the Claim above, this leads to a contradiction. □

Proposition 6.7.1 leads immediately to the following example.

Example 6.7.2 Let $G = \mathbb{Z}/6\mathbb{Z}$. Since G has elements of order 2 and 3, it follows from Proposition 6.7.1 that the canonical map $\operatorname{Typ} G^{\sqcup 0} \to \mathrm{V}(\Bbbk[G])$ is one-to-one but not surjective, for any division ring $\Bbbk$.

As witnessed by the following result, it is not so straightforward to find examples where the canonical map $\operatorname{Typ} S \to \mathrm{V}(\Bbbk\langle S\rangle)$ is not one-to-one.

Proposition 6.7.3 *The following statements hold, for any Boolean inverse semigroup S, any division ring $\Bbbk$, and the canonical map $f \colon \operatorname{Typ} S \to \mathrm{V}(\Bbbk\langle S\rangle)$:*

(1) *If S is locally matricial (cf. Definition* 5.1.1*), then f is an isomorphism.*
(2) *If S is semisimple (cf. Definition* 3.7.5*), then f is a coretraction (i.e., it has a left inverse for the composition of homomorphisms). In particular, f is one-to-one.*

Proof As a preliminary observation, we claim that the functors Typ, V, and $\Bbbk\langle\_\rangle$ all preserve directed colimits and finite direct products. We established this for the functor Typ in Proposition 4.1.9; for the functor V this is well known; and for the functor $\Bbbk\langle\_\rangle$ this is a straightforward categorical argument.

Now let S be a finite fundamental Boolean inverse semigroup. By Lawson [75, Theorem 4.18], S is a direct product of finitely many finite symmetric inverse semigroups. By the claim above, this case can be in turn reduced to the one where S is a finite symmetric inverse semigroup. Now if $S = \mathfrak{I}_n$, where n is a positive integer, then $(\operatorname{Typ} S, \operatorname{typ}_S(1)) \cong (\mathbb{Z}^+, n)$ (cf. Proposition 5.1.5). Further, from $\Bbbk\langle S\rangle \cong \mathrm{M}_n(\Bbbk)$ it follows that $(\mathrm{V}(\Bbbk\langle S\rangle), [1]_{\Bbbk\langle S\rangle})$ is isomorphic to $(\mathbb{Z}^+, n)$ as well. Since $f(\operatorname{typ}_S(1)) = [1]_{\Bbbk\langle S\rangle}$, it follows that f is an isomorphism.

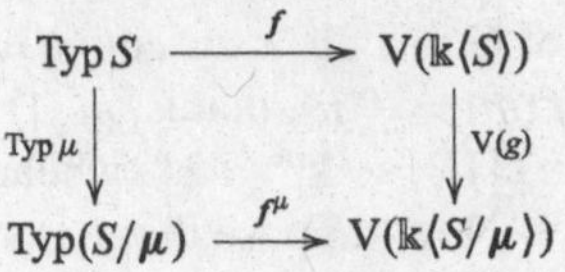

Fig. 6.2 A commutative diagram of commutative monoids

By the claim above, the conclusion (1) follows in a straightforward manner in case S is locally matricial. This holds, in particular, if S is a semisimple fundamental Boolean inverse semigroup, because in such a case, S is the directed union of all subsemigroups aSa where $a \in \operatorname{Idp} S$, and all the aSa are finite fundamental Boolean inverse semigroups.

(2) Now suppose that S is a semisimple Boolean inverse semigroup. Denote by $\boldsymbol{\mu}$ the largest idempotent-separating congruence of S. It follows from Howie [60, Theorem V.3.4] that $S/\boldsymbol{\mu}$ is fundamental; further, it follows from Proposition 3.4.5 that $S/\boldsymbol{\mu}$ is a Boolean inverse semigroup and that the canonical projection $\mu\colon S \twoheadrightarrow S/\boldsymbol{\mu}$ is an additive semigroup homomorphism. Further, it follows from Theorem 4.4.19 that μ is type-preserving. Since S is semisimple and S and $S/\boldsymbol{\mu}$ have the same idempotents, $S/\boldsymbol{\mu}$ is also semisimple.

By our result established above for semisimple fundamental Boolean inverse semigroups, the canonical map $\boldsymbol{f}^{\mu}\colon \operatorname{Typ}(S/\boldsymbol{\mu}) \to \mathrm{V}(\Bbbk\langle S/\boldsymbol{\mu}\rangle)$ is an isomorphism. Denote by $g\colon \Bbbk\langle S\rangle \twoheadrightarrow \Bbbk\langle S/\boldsymbol{\mu}\rangle$ the canonical map. We obtain the commutative diagram represented in Fig. 6.2. Since $\mathrm{V}(g) \circ \boldsymbol{f} = \boldsymbol{f}^{\mu} \circ \operatorname{Typ}\mu$ is an isomorphism, $\boldsymbol{f}$ is a coretraction. □

We shall now introduce a class of examples where $\boldsymbol{f}$ is not one-to-one. The following construction is a modification of the one of Example 4.9.3.

Notation 6.7.4 For a finite group G, of cardinality N, we set $\Omega_G = \mathbb{Z}^+ \times G$, and we set

$$\mathfrak{F}_G = \{X \subseteq \Omega_G \mid X \text{ is finite}\}\,,$$
$$\mathfrak{U}_G = \{X \subseteq \Omega_G \mid X \text{ is cofinite}\}\,,$$
$$\mathfrak{B}_G = \mathfrak{F}_G \cup \mathfrak{U}_G\,.$$

We denote by $\kappa(G)$ the set of all $x \in \mathfrak{I}_{\Omega_G}$, such that $\operatorname{dom}(x) \in \mathfrak{B}_G$, and such that if $\operatorname{dom}(x) \in \mathfrak{U}_G$, then there are $m \in \mathbb{Z}^+$ and $(k, g) \in \mathbb{Z} \times G$ such that

$$x(n,t) = (n + k, gt) \text{ for all } (n,t) \in [m, \infty) \times G\,. \tag{6.7.1}$$

For any $X \in \mathfrak{U}_G$, we denote by $\rho_G(X)$ the constant value of $\operatorname{card}\bigl(X \cap ([0,m) \times G)\bigr)$ modulo N, for large enough m. This defines a map $\rho_G\colon \mathfrak{U}_G \to \mathbb{Z}/N\mathbb{Z}$.

Lemma 6.7.5 *Let G be a finite group of cardinality N. Then $\kappa(G)$ is an additive Boolean inverse submonoid of $\mathfrak{I}_{\Omega_G}$. It is also a fundamental Boolean inverse meet-semigroup. Furthermore, for any $X, Y \in \mathfrak{B}_G$, if $\mathrm{id}_X \mathscr{D}_{\kappa(G)} \mathrm{id}_Y$, then $\rho_G(X) = \rho_G(Y)$.*

Note It is not hard to prove that $\mathrm{id}_X \mathscr{D}_{\kappa(G)} \mathrm{id}_Y$ follows from $\rho_G(X) = \rho_G(Y)$, for any $X, Y \in \mathfrak{U}_G$. We will not need that implication.

Proof The proofs of the statements about $\kappa(G)$ being an additive Boolean inverse submonoid of $\mathfrak{I}_{\Omega_G}$, together with a fundamental Boolean inverse meet-semigroup, are tedious but routine, and we leave them to the reader. Now let $X, Y \in \mathfrak{B}_G$ such that $\mathrm{id}_X \mathscr{D}_{\kappa(G)} \mathrm{id}_Y$. This means that there is a bijection $x: X \to Y$ belonging to $\kappa(G)$. We must prove that $\rho_G(X) = \rho_G(Y)$. If either X or Y is finite, then $\operatorname{card} X = \operatorname{card} Y$ and the desired conclusion is trivial. Suppose now that X and Y are both cofinite. Since X is cofinite, there are $m \in \mathbb{Z}^+$, $k \in \mathbb{Z}$, and $g \in G$ such that (6.7.1) holds. Furthermore, m can be taken large enough such that $[m, \infty) \times G \subseteq X$ and $[m + k, \infty) \times G \subseteq Y$. From (6.7.1) it follows that $x\big[[m, \infty) \times G\big] = [m + k, \infty) \times G$. Hence, forming set differences and since x is one-to-one, $x\big[X \cap ([0, m) \times G)\big] = Y \cap ([0, m + k) \times G)$. By evaluating the cardinalities of the two sides and reducing modulo N, we obtain $\rho_G(X) = \rho_G(Y)$. □

Lemma 6.7.6 *Let G be a finite group, of cardinality N, and let $\Bbbk$ be a division ring. If N is not a power of the characteristic of $\Bbbk$, then the canonical map $f \colon \operatorname{Typ} \kappa(G) \to \Bbbk\langle\kappa(G)\rangle$ is not one-to-one.*

Proof Throughout the proof we set $S = \kappa(G)$. By assumption, there are a prime number q, distinct from the characteristic of $\Bbbk$, and an element g of G of order q. In order to make it clear whether we are working in the Boolean inverse monoid S or in the $\Bbbk$-algebra $\Bbbk\langle S\rangle$, we shall denote by 1_X the canonical image of id_X in $\Bbbk\langle S\rangle$, for each finite or cofinite subset X of Ω_G. We shall also write 1 instead of 1_{Ω_G}.

The self-maps $\overline{g}$ and s of Ω_G defined by the rules $\overline{g}(n, t) = (n, gt)$ and $s(n, t) = (n + 1, t)$, for all $(n, t) \in \Omega_G$, both belong to S, and $s \circ \overline{g} = \overline{g} \circ s$.

Since q is distinct from the characteristic of $\Bbbk$, the element $e = (1/q) \sum_{k=0}^{q-1} \overline{g}^k$ is well defined, and it is an idempotent element of $\Bbbk\langle S\rangle$. Furthermore, from $s^{-1}s = 1$ it follows that e and ses^{-1} are Murray–von Neumann equivalent within $\Bbbk\langle S\rangle$; hence,

$$1 = e \oplus (1 - e) \sim_{\Bbbk\langle S\rangle} ses^{-1} \oplus (1 - e)\,. \tag{6.7.2}$$

Easy computations show that $ses^{-1} = e \cdot 1_{\mathbb{N} \times G} = e - e \cdot 1_{\{0\} \times G}$. Hence,

$$ses^{-1} \oplus (1 - e) = 1 - e \cdot 1_{\{0\} \times G}\,.$$

By writing $1 = 1_{\{0\} \times G} + 1_{\mathbb{N} \times G}$ and setting $w = 1_{\{0\} \times G} - e \cdot 1_{\{0\} \times G}$, we get

$$ses^{-1} \oplus (1 - e) = 1_{\mathbb{N} \times G} \oplus w\,. \tag{6.7.3}$$

Now the elements $\mathrm{id}_{\{0\} \times G}$ and $\overline{g}^k \circ \mathrm{id}_{\{0\} \times G}$ of S, for $0 \leq k < q$, are all finite orthogonal joins of the matrix units $e_{x,y}$, for $x, y \in G$, where $e_{x,y}$ denotes the unique element of S with domain $\{(0, y)\}$ and range $\{(0, x)\}$. It follows that $1_{\{0\} \times G}$ and $e \cdot 1_{\{0\} \times G}$ can both be represented as $N \times N$ matrices with entries in $\Bbbk$, and hence so can their difference, namely w. Since w is idempotent, and neither equal to 0 nor to $1_{\{0\}\times G}$, it is conjugate to $1_{\{0\}\times W}$ for some subset W of G distinct from $\varnothing$ and from G. (A closer look at the construction shows actually that the rank of w, which is equal to the cardinality of W, is equal to $(1 - 1/q)N$; however, we will need nothing more than $0 < \operatorname{card} W < N$.) Therefore, by (6.7.3), we obtain

$$ses^{-1} \oplus (1 - e) \sim_{\Bbbk\langle S\rangle} 1_W \oplus 1_{\mathbb{N}\times G} = 1_{W\sqcup(\mathbb{N}\times G)} \,.$$

By (6.7.2), it follows that, setting $X = W \sqcup (\mathbb{N} \times G)$,

$$1_{\Omega_G} \sim_{\Bbbk\langle G\rangle} 1_X \,,$$

that is, $f\bigl(\operatorname{typ}_S(\operatorname{id}_{\Omega_G})\bigr) = f\bigl(\operatorname{typ}_S(\operatorname{id}_X)\bigr)$. However, $\rho_G(\Omega_G) = 0$ and $\rho_G(X) = \operatorname{card} W \mod N \neq 0$, thus, by Lemma 6.7.5, $\operatorname{typ}_S(\operatorname{id}_{\Omega_G}) \neq \operatorname{typ}_S(\operatorname{id}_X)$. □

Remark 6.7.7 By mimicking the proof of Proposition 6.7.1, it is not hard to prove that under the assumptions of Lemma 6.7.6, the canonical map f is also not surjective.

Example 6.7.8 The order of the finite group $G = \mathbb{Z}/6\mathbb{Z}$ is not a prime power. By Lemmas 6.7.5 and 6.7.6, it follows that for the fundamental unital Boolean inverse meet-semigroup $\kappa(G)$, the canonical map $f\colon \operatorname{Typ}\kappa(G) \to \mathrm{V}(\Bbbk\langle\kappa(G)\rangle)$ is not one-to-one, for any division ring $\Bbbk$.

6.8 The Tensor Product of Two Boolean Inverse Semigroups

The first step of the construction of the tensor product of two Boolean inverse semigroups is given by the following lemma, whose straightforward proof we omit. Recall that the involutary semiring structure on $\mathrm{U}_{\mathrm{mon}}(S)$ is given by Proposition 6.2.2 while the involutary semiring structure on $\mathrm{U}_{\mathrm{mon}}(S) \otimes \mathrm{U}_{\mathrm{mon}}(T)$ is given by Proposition 6.2.3.

Lemma 6.8.1 *Let S and T be Boolean inverse semigroups. Then the set $S \otimes_0 T = \{x \otimes y \mid (x, y) \in S \times T\}$ is an inverse semigroup in the involutary semiring* $\mathrm{U}_{\mathrm{mon}}(S) \otimes \mathrm{U}_{\mathrm{mon}}(T)$.

Definition 6.8.2 Let S and T be Boolean inverse semigroups. We define $S \otimes T$ as the closure of $S \otimes_0 T$ under finite orthogonal sums, within $\mathrm{U}_{\mathrm{mon}}(S) \otimes \mathrm{U}_{\mathrm{mon}}(T)$.

Notation 6.8.3 Denote by $\mathfrak{J}_{S,T}$ the set of all elements of the free commutative monoid $\mathrm{F}_{\mathrm{mon}}\bigl(\mathrm{U}_{\mathrm{mon}}(S) \times \mathrm{U}_{\mathrm{mon}}(T)\bigr)$ of the form $\sum_{i=1}^{n}(x_i \bullet y_i)$, where each $(x_i, y_i) \in S \times T$ and the $x_i \otimes y_i$ are pairwise *orthogonal* (not only meet-orthogonal) within the involutary semiring $\mathrm{U}_{\mathrm{mon}}(S) \otimes \mathrm{U}_{\mathrm{mon}}(T)$.

Observe, in particular, that $S \otimes T = \{u/\leftrightarrows \mid u \in \mathfrak{J}_{S,T}\}$. By using the earlier result that every element of S has index at most 1 in $\mathrm{U}_{\mathrm{mon}}(S)$ (cf. Proposition 6.2.1), it is easy to obtain the following result, by aping the proof of the Claim in the proof of

Lemma 2.5.6. (The point is that whenever $z = x \oplus y$ in a Boolean inverse semigroup, x and y are orthogonal—not just meet-orthogonal.)

Lemma 6.8.4 *Whenever* $u \in \mathfrak{J}_{S,T}$ *and* $v \in \mathrm{F_{mon}}\big(\mathrm{U_{mon}}(S) \times \mathrm{U_{mon}}(T)\big)$, *if* $u \to^* v$, *then* $v \in \mathfrak{J}_{S,T}$.

Theorem 6.8.5 *Let* S *and* T *be Boolean inverse semigroups. Then* $S \otimes T$ *is an inverse semigroup in* $\mathrm{U_{mon}}(S) \otimes \mathrm{U_{mon}}(T)$. *Furthermore, it is Boolean inverse, and the properties* (1)–(4), *stated in Theorem* 6.1.6, *are all satisfied within the involutary semiring* $\mathrm{U_{mon}}(S) \otimes \mathrm{U_{mon}}(T)$. *Furthermore,* $\mathrm{U_{mon}}(S) \otimes \mathrm{U_{mon}}(T)$ *is additively generated by* $S \otimes T$.

Proof Since $\mathrm{U_{mon}}(S)$ and $\mathrm{U_{mon}}(T)$ are both conical refinement monoids, so is $\mathrm{U_{mon}}(S) \otimes \mathrm{U_{mon}}(T)$ (cf. Wehrung [121, Theorem 2.7]). In particular, by Lemma 6.1.5, $S \otimes T$ is an inverse semigroup in $\mathrm{U_{mon}}(S) \otimes \mathrm{U_{mon}}(T)$. Further, since, by Lemma 2.5.6, every pure tensor has index at most 1 in $\mathrm{U_{mon}}(S) \otimes \mathrm{U_{mon}}(T)$, it follows from Lemma 2.3.8 that this is also the case for every sum of pairwise meet-orthogonal pure tensors, thus, a fortiori, for every orthogonal sum of pure tensors. That is, every element of $S \otimes T$ has index at most 1 in $\mathrm{U_{mon}}(S) \otimes \mathrm{U_{mon}}(T)$.

Now we prove that $S \otimes T$ is a lower subset of $\mathrm{U_{mon}}(S) \otimes \mathrm{U_{mon}}(T)$.

Let $\boldsymbol{u} \in \mathrm{U_{mon}}(S) \otimes \mathrm{U_{mon}}(T)$ and $\boldsymbol{w} \in S \otimes T$ such that $\boldsymbol{u} \leq^+ \boldsymbol{w}$. We must prove that $\boldsymbol{u} \in S \otimes T$. There is $\boldsymbol{v} \in \mathrm{U_{mon}}(S) \otimes \mathrm{U_{mon}}(T)$ such that $\boldsymbol{w} = \boldsymbol{u} + \boldsymbol{v}$, and there is $w \in \mathfrak{J}_{S,T}$ such that $\boldsymbol{w} = w/\leftrightarrows$. Pick $u, v \in \mathrm{F_{mon}}(\mathrm{U_{mon}}(S) \times \mathrm{U_{mon}}(T))$ such that $\boldsymbol{u} = u/\leftrightarrows$ and $\boldsymbol{v} = v/\leftrightarrows$. Since $u + v \leftrightarrows w$, there is $w' \in \mathrm{F_{mon}}(\mathrm{U_{mon}}(S) \times \mathrm{U_{mon}}(T))$ such that $u + v \to^* w'$ and $w \to^* w'$. By Lemma 2.5.2, the first statement implies the existence of $u', v' \in \mathrm{F_{mon}}(\mathrm{U_{mon}}(S) \times \mathrm{U_{mon}}(T))$ such that $u \to^* u'$, $v \to^* v'$, and $w' = u' + v'$. Since $w \in \mathfrak{J}_{S,T}$ and by Lemma 6.8.4, $w' \in \mathfrak{J}_{S,T}$, thus $u' \in \mathfrak{J}_{S,T}$. Therefore, $\boldsymbol{u} = u'/\leftrightarrows \in S \otimes T$.

So far, we have verified Conditions (1) and (2) of Theorem 6.1.6, at the inverse semigroup $S \otimes T$ within the monoid $\mathrm{U_{mon}}(S) \otimes \mathrm{U_{mon}}(T)$. Further, (4) follows trivially from the definition of $S \otimes T$.

Finally, observe that $\mathrm{U_{mon}}(S)$ and $\mathrm{U_{mon}}(T)$ are both conical and proper (cf. Proposition 6.2.2), thus so is $\mathrm{U_{mon}}(S) \otimes \mathrm{U_{mon}}(T)$ (cf. Lemma 6.2.3). Every element $z \in S \otimes T$ can be written in the form $z = \sum_{i<n}(x_i \otimes y_i)$, where all $(x_i, y_i) \in S \otimes T$ and the $x_i \otimes y_i$ are pairwise orthogonal. It follows from that orthogonality assumption that $z^{-1}z = z^*z = \sum_{i<n}\big(\mathbf{d}(x_i) \otimes \mathbf{d}(y_i)\big)$. In particular, the idempotent elements of $S \otimes T$ are exactly the elements of the form $\boldsymbol{c} = \sum_{i<n}(a_i \otimes b_i)$, where $n \in \mathbb{Z}^+$, each $a_i \in \operatorname{Idp} S$, each $b_i \in \operatorname{Idp} T$, and the $a_i \otimes b_i$ are pairwise orthogonal. The elements $a = \bigvee_{i<n} a_i$ and $b = \bigvee_{i<n} b_i$ are idempotent in S and T, respectively. Since

$$a \otimes b = (a_i \otimes b_i) + \big((a \smallsetminus a_i) \otimes b_i\big) + \big(a_i \otimes (b \smallsetminus b_i)\big) + \big((a \smallsetminus a_i) \otimes (b \smallsetminus b_i)\big)$$

(within $\mathrm{U_{mon}}(S) \otimes \mathrm{U_{mon}}(T)$), we get $a_i \otimes b_i \leq^+ a \otimes b$, for each $i < n$. Since the $a_i \otimes b_i$ are pairwise orthogonal, they are, a fortiori, pairwise meet-orthogonal (cf. Lemmas 6.1.3 and 6.2.3). By Lemma 2.3.9, it follows that $\boldsymbol{c} \leq^+ a \otimes b$. Hence, the condition (3) of Theorem 6.1.6 follows.

Finally, every element of $\mathrm{U_{mon}}(S) \otimes \mathrm{U_{mon}}(T)$ is a finite sum of pure tensors; thus, a fortiori, it is a finite sum of elements of $S \otimes T$. □

In particular, $S \otimes T$, endowed with orthogonal addition, is a lower interval of $\mathrm{U_{mon}}(S) \otimes \mathrm{U_{mon}}(T)$. By applying Proposition 2.2.4, we get the following.

Corollary 6.8.6 *For any Boolean inverse monoids S and T, there is an isomorphism from $\mathrm{U_{mon}}(S) \otimes \mathrm{U_{mon}}(T)$ onto $\mathrm{U_{mon}}(S \otimes T)$ that fixes all elements of $S \otimes T$. In particular, $\mathrm{U_{mon}}(S \otimes T) \cong \mathrm{U_{mon}}(S) \otimes \mathrm{U_{mon}}(T)$.*

Despite the apparent complexity of the construction of the tensor product of two Boolean inverse semigroups, we shall now see that it is the solution of a rather easily formulated universal property, similar to the one for the tensor product of rings.

Definition 6.8.7 Let S, T, and Z be Boolean inverse semigroups. A map $f: S \times T \to Z$ is *bi-additive* if the following conditions hold:

(1) $f(x, y)f(x', y') = f(xx', yy')$, whenever $x, x' \in S$ and $y, y' \in T$;
(2) $f(x, y)^{-1} = f(x^{-1}, y^{-1})$, whenever $x \in S$ and $y \in T$;
(3) $f(x_0 \oplus x_1, y) = f(x_0, y) \oplus f(x_1, y)$, whenever $x_0, x_1 \in S$, $y \in T$, and $x_0 \perp x_1$;
(4) $f(x, y_0 \oplus y_1) = f(x, y_0) \oplus f(x, y_1)$, whenever $x \in S$, $y_0, y_1 \in T$, and $y_0 \perp y_1$.

Observe, in particular, that by the last two axioms above, $f(0, y) = 0$ for each $y \in T$, and $f(x, 0) = 0$ for each $x \in S$.

The proof of the following lemma is routine application of our earlier results, and we omit it.

Lemma 6.8.8 *Let S and T be Boolean inverse semigroups. Then the canonical map $S \times T \to S \otimes T$, $(x, y) \mapsto x \otimes y$ is bi-additive.*

Lemma 6.8.9 *Let S, T, Z be Boolean inverse semigroups and let $f: S \times T \to Z$ be bi-additive. Then $x_0 \otimes y_0 \perp x_1 \otimes y_1$ implies that $f(x_0, y_0) \perp f(x_1, y_1)$, for all $x_0, x_1 \in S$ and all $y_0, y_1 \in T$.*

Proof Our assumption $x_0 \otimes y_0 \perp x_1 \otimes y_1$ means that

$$x_0^{-1}x_1 \otimes y_0^{-1}y_1 = x_0x_1^{-1} \otimes y_0y_1^{-1} = 0\,,$$

that is,

$$(\text{either } x_0^{-1}x_1 = 0 \text{ or } y_0^{-1}y_1 = 0) \quad \text{and} \quad (\text{either } x_0x_1^{-1} = 0 \text{ or } y_0y_1^{-1} = 0)\,.$$

The element $f(x_0, y_0)^{-1}f(x_1, y_1) = f(x_0^{-1}x_1, y_0^{-1}y_1)$ vanishes if either $x_0^{-1}x_1 = 0$ or $y_0^{-1}y_1 = 0$, thus in any case. Symmetrically, the element $f(x_0, y_0)f(x_1, y_1)^{-1}$ vanishes if either $x_0x_1^{-1} = 0$ or $y_0y_1^{-1} = 0$, thus in any case. □

Now we can state the universal property of the tensor product of two Boolean inverse semigroups.

Theorem 6.8.10 *Let S and T be Boolean inverse semigroups. Then the canonical map $S \times T \to S \otimes T$ is universal among all the bi-additive maps from $S \times T$ to a Boolean inverse semigroup.*

Proof The easy direction is provided by Lemma 6.8.8.

Now let Z be a Boolean inverse semigroup and let $f\colon S \times T \to Z$ be bi-additive. We need to prove that there is a unique additive semigroup homomorphism $\overline{f}\colon S \otimes T \to Z$ such that $f(x, y) = \overline{f}(x \otimes y)$ for all $(x, y) \in S \times T$. Since every element of $S \otimes T$ is a finite orthogonal sum of elements of the form $x \otimes y$, the uniqueness statement follows easily from Lemma 6.8.9.

Now we deal with the existence part. Standard arguments, involving the universal properties defining $\mathrm{U_{mon}}(S)$ and $\mathrm{U_{mon}}(T)$, show that f extends to a (monoid) bimorphism $f_1\colon \mathrm{U_{mon}}(S) \times \mathrm{U_{mon}}(T) \to \mathrm{U_{mon}}(Z)$. By the universal property of the (monoid) tensor product, there is a unique monoid homomorphism

$$f_2\colon \mathrm{U_{mon}}(S) \otimes \mathrm{U_{mon}}(T) \to \mathrm{U_{mon}}(Z)$$

such that

$$f_2(x \otimes y) = f_1(x, y) \text{ for all } (x, y) \in \mathrm{U_{mon}}(S) \times \mathrm{U_{mon}}(T)\,.$$

Now $f_2(x \otimes y) = f(x, y)$ belongs to Z whenever $(x, y) \in S \times T$. By Lemma 6.8.9, it follows that $f_2[S \otimes T]$ is contained in Z. Accordingly, we can define the domain-range restriction $\overline{f}$ of f_2 from $S \otimes T$ to Z. By Lemma 6.8.9, $\overline{f}$ preserves all finite orthogonal sums of pure tensors. Thus, $\overline{f}$ preserves finite orthogonal sums. By the first two axioms defining bi-additive maps, the proof that $\overline{f}$ is a homomorphism of inverse semigroups is routine. □

6.9 The Type Monoid of a Tensor Product

For a Boolean inverse semigroup S, Green's relation $\mathscr{D}_S$ is an additive, conical V-equivalence on S (cf. Lemma 4.1.1), thus, by Theorem 2.4.6, it extends to a unique additive V-equivalence $\mathscr{D}_S^+ = \mathrm{U_{mon}}(\mathscr{D}_S)$ on $\mathrm{U_{mon}}(S)$. Moreover, still by using Theorem 2.4.6,

$$\operatorname{Typ} S = \mathrm{U_{mon}}(S/\mathscr{D}_S) \cong \mathrm{U_{mon}}(S)/\mathscr{D}_S^+\,,$$

thus expressing $\operatorname{Typ} S$ as the quotient of a conical refinement monoid under a conical V-congruence.

We shall now pursue this idea on tensor products of Boolean inverse semigroups. Recall that the tensor product of two congruences (of commutative monoids) is introduced in Notation 2.6.3. For Boolean inverse semigroups S and T, the enveloping monoids $\mathrm{U_{mon}}(S)$ and $\mathrm{U_{mon}}(T)$ are both conical refinement monoids. Hence, the binary relation $\boldsymbol{\gamma} = \mathscr{D}_S^+ \otimes \mathscr{D}_T^+$ is, by Corollary 2.6.4, a conical V-congruence of the monoid $\mathrm{U_{mon}}(S) \otimes \mathrm{U_{mon}}(T)$. Since the tensor product $S \otimes T$ (of Boolean inverse semigroups) is constructed as a subset of the monoid $\mathrm{U_{mon}}(S) \otimes$

$\mathrm{U_{mon}}(T)$, the restriction of $\boldsymbol{\gamma}$ to $S \otimes T$ is well defined. It turns out that much more is true.

Lemma 6.9.1 *The restriction of $\boldsymbol{\gamma}$ to $S \otimes T$ is equal to $\mathscr{D}_{S\otimes T}$.*

Proof Throughout the proof we shall set $M = \mathrm{U_{mon}}(S)$, $N = \mathrm{U_{mon}}(T)$, $\boldsymbol{\alpha} = \mathscr{D}_S^+$, $\boldsymbol{\beta} = \mathscr{D}_T^+$, and we shall denote by $\alpha\colon M \twoheadrightarrow M/\boldsymbol{\alpha}$ and $\beta\colon N \twoheadrightarrow N/\boldsymbol{\beta}$ the associated canonical projections. We shall use the notation of Sects. 2.5 and 2.6, in particular viewing the elements of $M \otimes N$ as equivalence classes of elements of $\mathrm{F_{mon}}(M \times N)$ with respect to the binary relation denoted there by $\leftrightarrows$. We shall also use the set $\mathfrak{I}_{S,T}$ introduced in Notation 6.8.3.

Let $\boldsymbol{u}, \boldsymbol{v} \in S \otimes T$, we must prove that $\boldsymbol{u} \equiv_{\boldsymbol{\gamma}} \boldsymbol{v}$ iff $\boldsymbol{u}\ \mathscr{D}_{S\otimes T}\ \boldsymbol{v}$.

Suppose first that $(\boldsymbol{u}, \boldsymbol{v}) \in \boldsymbol{\gamma}$, that is, $(\boldsymbol{u}, \boldsymbol{v}) \in \boldsymbol{\alpha} \otimes \boldsymbol{\beta}$. Pick $u, v \in \mathfrak{I}_{S,T}$ such that $\boldsymbol{u} = u/{\leftrightarrows}$ and $\boldsymbol{v} = v/{\leftrightarrows}$. Our assumption means that $(\alpha \bullet \beta)(u) \leftrightarrows (\alpha \bullet \beta)(v)$, that is, there is $w \in \mathrm{F_{mon}}\big((M/\boldsymbol{\alpha}) \times (N/\boldsymbol{\beta})\big)$ such that $(\alpha\bullet\beta)(u) \rightarrow^* w$ and $(\alpha\bullet\beta)(v) \rightarrow^* w$. By Lemma 2.6.1, there are $u', v' \in \mathrm{F_{mon}}(M \times N)$ such that $u \rightarrow^* u'$, $v \rightarrow^* v'$, and $(\alpha \bullet \beta)(u') = (\alpha \bullet \beta)(v') = w$. By Lemma 6.8.4, u' and v' both belong to $\mathfrak{I}_{S,T}$. Writing $u' = \sum_{i=1}^{n}(x_i \bullet y_i)$ and $v' = \sum_{i=1}^{n'}(x_i' \bullet y_i')$, both orthogonal sums, it follows that

$$w = \sum_{i=1}^{n}\big(\alpha(x_i) \bullet \beta(y_i)\big) = \sum_{i=1}^{n'}\big(\alpha(x_i') \bullet \beta(y_i')\big),$$

thus $n = n'$ and there is a permutation σ of $[n]$ such that

$$\big(\alpha(x_i'), \beta(y_i')\big) = \big(\alpha(x_{\sigma(i)}), \beta(y_{\sigma(i)})\big) \quad \text{for all } i \in [n],$$

that is,

$$x_i'\ \mathscr{D}_S\ x_{\sigma(i)} \text{ and } y_i'\ \mathscr{D}_T\ y_{\sigma(i)}, \quad \text{for all } i \in [n].$$

Picking $s_i \in S$ and $t_i \in T$ such that $\mathbf{d}(s_i) = x_i'$, $\mathbf{r}(s_i) = x_{\sigma(i)}$, $\mathbf{d}(t_i) = y_i'$, and $\mathbf{r}(t_i) = y_{\sigma(i)}$, it follows that $\mathbf{d}(s_i \otimes t_i) = x_i' \otimes y_i'$ and $\mathbf{r}(s_i \otimes t_i) = x_{\sigma(i)} \otimes y_{\sigma(i)}$, whence

$$(x_i' \otimes y_i')\ \mathscr{D}_{S\otimes T}\ (x_{\sigma(i)} \otimes y_{\sigma(i)}), \quad \text{for all } i \in [n].$$

By forming the (orthogonal) joins of all those relations in $S \otimes T$ and observing that $\boldsymbol{u} = u'/{\leftrightarrows}$ and $\boldsymbol{v} = v'/{\leftrightarrows}$, it follows that $\boldsymbol{u}\ \mathscr{D}_{S\otimes T}\ \boldsymbol{v}$.

Let, conversely, $(\boldsymbol{u}, \boldsymbol{v}) \in \mathscr{D}_{S\otimes T}$. This means that there is $\boldsymbol{z} \in S \otimes T$ such that $\boldsymbol{u} = \mathbf{d}(\boldsymbol{z})$ and $\boldsymbol{v} = \mathbf{r}(\boldsymbol{z})$. Write $\boldsymbol{z} = z/{\leftrightarrows}$, with $z = \sum_{i=1}^{n}(x_i \bullet y_i) \in \mathfrak{I}_{S,T}$. Since the elements $x_i \otimes y_i$ are pairwise orthogonal, we get $\boldsymbol{u} = \bigoplus_{i=1}^{n}\big(\mathbf{d}(x_i) \otimes \mathbf{d}(y_i)\big)$ and $\boldsymbol{v} = \bigoplus_{i=1}^{n}\big(\mathbf{r}(x_i) \otimes \mathbf{r}(y_i)\big)$. By the definition of $\boldsymbol{\alpha}$ and $\boldsymbol{\beta}$, $\mathbf{d}(x_i) \equiv_{\boldsymbol{\alpha}} \mathbf{r}(x_i)$ and $\mathbf{d}(y_i) \equiv_{\boldsymbol{\beta}} \mathbf{r}(y_i)$ for each $i \in [n]$, thus $\mathbf{d}(x_i) \otimes \mathbf{d}(y_i) \equiv_{\boldsymbol{\gamma}} \mathbf{r}(x_i) \otimes \mathbf{r}(y_i)$, and thus, forming the (orthogonal) joins of those relations, $\boldsymbol{u} \equiv_{\boldsymbol{\gamma}} \boldsymbol{v}$. □

Theorem 6.9.2 *Let S and T be Boolean inverse semigroups. There is a unique monoid isomorphism* $\eta\colon (\mathrm{Typ}\, S) \otimes (\mathrm{Typ}\, T) \to \mathrm{Typ}(S \otimes T)$ *such that*

$$\eta\big(\mathrm{typ}_S(x) \otimes \mathrm{typ}_T(y)\big) = \mathrm{typ}_{S\otimes T}(x \otimes y) \text{ for every } (x, y) \in S \times T\,. \tag{6.9.1}$$

Proof Since $(\mathrm{Typ}\, S) \otimes (\mathrm{Typ}\, T)$ is additively generated by the elements of the form $\mathrm{typ}_S(x) \otimes \mathrm{typ}_T(y)$, the uniqueness of η is trivial. Let us deal with existence.

Throughout the proof we shall identify the monoids $\mathrm{U}_{\mathrm{mon}}(S) \otimes \mathrm{U}_{\mathrm{mon}}(T)$ and $\mathrm{U}_{\mathrm{mon}}(S \otimes T)$, via Corollary 6.8.6.

The binary relation $\boldsymbol{\gamma} = \mathscr{D}_S^+ \otimes \mathscr{D}_T^+$ is a conical V-congruence of the commutative monoid $\mathrm{U}_{\mathrm{mon}}(S) \otimes \mathrm{U}_{\mathrm{mon}}(T)$, and further, by using the definition of the tensor product of two congruences,

$$\begin{aligned}(\mathrm{Typ}\, S) \otimes (\mathrm{Typ}\, T) &= \big(\mathrm{U}_{\mathrm{mon}}(S)/\mathscr{D}_S^+\big) \otimes \big(\mathrm{U}_{\mathrm{mon}}(T)/\mathscr{D}_T^+\big)\\ &\cong \mathrm{U}_{\mathrm{mon}}(S) \otimes \mathrm{U}_{\mathrm{mon}}(T)/\boldsymbol{\gamma} = \mathrm{U}_{\mathrm{mon}}(S \otimes T)/\boldsymbol{\gamma}\,. \end{aligned} \tag{6.9.2}$$

Since $S\otimes T$, endowed with its orthogonal addition, is a lower interval of $\mathrm{U}_{\mathrm{mon}}(S\otimes T)$, it follows from Lemma 6.9.1, together with Theorem 2.4.6, that $\boldsymbol{\gamma} = \mathrm{U}_{\mathrm{mon}}(\mathscr{D}_{S\otimes T})$. By using again Theorem 2.4.6, it follows that

$$\mathrm{U}_{\mathrm{mon}}(S \otimes T)/\boldsymbol{\gamma} \cong \mathrm{U}_{\mathrm{mon}}\big(S \otimes T/\mathscr{D}_{S\otimes T}\big) = \mathrm{Typ}(S \otimes T)\,. \tag{6.9.3}$$

Further, it is straightforward to verify that the composition η of the isomorphisms given by (6.9.2) and (6.9.3) satisfies (6.9.1). □

Chapter 7
Discussion

Problem 1 Let M be a countable conical refinement monoid and let $\Bbbk$ be a countable field. Is there a countable fundamental Boolean inverse semigroup S such that $\operatorname{Typ} S \cong \mathrm{V}(\Bbbk\langle S\rangle) \cong M$?

By Theorem 4.8.9, there is a countable fundamental Boolean inverse semigroup S such that $\operatorname{Typ} S \cong M$. However, we have seen with Example 6.7.8 that the canonical map $\operatorname{Typ} S \to \mathrm{V}(\Bbbk\langle S\rangle)$ may not be one-to-one.

Problem 2 Let S be a Boolean inverse semigroup. Is the canonical map $\operatorname{Typ} S \to \mathrm{V}(\mathbb{Z}\langle S\rangle)$ an isomorphism?

By Examples 6.7.2 and 6.7.8, the canonical map $\operatorname{Typ} S \to \mathrm{V}(\Bbbk\langle S\rangle)$ (cf. Sect. 6.7) may be neither one-to-one, nor surjective, for any division ring $\Bbbk$. However, the constructions of the idempotents introduced to establish this, in the proofs of Proposition 6.7.1 and Lemma 6.7.6, depend of the characteristic of $\Bbbk$. For a convenient description of $\mathbb{Z}\langle S\rangle$, see Example 6.3.10.

It follows immediately from Passman [92, Theorem 22.7] (which is also mentioned without proof in Kaplansky [66, p. 123]) that the group ring $\mathbb{Z}[G]$ has no nontrivial idempotents. (Any nontrivial idempotent of $\mathbb{Z}[G]$ is a nontrivial idempotent of $\mathbb{C}[G]$, thus its trace lies strictly between 0 and 1; but that trace is also an integer.) This suggests, without proving it (matrix rings over $\mathbb{Z}[G]$ may have nontrivial idempotents), that the specialization of Problem 2 to the case where $S = G^{\sqcup 0}$, for a group G (cf. Definition 1.5.1), may indeed have a positive solution. However, even this special case does not seem to be known.

Problem 3 Is every conical refinement monoid M of cardinality $\aleph_1$ group-measurable? That is, is there a Boolean ring B, with an action of a group G, such that $M \cong \mathbb{Z}^+\langle B\rangle /\!\!/ G$? Equivalently, does $M \cong \operatorname{Typ} S$ for some Boolean inverse semigroup S?

Some evidence for a positive solution to Problem 3 is the following. By Theorem 4.8.9, every countable conical refinement monoid is group-measurable. Also, by Dobbertin's Theorem (cf. Theorem 4.6.7), every conical refinement monoid of cardinality at most $\aleph_1$ is V-measurable. Nevertheless, not every V-measure is

F. Wehrung, *Refinement Monoids, Equidecomposability Types, and Boolean Inverse Semigroups*, Lecture Notes in Mathematics 2188,
DOI 10.1007/978-3-319-61599-8_7

group-induced, or even groupoid-induced (cf. Examples 4.7.8 and 5.4.3). Also, there are conical refinement monoids of cardinality $\aleph_2$ that are not group-measurable (cf. Theorem 4.6.9).

Problem 4 Let G be an Abelian lattice-ordered group. Is there a fundamental Boolean inverse meet-semigroup S such that $\operatorname{Typ} S \cong G^+$?

By Theorems 5.1.10 and 5.2.9, any counterexample to Problem 4 would need to be non-projectable with at least $\aleph_2$ elements. A countable conical refinement monoid, not isomorphic to the type monoid of any fundamental Boolean inverse meet-semigroup, is given in Example 4.9.4.

We may state, similarly, the following problem.

Problem 5 Let G be an Abelian lattice-ordered group. Is there a locally matricial inverse semigroup S such that $G^+ \cong \operatorname{Typ} S$?

By Theorem 5.1.10, any counterexample to Problem 5 would need to have at least $\aleph_2$ elements. The group action constructed in Theorem 5.2.9 has underlying group G, which is not locally finite unless G is trivial. A related problem, stated in Goodearl [50, Problem 31] and credited there to Handelman, asks whether there exists a locally matricial algebra R, over a field, such that $K_0(R) \cong \mathbb{R}$ (as an ordered group). Due to Goodearl and Handelman [52], the answer to that question is positive if the Continuum Hypothesis holds.

The apparent conflicts between Theorem 5.4.1, Example 5.4.3, and Example 5.4.4 suggest the following problem.

Problem 6 Let M be a conical commutative monoid. Can M be embedded into a conical refinement monoid N such that there are a complete atomic Boolean ring B and a V-measure $\mu: B \to N$ with range generating N as a submonoid?

Some evidence for a positive solution of Problem 6 is given in Theorems 5.4.7 and 5.4.8.

For the statement of the following problem, we first recall that an embedding $f: M \hookrightarrow N$ of monoids is *pure* if every finite equation system, with parameters from $f[M]$, which has a solution in N, also has a solution in $f[M]$.

Problem 7 Does every refinement algebra have a pure embedding into the commutative monoid $\mathbb{Z}^+\langle B\rangle /\!\!/ G$, for some action of a group G on a complete Boolean ring B?

By Theorem 5.4.10, if B is complete, then $\mathbb{Z}^+\langle B\rangle /\!\!/ G$ is a refinement algebra. Moreover, it is obvious that every pure submonoid of a refinement algebra is a refinement algebra.

The following problem is motivated by the realization problem of conical refinement monoids, with at most $\aleph_1$ elements, as $\mathrm{V}(R)$ for rings R satisfying various conditions (e.g., regular, exchange, C*-algebra of real rank zero). For a survey about some of those problems, see Ara [6].

Problem 8 Find sufficient conditions, as general as possible, ensuring that the ring $K\langle S\rangle$ is an exchange ring (resp., a regular ring), for a unital ring K and a Boolean inverse semigroup S. Similarly, find sufficient conditions for the C*-algebra $\mathrm{C}^*\langle S\rangle$ to have real rank zero. Same question for $\mathrm{C}^*_{\mathrm{red}}\langle S\rangle$ (cf. Remark 6.4.12).

Recall that $\mathbb{C}\langle S\rangle$ is isomorphic to a dense involutive subalgebra of both $C^*\langle S\rangle$ and $C^*_{red}\langle S\rangle$ (cf. Theorem 6.4.11 and Remark 6.4.12).

Problem 9 Is the category of all distributive inverse semigroups, with semigroup homomorphisms preserving all finite compatible joins, a finitary variety in the sense of Adámek and Rosický [3, Chap. 3]?

We have seen in Sect. 3.2 that the concrete category of all Boolean inverse semigroups, with additive semigroup homomorphisms, is a variety, namely the variety of all biases. A similar result holds for Boolean inverse meet-semigroups with meet-preserving additive semigroup homomorphisms: just add to the axioms defining biases (cf. Definition 3.2.3) the identities stating that $(S, \wedge)$ is a semilattice, together with the identities

$$\mathsf{x} \wedge \mathsf{y} = \mathsf{x}\,\mathbf{d}(\mathsf{x} \wedge \mathsf{y})\,; \quad \mathsf{x}\,\mathbf{d}(\mathsf{y}) = \mathsf{x}\,\mathbf{d}(\mathsf{y}) \wedge \mathsf{x}\,.$$

On the other hand, the analogue of that result for *fundamental* Boolean inverse semigroups fails. Indeed, consider the construction of Example 3.7.13 and pick any additive semigroup embedding $g\colon \mathbb{Z}^{\sqcup 0} \hookrightarrow \mathfrak{I}_{\mathbb{Z}^+}$. Then $g \circ f$ is an additive semigroup homomorphism from the fundamental Boolean inverse semigroup S to the symmetric Boolean inverse semigroup $\mathfrak{I}_{\mathbb{Z}^+}$. Nevertheless, the image of $g \circ f$, which is isomorphic to $\mathbb{Z}^{\sqcup 0}$, is not fundamental. Hence there is no analogue of the concept of bias for fundamental Boolean inverse semigroups.

The situation for distributive inverse semigroups is similar. The following example was constructed by Ganna Kudryavtseva and the author, with the help of the `Mace4` program (cf. McCune [81]).

Example 7.0.1 A finite distributive inverse monoid S, satisfying the identity $\mathsf{x}^3 = \mathsf{x}$, with a monoid homomorphism $g\colon S \to \mathfrak{I}_3$ preserving zero, all compatible joins, and all meets, such that the inverse monoid $g[S]$ is not distributive.

Proof Consider the following distributive inverse monoid, with zero element 0 and unit 6, whose table is represented in Table 7.1.

In particular, $\operatorname{Idp} S = \{0, 1, 3, 5, 6\}$. Observe that $x^3 = x$ for every $x \in S$. The equivalence relation $\boldsymbol{\theta}$ on S, whose only nontrivial block is $\{0, 4, 5\}$, is a semigroup congruence on S. It is also a congruence for the meet operation on S.

Table 7.1 Table of the inverse semigroup S

·	0	1	2	3	4	5	6
0	0	0	0	0	0	0	0
1	0	1	4	5	4	5	1
2	0	4	3	2	5	4	2
3	0	5	2	3	4	5	3
4	0	4	5	4	5	4	4
5	0	5	4	5	4	5	5
6	0	1	2	3	4	5	6

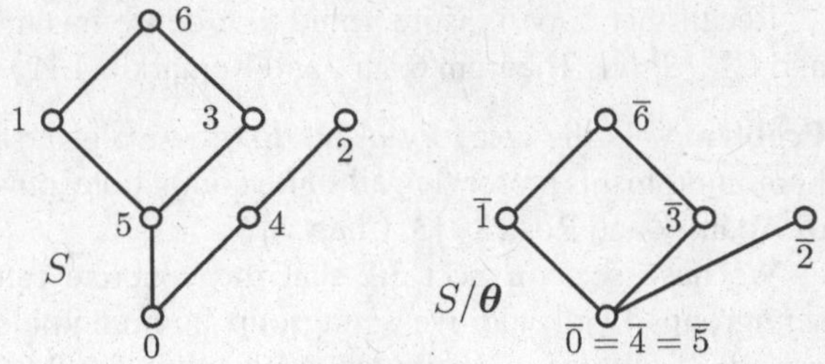

Fig. 7.1 The natural partial orderings on S and S/θ

The natural orderings on S and $S/\boldsymbol{\theta}$ are represented in Fig. 7.1 (we set $\overline{x} = x/\boldsymbol{\theta}$, for every $x \in S$; hence $\overline{0} = \overline{4} = \overline{5}$). Two elements x and y of $S \setminus \{0\}$ are compatible iff either $\{x, y\} \subseteq \{1, 3, 5, 6\}$ or $\{x, y\} \subseteq \{2, 4\}$. Thus, the only nontrivial compatible join in S is $6 = 1 \vee 3$.

Now we let $f\colon S/\boldsymbol{\theta} \to \mathfrak{I}_3$ defined by $f(\overline{0}) = \varnothing$, $f(\overline{1}) = \mathrm{id}_{\{1\}}$, $f(\overline{2}) = \begin{pmatrix} 2 & 3 \\ 3 & 2 \end{pmatrix}$, $f(\overline{3}) = \mathrm{id}_{\{2,3\}}$, $f(\overline{6}) = \mathrm{id}_{\{1,2,3\}}$. It is straightforward to verify that f is a monoid embedding from $S/\boldsymbol{\theta}$ into $\mathfrak{I}_3$.

Denote by $\theta\colon S \twoheadrightarrow S/\boldsymbol{\theta}$ the canonical projection. The composite $g = f \circ \theta$ is a 0-preserving monoid homomorphism $S \to \mathfrak{I}_3$. Furthermore,

$$g(1) \vee g(3) = \mathrm{id}_{\{1\}} \vee \mathrm{id}_{\{2,3\}} = \mathrm{id}_{\{1,2,3\}} = g(6)\,,$$

so g preserves all compatible joins.

The elements $g(1) = \mathrm{id}_{\{1\}}$ and $g(2) = \begin{pmatrix} 2 & 3 \\ 3 & 2 \end{pmatrix}$ are compatible in the image of g: indeed, they have a join in $\mathfrak{I}_3$, namely $\begin{pmatrix} 1 & 2 & 3 \\ 1 & 3 & 2 \end{pmatrix}$. Nevertheless, $\{\overline{1}, \overline{2}\}$ has no upper bound in $S/\boldsymbol{\theta}$. Since f is an embedding, it follows that $\{g(1), g(2)\}$ has no upper bound in the inverse semigroup $f[S/\boldsymbol{\theta}] = g[S]$. Therefore, the image of g is not a distributive inverse semigroup. □

It follows that the *concrete* category of all distributive inverse semigroups, with semigroup homomorphisms preserving all finite compatible joins, is not a variety of algebras.

Problem 10 Let H be a (real or complex) Hilbert space. Is the enveloping monoid $\mathrm{U_{mon}}(\mathrm{Sub}\,H)$ of $\mathrm{Sub}\,H$ cancellative?

As observed in Example 2.7.15, the partial commutative monoid $(\mathrm{Sub}\,H, \oplus, \{0\})$ embeds into the full cancellative monoid $\mathrm{Proj}^+ H$ of all operators on H that can be expressed as finite sums of projections of H. Nevertheless, it is also observed there that if H is at least two-dimensional, then $\mathrm{Proj}^+ H$ is a proper quotient of the enveloping monoid $\mathrm{U_{mon}}(\mathrm{Sub}\,H)$. Further, we observe, at the end of Example 2.7.15, that Problem 10 has a positive solution in case H is two-dimensional, so this problem is interesting only in case H is at least three-dimensional.

Bibliography

1. Abrams, G., Aranda Pino, G.: The Leavitt path algebra of a graph. J. Algebra **293**(2), 319–334 (2005). MR 2172342 (2007b:46085)
2. Abrams, G., Aranda Pino, G.: The Leavitt path algebras of arbitrary graphs. Houst. J. Math. **34**(2), 423–442 (2008). MR 2417402 (2009h:16043)
3. Adámek, J., Rosický, J.: Locally Presentable and Accessible Categories. London Mathematical Society Lecture Note Series, vol. 189. Cambridge University Press, Cambridge (1994). MR 1294136 (95j:18001)
4. Anderson, M., Feil, T.: Lattice-Ordered Groups: An introduction. Reidel Texts in the Mathematical Sciences. D. Reidel Publishing Co., Dordrecht (1988). MR 937703 (90b:06001)
5. Ara, P.: Extensions of exchange rings. J. Algebra **197**(2), 409–423 (1997). MR 1483771 (98j:16021)
6. Ara, P.: The realization problem for von Neumann regular rings. In: Ring Theory 2007, pp. 21–37. World Scientific Publishing, Hackensack, NJ (2009). MR 2513205 (2010k:16020)
7. Ara, P., Exel, R.: Dynamical systems associated to separated graphs, graph algebras, and paradoxical decompositions. Adv. Math. **252**, 748–804 (2014). MR 3144248
8. Ara, P., Facchini, A.: Direct sum decompositions of modules, almost trace ideals, and pullbacks of monoids. Forum Math. **18**(3), 365–389 (2006). MR 2237927 (2007d:16012)
9. Ara, P., Goodearl, K.R.: Leavitt path algebras of separated graphs. J. Reine Angew. Math. **669**, 165–224 (2012). MR 2980456
10. Ara, P., Goodearl, K.R.: Tame and wild refinement monoids. Semigroup Forum **91**(1), 1–27 (2015). MR 3369375
11. Ara, P., Pardo, E.: Refinement monoids with weak comparability and applications to regular rings and C^*-algebras. Proc. Am. Math. Soc. **124**(3), 715–720 (1996). MR 1301484 (96f:46124)
12. Ara, P., Goodearl, K.R., Pardo, E., Tyukavkin, D.V.: K-theoretically simple von Neumann regular rings. J. Algebr. **174**(2), 659–677 (1995). MR 1334230 (96g:16012)
13. Ara, P., Goodearl, K.R., O'Meara, K.C., Pardo, E.: Separative cancellation for projective modules over exchange rings. Israel J. Math. **105**, 105–137 (1998). MR 1639739 (99g:16006)
14. Ara, P., Moreno, M.A., Pardo, E.: Nonstable K-theory for graph algebras. Algebr. Represent. Theory **10**(2), 157–178 (2007). MR 2310414 (2008b:46094)
15. Armstrong, T.E.: Invariance of full conditional probabilities under group actions. In: Measure and Measurable Dynamics (Rochester, NY, 1987). Contemporary Mathematics, vol. 94, pp. 1–21. American Mathematical Society, Providence, RI (1989). MR 1012973 (91d:28007)
16. Ash, C.J., Hall, T.E.: Inverse semigroups on graphs. Semigroup Forum **11**(2), 140–145 (1975/1976). MR 0387449 (52 #8292)

F. Wehrung, *Refinement Monoids, Equidecomposability Types, and Boolean Inverse Semigroups*, Lecture Notes in Mathematics 2188,
DOI 10.1007/978-3-319-61599-8

17. Banach, S.: Un théorème sur les transformations biunivoques. Fund. Math. **6**(1), 236–239 (1924) (French)
18. Banach, S., Tarski, A.: Sur la décomposition des ensembles de points en parties respectivement congruentes. Fund. Math. **6**(1), 244–277 (1924) (French)
19. Barwise, J.: Admissible Sets and Structures: An Approach to Definability Theory. Perspectives in Mathematical Logic. Springer, Berlin/New York (1975). MR 0424560 (54 #12519)
20. Bergman, G.M.: Coproducts and some universal ring constructions. Trans. Am. Math. Soc. **200**, 33–88 (1974). MR 0357503 (50 #9971)
21. Bergman, G.M., Dicks, W.: Universal derivations and universal ring constructions. Pac. J. Math. **79**(2), 293–337 (1978). MR 531320 (81b:16024)
22. Bigard, A., Keimel, K., Wolfenstein, S.: Groupes et Anneaux Réticulés. Lecture Notes in Mathematics, vol. 608. Springer, Berlin/New York (1977). MR 0552653 (58 #27688)
23. Birkhoff, G.: On the combination of subalgebras. Math. Proc. Camb. Philos. Soc. **29**(4), 441–464 (1933)
24. Blackadar, B.: Rational C^*-algebras and nonstable K-theory. In: Proceedings of the 7th Great Plains Operator Theory Seminar (Lawrence, KS, 1987), vol. 20. London Mathematical Society Lecture Note Series, vol. 2, pp. 285–316 (1990). MR 1065831 (92e:46135)
25. Brookfield, G.J.: Monoids and categories of Nœtherian modules. ProQuest LLC, Ann Arbor, MI, Thesis (Ph.D.). University of California, Santa Barbara (1997). MR 2696168
26. Ceccherini-Silberstein, T., Grigorčuk, R.I., de la Harpe, P.: Amenability and paradoxical decompositions for pseudogroups and discrete metric spaces. Tr. Mat. Inst. Steklova **224**, 68–111 (1999). Algebra. Topol. Differ. Uravn. i ikh Prilozh., English translation in Proc. Steklov Inst. Math. 1999 **224**(1), 57–97. MR 1721355 (2001h:43001)
27. Chang, C.C.: Algebraic analysis of many valued logics. Trans. Am. Math. Soc. **88**, 467–490 (1958). MR 0094302 (20 #821)
28. Chen, H.: On separative refinement monoids. Bull. Korean Math. Soc. **46**(3), 489–498 (2009). MR 2522861 (2010f:20060)
29. Cignoli, R., D'Ottaviano, I.M.L., Mundici, D.: Algebraic Foundations of Many-Valued Reasoning. Trends in Logic—Studia Logica Library, vol. 7. Kluwer Academic Publishers, Dordrecht (2000). MR 1786097 (2001j:03114)
30. Clifford, A.H., Preston, G.B.: The Algebraic Theory of Semigroups. Vol. II. Mathematical Surveys, vol. 7. American Mathematical Society, Providence, RI (1967). MR 0218472 (36 #1558)
31. Cohn, P.M.: On the free product of associative rings. Math. Z. **71**, 380–398 (1959). MR 0106918 (21 #5648)
32. Dobbertin, H.: On Vaught's criterion for isomorphisms of countable Boolean algebras. Algebra Univers. **15**(1), 95–114 (1982). MR 663956 (83m:06017)
33. Dobbertin, H.: Refinement monoids, Vaught monoids, and Boolean algebras. Math. Ann. **265**(4), 473–487 (1983). MR 721882 (85e:06016)
34. Dobbertin, H.: Measurable refinement monoids and applications to distributive semilattices, Heyting algebras, and Stone spaces. Math. Z. **187**(1), 13–21 (1984). MR 753415 (85h:20072)
35. Dobbertin, H.: Vaught measures and their applications in lattice theory. J. Pure Appl. Algebra **43**(1), 27–51 (1986). MR 862871 (87k:06032)
36. Duncan, J., Paterson, A.L.T.: C^*-algebras of inverse semigroups. Proc. Edinb. Math. Soc. (2) **28**(1), 41–58 (1985). MR 785726 (86h:46090)
37. Dvurečenskij, A., Pulmannová, S.: New Trends in Quantum Structures. Mathematics and its Applications, vol. 516. Kluwer Academic Publishers/Ister Science, Dordrecht/Bratislava (2000). MR 1861369 (2002h:81021)
38. Effros, E.G., Handelman, D.E., Shen, C.L.: Dimension groups and their affine representations. Am. J. Math. **102**(2), 385–407 (1980). MR 564479 (83g:46061)
39. Elliott, G.A.: On the classification of inductive limits of sequences of semisimple finite-dimensional algebras. J. Algebra **38**(1), 29–44 (1976). MR 0397420 (53 #1279)
40. Exel, R.: Inverse semigroups and combinatorial C^*-algebras. Bull. Braz. Math. Soc. (N.S.) **39**(2), 191–313 (2008). MR 2419901

41. Exel, R.: Tight representations of semilattices and inverse semigroups. Semigroup Forum **79**(1), 159–182 (2009). MR 2534230
42. Exel, R.: Reconstructing a totally disconnected groupoid from its ample semigroup. Proc. Am. Math. Soc. **138**(8), 2991–3001 (2010). MR 2644910
43. Exel, R., Gonçalves, D., Starling, C.: The tiling C*-algebra viewed as a tight inverse semigroup algebra. Semigroup Forum **84**(2), 229–240 (2012). MR 2898758
44. Fekete, M.: Über die Verteilung der Wurzeln bei gewissen algebraischen Gleichungen mit ganzzahligen Koeffizienten. Math. Z. **17**(1), 228–249 (1923). MR 1544613
45. Fillmore, P.A.: On sums of projections. J. Funct. Anal. **4**, 146–152 (1969). MR 0246150 (39 #7455)
46. Foster, A.L.: The idempotent elements of a commutative ring form a Boolean algebra; ring-duality and transformation theory. Duke Math. J. **12**, 143–152 (1945). MR 0012264 (7,1c)
47. Foulis, D.J., Bennett, M.K.: Effect algebras and unsharp quantum logics. Found. Phys. **24**(10), 1331–1352 (1994). Special issue dedicated to Constantin Piron on the occasion of his sixtieth birthday. MR 1304942 (95k:06020)
48. Freese, R., Nation, J.B.: Congruence lattices of semilattices. Pac. J. Math. **49**, 51–58 (1973). MR 0332590 (48 #10916)
49. Goodearl, K.R.: Partially Ordered Abelian Groups with Interpolation. Mathematical Surveys and Monographs, vol. 20. American Mathematical Society, Providence, RI (1986). MR 845783 (88f:06013)
50. Goodearl, K.R.: von Neumann Regular Rings, 2nd edn. Robert E. Krieger Publishing Co., Inc., Malabar, FL (1991). MR 1150975 (93m:16006)
51. Goodearl, K.R.: von Neumann regular rings and direct sum decomposition problems. In: Abelian Groups and Modules (Padova, 1994). Mathematics and Its Applications, vol. 343, pp. 249–255. Kluwer Academic Publishers, Dordrecht (1995). MR 1378203
52. Goodearl, K.R., Handelman, D.E.: Tensor products of dimension groups and K_0 of unit-regular rings. Can. J. Math. **38**(3), 633–658 (1986). MR 845669 (87i:16043)
53. Grätzer, G.: Lattice Theory: Foundation. Birkhäuser/Springer, Basel AG, Basel (2011). MR 2768581 (2012f:06001)
54. Greechie, R.J.: Orthomodular lattices admitting no states. J. Combin. Theory Ser. A **10**, 119–132 (1971). MR 0274355 (43 #120)
55. Grigorčuk, R.I.: On Milnor's problem of group growth. Sov. Math. Dokl. **28**, 23–26 (1983) (English. Russian original). MR 0712546 (85g:20042)
56. Grillet, P.-A.: Interpolation properties and tensor product of semigroups. Semigroup Forum **1**(2), 162–168 (1970). MR 0267022 (42 #1924)
57. Grillet, P.-A.: Directed colimits of free commutative semigroups. J. Pure Appl. Algebra **9**(1), 73–87 (1976/1977). MR 0422461 (54 #10450)
58. Hanf, W.: Primitive Boolean algebras. In: Proceedings of the Tarski Symposium (Proceedings of the Symposia Pure Mathematics, University of California, Berkeley, CA, 1971), vol. 25, pp. 75–90. American Mathematical Society, Providence, RI (1974). MR 0379182 (52 #88)
59. Hewitt, E., Zuckerman, H.S.: The l_1-algebra of a commutative semigroup. Trans. Am. Math. Soc. **83**(1), 70–97 (1956). MR 0081908 (18,465b)
60. Howie, J.M.: An Introduction to Semigroup Theory. London Mathematical Society Monographs, vol. 7. Academic/Harcourt Brace Jovanovich Publishers, London/New York (1976). MR 0466355 (57 #6235)
61. Jech, T.: The Axiom of Choice. Studies in Logic and the Foundations of Mathematics, vol. 75. North-Holland Publishing Co./American Elsevier Publishing Co., Inc., Amsterdam, London/New York (1973). MR 0396271
62. Jones, D.G., Lawson, M.V.: Graph inverse semigroups: their characterization and completion. J. Algebra **409**, 444–473 (2014). MR 3198850
63. Jónsson, B.: On the representation of lattices. Math. Scand. **1**, 193–206 (1953). MR 0058567 (15,389d)
64. Kado, J.: Unit-regular rings and simple self-injective rings. Osaka J. Math. **18**(1), 55–61 (1981). MR 609977 (82h:16008)

65. Kalmbach, G.: Orthomodular Lattices. London Mathematical Society Monographs, vol. 18. Academic/Harcourt Brace Jovanovich Publishers, London/New York (1983). MR 716496 (85f:06012)
66. Kaplansky, I.: Fields and Rings. The University of Chicago Press, Chicago, IL, London (1969). MR 0269449 (42 #4345)
67. Karp, C.R.: Finite-quantifier equivalence. In: Theory of Models (Proceedings of the 1963 International Symposium at Berkeley), pp. 407–412. North-Holland, Amsterdam (1965). MR 0209132 (35 #36)
68. Kerr, D.: C*-Algebras and Topological Dynamics: Finite Approximation and Paradoxicality. CRM Advanced Course Books, Birkhäuser (to appear)
69. Kerr, D., Nowak, P.W.: Residually finite actions and crossed products. Ergod. Theory Dyn. Syst. **32**(5), 1585–1614 (2012). MR 2974211
70. Ketonen, J.: The structure of countable Boolean algebras. Ann. Math. (2) **108**(1), 41–89 (1978). MR 0491391
71. Kudryavtseva, G., Lawson, M.V., Lenz, D.H., Resende, P.: Invariant means on Boolean inverse monoids. Semigroup Forum **92**(1), 77–101 (2016). MR 3448402
72. Lawson, M.V.: Enlargements of regular semigroups. Proc. Edinb. Math. Soc. (2) **39**(3), 425–460 (1996). MR 1417688 (97k:20104)
73. Lawson, M.V.: Inverse Semigroups: The Theory of Partial Symmetries. World Scientific Publishing Co., Inc., River Edge, NJ (1998). MR 1694900 (2000g:20123)
74. Lawson, M.V.: A noncommutative generalization of Stone duality. J. Aust. Math. Soc. **88**(3), 385–404 (2010). MR 2827424 (2012h:20141)
75. Lawson, M.V.: Non-commutative Stone duality: inverse semigroups, topological groupoids and C^*-algebras. Int. J. Algebra Comput. **22**(6), 1250058, 47 pp. (2012). MR 2974110
76. Lawson, M.V., Lenz, D.H.: Pseudogroups and their étale groupoids. Adv. Math. **244**, 117–170 (2013). MR 3077869
77. Lawson, M.V., Scott, P.: AF inverse monoids and the structure of countable MV-algebras. J. Pure Appl. Algebra **221**(1), 45–74 (2017). MR 3531463
78. Leech, J.: Skew Boolean algebras. Algebra Univers. **27**(4), 497–506 (1990). MR 1387897 (97a:06018)
79. Leech, J.: Inverse monoids with a natural semilattice ordering. Proc. Lond. Math. Soc. (3) **70**(1), 146–182 (1995). MR 1300843
80. Mal'cev, A.I.: On the general theory of algebraic systems. Mat. Sb. N.S. **35**(77), 3–20 (1954). MR 0065533 (16,440e)
81. McCune, W.: Prover9 and Mace4 [computer software], 2005–2010
82. McKenzie, R.N., McNulty, G.F., Taylor, W.F.: Algebras, Lattices, Varieties. Vol. I. The Wadsworth & Brooks/Cole Mathematics Series. Wadsworth & Brooks/Cole Advanced Books & Software, Monterey, CA (1987). MR 883644 (88e:08001)
83. Mesyan, Z., Mitchell, J.D., Morayne, M., Péresse, Y.H.: Topological graph inverse semigroups (2013). arXiv 1306.5388, version 1
84. Moreira Dos Santos, C.: Decomposition of strongly separative monoids. J. Pure Appl. Algebra **172**(1), 25–47 (2002). MR 1904228 (2003d:06020)
85. Moreira Dos Santos, C.: A refinement monoid whose maximal antisymmetric quotient is not a refinement monoid. Semigroup Forum **65**(2), 249–263 (2002). MR 1911728 (2003c:20074)
86. Mundici, D.: Interpretation of AF C^*-algebras in Łukasiewicz sentential calculus. J. Funct. Anal. **65**(1), 15–63 (1986). MR 819173 (87k:46146)
87. Munn, W.D.: Some recent results on the structure of inverse semigroups. In: Semigroups (Proceedings of a Symposium, Wayne State Univeristy, Detroit, MI, 1968), pp. 107–123. Academic, New York (1969). MR 0262404 (41 #7012)
88. Nambooripad, K.S.S.: The natural partial order on a regular semigroup. Proc. Edinb. Math. Soc. (2) **23**(3), 249–260 (1980). MR 620922 (82g:20092)
89. Navara, M.: An orthomodular lattice admitting no group-valued measure. Proc. Am. Math. Soc. **122**(1), 7–12 (1994). MR 1191871 (94k:06007)

90. Ol'šanskiĭ, A.J.: On the question of the existence of an invariant mean on a group. Uspekhi Mat. Nauk **35**(4)(214), 199–200 (1980). MR 586204 (82b:43002)
91. O'Meara, K.C.: The exchange property for row and column-finite matrix rings. J. Algebra **268**(2), 744–749 (2003). MR 2009331 (2004i:16040)
92. Passman, D.S.: Infinite Group Rings. Pure and Applied Mathematics, vol. 6. Marcel Dekker, Inc., New York (1971). MR 0314951 (47 #3500)
93. Paterson, A.L.T.: Groupoids, Inverse Semigroups, and Their Operator Algebras. Progress in Mathematics, vol. 170. Birkhäuser Boston, Inc., Boston, MA (1999). MR 1724106 (2001a:22003)
94. Pierce, R.S.: Countable Boolean algebras. In: Handbook of Boolean Algebras, vol. 3, pp. 775–876. North-Holland, Amsterdam (1989). MR 991610
95. Rabanovich, V.I., Samoĭlenko, Y.S.: Scalar operators representable as a sum of projectors. Ukraïn. Mat. Zh. **53**(7), 939–952 (2001). MR 2031291 (2004i:47073)
96. Rainone, T.: Finiteness and paradoxical decompositions in C*-dynamical systems (2015). arXiv 1502.06153, version 1
97. Renault, J.: A Groupoid Approach to C^*-Algebras. Lecture Notes in Mathematics, vol. 793. Springer, Berlin (1980). MR 584266 (82h:46075)
98. Resende, P.: A note on infinitely distributive inverse semigroups. Semigroup Forum **73**(1), 156–158 (2006). MR 2277324 (2007j:20094)
99. Rørdam, M., Sierakowski, A.: Purely infinite C^*-algebras arising from crossed products. Ergodic Theory Dyn. Syst. **32**(1), 273–293 (2012). MR 2873171 (2012m:46063)
100. Rosenblatt, J.M.: Invariant measures and growth conditions. Trans. Am. Math. Soc. **193**, 33–53 (1974). MR 0342955 (49 #7699)
101. Schein, B.M.: On the theory of generalized groups (Russian). Dokl. Akad. Nauk SSSR **153**, 296–299 (1963). English translation in Soviet Math. Dokl. **4**, 1680–1683 (1963). MR 0170966 (30 #1200)
102. Schein, B.M.: Completions, translational hulls and ideal extensions of inverse semigroups. Czechoslov. Math. J. **23**(98), 575–610 (1973). MR 0325820 (48 #4166)
103. Solomon, L.: Representations of the rook monoid. J. Algebra **256**(2), 309–342 (2002). MR 1939108 (2003m:20091)
104. Steinberg, B.: A groupoid approach to discrete inverse semigroup algebras. Adv. Math. **223**(2), 689–727 (2010). MR 2565546 (2010k:20113)
105. Steinberg, B.: Simplicity, primitivity and semiprimitivity of étale groupoid algebras with applications to inverse semigroup algebras (2014). arXiv 1408.6014, version 1
106. Stone, M.H.: The theory of representations for Boolean algebras. Trans. Am. Math. Soc. **40**(1), 37–111 (1936). MR 1501865
107. Stone, M.H.: Applications of the theory of Boolean rings to general topology. Trans. Am. Math. Soc. **41**(3), 375–481 (1937). MR 1501905
108. Tarski, A.: Cancellation laws in the arithmetic of cardinals. Fund. Math. **36**, 77–92 (1949). MR 0032710
109. Tarski, A.: Cardinal Algebras. With an Appendix: Cardinal Products of Isomorphism Types, by Bjarni Jónsson and Alfred Tarski. Oxford University Press, New York, NY (1949). MR 0029954 (10,686f)
110. Tarski, A.: Ordinal Algebras. With Appendices by Chen-Chung Chang and Bjarni Jónsson. North-Holland Publishing Co., Amsterdam (1956). MR 0082935
111. Truss, J.K.: The failure of cancellation laws for equidecomposability types. Can. J. Math. **42**(4), 590–606 (1990). MR 1074225 (91k:03147)
112. Vagner, V.V.: On the theory of antigroups. Izv. Vysš. Učebn. Zaved. Matematika **4**(107), 3–15 (1971). MR 0294544 (45 #3614)
113. Vagner, V.V.: t-Simple representations of antigroups. Izv. Vysš. Učebn. Zaved. Matematika **9**(112), 18–29 (1971). MR 0289678 (44 #6866)
114. Vaught, R.L.: Topics in the theory of arithmetical classes and Boolean algebras. ProQuest LLC, Ann Arbor, MI, Thesis (Ph.D.). University of California, Berkeley (1955). MR 2938637

115. Wagon, S.: The Banach-Tarski Paradox. Encyclopedia of Mathematics and Its Applications, vol. 24. Cambridge University Press, Cambridge (1985). With a foreword by Jan Mycielski. MR 803509 (87e:04007)
116. Wallis, A.R.: Semigroup and category-theoretic approaches to partial symmetry, Ph.D. thesis. Heriot-Watt University, Edinburgh (2013)
117. Weber, H.: There are orthomodular lattices without nontrivial group-valued states: a computer-based construction. J. Math. Anal. Appl. **183**(1), 89–93 (1994). MR 1273434 (95e:06026)
118. Wehrung, F.: Injective positively ordered monoids. I. J. Pure Appl. Algebra **83**(1), 43–82 (1992). MR 1190444 (93k:06023)
119. Wehrung, F.: Injective positively ordered monoids. II. J. Pure Appl. Algebra **83**(1), 83–100 (1992). MR 1190444 (93k:06023)
120. Wehrung, F.: The universal theory of ordered equidecomposability types semigroups. Can. J. Math. **46**(5), 1093–1120 (1994). MR 1295133 (95i:06025)
121. Wehrung, F.: Tensor products of structures with interpolation. Pac. J. Math. **176**(1), 267–285 (1996). MR 1433994 (98e:06010)
122. Wehrung, F.: The dimension monoid of a lattice. Algebra Univers. **40**(3), 247–411 (1998). MR 1668068 (2000i:06014)
123. Wehrung, F.: Embedding simple commutative monoids into simple refinement monoids. Semigroup Forum **56**(1), 104–129 (1998). MR 1490558 (99b:20092)
124. Wehrung, F.: Non-measurability properties of interpolation vector spaces. Israel J. Math. **103**, 177–206 (1998). MR 1613568 (99g:06023)
125. Wehrung, F.: A K_0-avoiding dimension group with an order-unit of index two. J. Algebra **301**(2), 728–747 (2006). MR 2236765 (2007e:16011)
126. Zhitomirskiy, G.I.: Inverse semigroups and fiberings. In: Semigroups (Luino, 1992), pp. 311–321. World Scientific Publishing, River Edge, NJ (1993). MR 1647275
127. Zhitomirskiy, G.I.: Topologically complete representations of inverse semigroups. Semigroup Forum **66**(1), 121–130 (2003). MR 1939670 (2003h:20113)

Author Index

F. Wehrung, *Refinement Monoids, Equidecomposability Types, and Boolean Inverse Semigroups*, Lecture Notes in Mathematics 2188,
DOI 10.1007/978-3-319-61599-8

Glossary of Notation

0_P, 1_P, *11*

$a \xrightarrow{x} b$, *110*
$|x|$ (in lattice-ordered groups), *162*
ad_g (inner endomorphism), *106*
$a^{\perp}$ (in lattice-ordered groups), *162*
$x \approx_N y$, *16*
$\rightarrow$ (on $\mathrm{F}_{\mathrm{mon}}(P)$), *26*
$\rightarrow$ (on $\mathrm{F}_{\mathrm{mon}}(M \times N)$), *40*
$\rightarrow^{*}$ (on $\mathrm{F}_{\mathrm{mon}}(M \times N)$), *40*
$\leftrightarrows$ (on $\mathrm{F}_{\mathrm{mon}}(M \times N)$), *40*
$\mathrm{At}\,P$, *11*
$\mathrm{Aut}(B)$, *134*
$\mathrm{Aut}(B, \mu)$, *134*

$\mathbf{B}(\Omega)$, *203*
$\boldsymbol{I}\colon A \cong_{\mathrm{p}} B, A \approxeq_{\mathrm{p}} B$ (back-and-forth system), *129*
$\mathrm{B}_{\Omega}^{\oplus}(S)$ (generalized rook matrices), *96*
$\oslash$ (skew difference), 7, *82*
∇ (skew join), 7, *82*
$x \bullet y$ (on monoid elements), *40*
$f \bullet g$ (on monoid homomorphisms), *43*
$c = a \boxplus b$ (disjunctive addition), *32*
$\mathrm{BR}(D)$ (enveloping Boolean ring), *158*
$B \downarrow \neg c$, *159*
$\|f = g\|$, *121*

$\mathbb{C}$, *16*
$\mathrm{comp}(a : b)$ (weak comparability set), *20*
$R \rtimes G$ (R ring), *53*
$S \rtimes G$ (S Boolean inverse semigroup), *100*

$\mathrm{C}^{*}_{\mathrm{red}}\langle S \rangle$ (reduced additive enveloping C*-algebra), *203*
$\mathrm{C}^{*}\langle S \rangle$ (additive enveloping C*-algebra), *202*

$\mathbf{d}(x)$ (domain), *72*
$\Delta(x, y)$ (generators of the dimension monoid), *160*
$\delta_{x,y}$ (Kronecker symbol), *77*
$D^{[2]}$, *159*
$\mathscr{D}$ (Green's relation), 6, *73*
$y \smallsetminus x$ (in a generalized Boolean algebra), *14*
$y \smallsetminus x$ (in a Boolean inverse semigroup), *78*
$\mathrm{Dim}\, G$ (dimension monoid), *160*
$\mathrm{dom}\, f$ (domain), *10*
$\dot{x}$ (in $\mathrm{F}_{\mathrm{mon}}(P)$), *26*
$\mathscr{D}_S^{+}$, *217*

$E_{G,n}$, E_G, *61*
ε_P (embedding $P \hookrightarrow \mathrm{U}_{\mathrm{mon}}(P)$), *26*
$[u]$ (in $\mathrm{F}_{\mathrm{mon}}(P)$), *26*
$a \simeq_{\mu}^{\mathrm{gp}} b$, *134*
$a \sim_{\mu}^{\mathrm{gp}} b$, *134*
$a \sim_{\mu}^{\mathrm{gpd}} b$, *134*
$\equiv$ (on $\mathrm{F}_{\mathrm{mon}}(P)$), *26*
$\circeq$ (on $\mathrm{F}_{\mathrm{mon}}(P)$), *26*

$f[X]$, *10*
$f^{-1}[X]$, *10*
$\langle g_1, g_2 \rangle^{-n}(c)$, *170*
$\mathrm{F}_{\mathrm{mon}}(X)$ (free commutative monoid), *26*
$f \restriction_X$ (restriction), *10*

F. Wehrung, *Refinement Monoids, Equidecomposability Types, and Boolean Inverse Semigroups*, Lecture Notes in Mathematics 2188,
DOI 10.1007/978-3-319-61599-8

Index

F. Wehrung, *Refinement Monoids, Equidecomposability Types, and Boolean Inverse Semigroups*, Lecture Notes in Mathematics 2188,
DOI 10.1007/978-3-319-61599-8

LECTURE NOTES IN MATHEMATICS **Springer**

Editorial Policy

1. Lecture Notes aim to report new developments in all areas of mathematics and their applications – quickly, informally and at a high level. Mathematical texts analysing new developments in modelling and numerical simulation are welcome.

   Manuscripts should be reasonably self-contained and rounded off. Thus they may, and often will, present not only results of the author but also related work by other people. They may be based on specialised lecture courses. Furthermore, the manuscripts should provide sufficient motivation, examples and applications. This clearly distinguishes Lecture Notes from journal articles or technical reports which normally are very concise. Articles intended for a journal but too long to be accepted by most journals, usually do not have this "lecture notes" character. For similar reasons it is unusual for doctoral theses to be accepted for the Lecture Notes series, though habilitation theses may be appropriate.

2. Besides monographs, multi-author manuscripts resulting from SUMMER SCHOOLS or similar INTENSIVE COURSES are welcome, provided their objective was held to present an active mathematical topic to an audience at the beginning or intermediate graduate level (a list of participants should be provided).

   The resulting manuscript should not be just a collection of course notes, but should require advance planning and coordination among the main lecturers. The subject matter should dictate the structure of the book. This structure should be motivated and explained in a scientific introduction, and the notation, references, index and formulation of results should be, if possible, unified by the editors. Each contribution should have an abstract and an introduction referring to the other contributions. In other words, more preparatory work must go into a multi-authored volume than simply assembling a disparate collection of papers, communicated at the event.

3. Manuscripts should be submitted either online at www.editorialmanager.com/lnm to Springer's mathematics editorial in Heidelberg, or electronically to one of the series editors. Authors should be aware that incomplete or insufficiently close-to-final manuscripts almost always result in longer refereeing times and nevertheless unclear referees' recommendations, making further refereeing of a final draft necessary. The strict minimum amount of material that will be considered should include a detailed outline describing the planned contents of each chapter, a bibliography and several sample chapters. Parallel submission of a manuscript to another publisher while under consideration for LNM is not acceptable and can lead to rejection.

4. In general, **monographs** will be sent out to at least 2 external referees for evaluation.

   A final decision to publish can be made only on the basis of the complete manuscript, however a refereeing process leading to a preliminary decision can be based on a pre-final or incomplete manuscript.

   Volume Editors of **multi-author works** are expected to arrange for the refereeing, to the usual scientific standards, of the individual contributions. If the resulting reports can be forwarded to the LNM Editorial Board, this is very helpful. If no reports are forwarded or if other questions remain unclear in respect of homogeneity etc, the series editors may wish to consult external referees for an overall evaluation of the volume.

5. Manuscripts should in general be submitted in English. Final manuscripts should contain at least 100 pages of mathematical text and should always include

   - a table of contents;
   - an informative introduction, with adequate motivation and perhaps some historical remarks: it should be accessible to a reader not intimately familiar with the topic treated;
   - a subject index: as a rule this is genuinely helpful for the reader.
   - For evaluation purposes, manuscripts should be submitted as pdf files.

6. Careful preparation of the manuscripts will help keep production time short besides ensuring satisfactory appearance of the finished book in print and online. After acceptance of the manuscript authors will be asked to prepare the final LaTeX source files (see LaTeX templates online: https://www.springer.com/gb/authors-editors/book-authors-editors/manuscriptpreparation/5636) plus the corresponding pdf- or zipped ps-file. The LaTeX source files are essential for producing the full-text online version of the book, see http://link.springer.com/bookseries/304 for the existing online volumes of LNM). The technical production of a Lecture Notes volume takes approximately 12 weeks. Additional instructions, if necessary, are available on request from lnm@springer.com.

7. Authors receive a total of 30 free copies of their volume and free access to their book on SpringerLink, but no royalties. They are entitled to a discount of 33.3 % on the price of Springer books purchased for their personal use, if ordering directly from Springer.

8. Commitment to publish is made by a *Publishing Agreement*; contributing authors of multiauthor books are requested to sign a *Consent to Publish form*. Springer-Verlag registers the copyright for each volume. Authors are free to reuse material contained in their LNM volumes in later publications: a brief written (or e-mail) request for formal permission is sufficient.

Addresses:
Professor Jean-Michel Morel, CMLA, École Normale Supérieure de Cachan, France
E-mail: moreljeanmichel@gmail.com

Professor Bernard Teissier, Equipe Géométrie et Dynamique,
Institut de Mathématiques de Jussieu – Paris Rive Gauche, Paris, France
E-mail: bernard.teissier@imj-prg.fr

Springer: Ute McCrory, Mathematics, Heidelberg, Germany,
E-mail: lnm@springer.com

Printed in the United States
By Bookmasters